X-Ray Diffraction and the Identification and Analysis of Clay Minerals

DATE DUE

X-Ray Diffraction and
the Identification and Analysis
of Clay Minerals

SECOND EDITION

DUANE M. MOORE
Illinois State Geological Survey

ROBERT C. REYNOLDS, JR.
Dartmouth College

Oxford New York
OXFORD UNIVERSITY PRESS
1997

University Press

d New York
Athens Auckland Bangkok Bombay
Calcutta Cape Town Dar es Salaam Delhi
Florence Hong Kong Istanbul Karachi
Kuala Lumpur Madras Madrid Melbourne
Mexico City Nairobi Paris Singapore
Taipei Tokyo Toronto

and associated companies in

Berlin Ibadan

Published by Oxford University Press, Inc.
198 Madison Avenue, New York, New York 10016

Oxford is a registered trademark of Oxford University Press

Library of Congress Cataloging-in-Publication Data

Moore, Duane Milton, 1933-
X-ray diffraction and the identification and analysis of clay minerals / Duane M. Moore,
Robert C. Reynolds, Jr. -- 2nd ed.
p. cm.
Includes bibliographical references (p. -) and index.
ISBN 0-19-508713-5 (pbk.)
1. Clay--Analysis. 2. Clay minerals. 3. X-ray crystallography.
I. Reynolds, Robert C., Jr., 1927- . II. Title
QE471.3.M66 1996
549'.6--dc20 96-22767

1 3 5 7 9 8 6 4 2
Printed in the United States of America
on acid-free paper

To John Hower, our friend.

Contents

Preface to the Second Edition

As we look back over what we've done for this new edition, it is apparent that our coverage has broadened so that the title of this book is no longer accurate. It has seemed necessary for us to add comments about the origins of the various clay minerals, to add emphasis on soils, and to at least try to introduce some of the organic chemistry associated with the study and application of clay minerals. Perhaps we may only be more brazen, not necessarily smarter. Our biggest surprise was how much new information has been generated since the publication of the first edition and how much harder it has been to smoothly (we hope) incorporate this into the second edition than it was to write the first edition. We hope that our efforts to prepare photo-ready copy to help keep the price of the book as low as possible will carry through the entire publication process.

We leave the Preface to the first edition because it still defines what we are trying to do. From the first edition, Chapter 4 has been split into two chapters, 4 and 5. Chapter 10 is brand new and is an introduction to using X-ray diffraction tracings from random powders.

We would like to add to the acknowledgments from the first edition that Randy Hughes and Georg Grathoff read and commented on all chapters, Don Henderson helped greatly with graphics and with formatting and layout, Herb Glass offered insightful comments on Chapter 10, Sam Savin and his students provided some helpful feedback, and Tom McGreary, Philip Zodhiates, and an anonymous editor from Oxford University Press edited the manuscript for clarity and consistency. Support of the Illinois State Geological Survey was shown in many ways and is greatly appreciated.

D. M. Moore and Robert C. Reynolds, Jr.
July, 1996

Preface to the First Edition

This book is for students; it is an instrument for teaching. Introductory geology, introductory chemistry, and a course in mineralogy are the only prerequisites needed. We do use some physical chemistry, some physics, some geochemistry, some mathematics but you will have encountered most of the terms and concepts we use. Our hope is that one of the things this book will do is to give you motive and purpose for further study in these disciplines that are so important for understanding minerals.

This book is a combination text and lab manual. This is reflected in the format. The wire binding allows it to lay flat on the lab bench. Although our topic is clay mineralogy, it is only a convenient vehicle for instruction in (1) analytical thinking, (2) the use of electronic gadgets to augment such thinking, and (3) the art of experimenting (or directed messing around). Our primary audience is potential chemists, engineers, geologists, and soil scientists, although professionals will find explanations, examples, and methods helpful to them. But we have not include them in our primary audience. Our expertise is in the clay mineralogy of sedimentary rocks. Subject matter as closely related as the clay minerals of soils, of hydrothermal systems, and of failed slopes is complex enough for us that we defer to others for examples and general statements about them. However, you will find here a vocabulary that will introduce you to these closely related topics.

Our assumptions and goals are the following. We think it is crucial to achieve historical perspective, to understand electronic instruments as extensions of our senses, to understand the interaction of, in this case X-rays, and crystalline materials, and to understand the applications and limitations of the data produced by this interaction. The organization of the book reflects these assumptions and the goals they imply: Chapter 1 should say to you that historical context is as important to science as to any other discipline and that scientists are people. Chapters 2 and 3 emphasize that we necd to understand what goes on inside of the electronic gadgetry and in the interaction of X-rays and crystalline materials. Such understandings are important for interpreting the data produced by the gadgets and to relate them to the details of the material being studied. The remaining chapters show that, as an end-product of collecting and analyzing data, an understanding of clay minerals is developing. This understanding provides clues for an interpretation of agricultural, engineering, geological, and industrial problems, with it clearly implied that careful study of any aspect of nature can yield insight and interpretation. In addition, we hope it is evident to every reader that there are many unsolved, inviting problems; many gaps in the understanding developed to date.

Further clues to what we consider important are the voice we have tried to maintain, our emphasis on use of older, simpler X-ray machines, and our mixing of theoretical discussion and recipe-like directions. For the voice, we have tried to keep the text unintimidating, and informal, but nonetheless rigorous. The book is probably a bit more colloquial than most textbooks because we want to communicate our delight and enthusiasm for the subject matter, and from a conviction that we need to be deliberate about not taking ourselves too seriously. We explain some equations in more detail than is common and have introduced others with the expectation that they can be accepted on faith. Questions are scattered through the text to remind you to read with a questioning mind: "Am I understanding this?" In some cases the answer follows; in others it does not. We want you to form the habit of asking yourself questions without explicit signals to do so. We make some points in the book that are deliberately repeated and then tied together with page numbers, *e.g.*, the importance of Bragg's law. Also implied is a sense that generalizations and facts are not final and absolute but are only the best we can manage at the time. They should be questioned; accepted only tentatively. There are instances in which ambivalence is respectable, even proper.

We emphasize the use of older, simpler X-ray machines for one, compelling reason: They make far more sense for teaching than do automated machines. Some suppliers of X-ray equipment offer what they call tabletop models, which perhaps could meet the goals of teaching the geometry of X-ray diffraction and the manipulation of parts of the apparatus in order to obtain optimal results. However, between what is sealed up for safety reasons and what is automated, not much can be adjusted, manipulated, or watched. Automated X-ray diffraction machines are for research and for production control. They are excellent when large numbers of very similar samples need to be run in the shortest possible time. These are not the criteria for teaching, however; they are anathema to teaching. Automated machines have computing facilities that may be used for everything from aligning θ - 2θ to quantitative phase analysis to peak stripping to data management. The programs are proprietary and therefore usually unavailable for inspection. The assumptions built into the programs remain unknown. The only way to train people who can create, design, or modify, let alone run, automated machines or to write the programs for them, is to have them start with the geometry of the machine exposed —to allow them access to a machine that can be manipulated, one with which they can experiment.

Mixing of recipe-like instructions for preparing a slide and interpreting an X-ray diffractogram with a theoretical consideration of the influence of Fe in the octahedral sheet on the structure factor F of a chlorite, reflects our conviction that the details of laboratory preparation and of theoretical considerations must mutually inform and reinforce one another. They make much less sense when separated from one another.

Another goal of this book is to have it serve as a bridge from Ralph Grim's and Dorothy Carroll's books to literature such as Brindley and Brown (1980), James (1965), Klug and Alexander (1974), Newman (1987),Velde (1985), and Wilson (1987).

This book had at its conception, a plan for a chapter on the origin, or general petrology, of clay minerals, but it rapidly became too bulky and began to carry us off the track of our original goal. Small parts of this have been added to Chapter 4. Perhaps another time.

ACKNOWLEDGMENTS

Our first acknowledgment must go to those who have gone before us and from whom we have taken so much. We have tried to recognize the source of the facts and ideas we have used and to acknowledge them in the text. Almost certainly some facts and ideas are unacknowledged, and are unacknowledged because they have become so much a part of our thinking; perhaps the ultimate acknowledgment.

This book began from two sources. One source was a set of notes for a short course given in Pakistan while one of us (DMM) was supported by a Fulbright Lectureship, 1984-85. The other was a set of notes made in contemplation of writing a book (RCR). Dartmouth College, the Illinois State Geological Survey, and Knox College provided institutional support. Copy editor Tom Whipple did considerable cleaning up of the manuscript. The manuscript was prepared camera ready in Times® font by using Macintosh® computers, Word® version 3.01, and Laserwriter® printers. We thank friends and organizations for their permissions to use figures and photographs. They are individually acknowledged in captions. Hosts of colleagues and students, and students now colleagues, provided information, advice, emotional and logistical support, and even ridicule when it was appropriate. Stephen Altaner and Ralph Grim read the entire manuscript. They offered genuine help and insight in terms of the history, the science, and the clarity of our subject matter. The following read parts of the manuscript and offered helpful discussion and criticism: James Drever, Herbert Glass, Randall Hughes, Blair Jones, Robert Lander, James Matthews, Herman Roberson and his students, Shelley Roberts, Russell Sutton, Norma Vergo, and Douglas Wilson. And finally, we thank Roseann Reynolds for putting up with the confusion and general disruption of her household during the gestation of this book.

D. M. Moore and Robert C. Reynolds, Jr.

REFERENCES

Brindley, G. W., and Brown, G., editors, (1980) *Crystal Structures of Clay Minerals and Their X-Ray Identification*: Mineralogical Society, London, 495 p.

James, R. W. (1965) *The Optical Principles of the Diffraction of X-Rays*: Vol. II of *The Crystalline State*: Series edited by Sir Lawrence Bragg, Cornell University Press, Ithaca, 664 p.

Klug, H. P., and Alexander, L. E. (1974) *X-Ray Diffraction Procedures*, 2nd ed.: Wiley, New York, 966 p.

Newman, A. C. D., editor (1987) *Chemistry of Clays and Clay Minerals*: Monograph No. 6, Mineralogical Society, Wiley, New York, 480 p.

Velde, B. (1985) *Clay Minerals: a Physico-Chemical Explanation of Their Occurrence*: Elsevier, Amsterdam, 427 p.

Wilson, M. J., editor (1987) *A Handbook of Determinative Methods in Clay Mineralogy*: Blackie, Glasgow, 320 p.

X-Ray Diffraction and the Identification and Analysis of Clay Minerals

Chapter 1
Introduction and Historical Background

Clay minerals, crystallites the size of colloidal particles, viruses, or the particles in smoke from a cigarette, are the most abundant minerals at the surface of the earth. They are major components of soils, of sedimentary rocks, the outcrops of which cover about 75% of the earth's land surface, and of the pelagic oozes blanketing the ocean basins. These hydrous alumino silicates, which form about one-third of all sedimentary rocks (or 1×10^{25} grams), represent the weathering of an enormous volume of other silicate minerals. They are most often the products of chemical weathering of minerals formed deep within the earth, feldspars and other silicate minerals; weathering is initiated when they meet the atmosphere and react with it. This environment in which clay minerals form also is that occupied by humans and all other forms of life—reason enough perhaps that we should try to understand them.

To geologists, clay is a term for a rock made of at least 50% material less than (<) 2 μm (= micrometer, formerly called micron, = 0.001 mm, or one-millionth of a meter, 10^{-6} m, and is used in the sense of equivalent spherical diameter; see Stoke's law, Chapter 6, p. 213). Engineers and soil scientists use clay as a size term calling it material < 4 μm. To them, clay carries no implications about composition or mineralogy. In this book, we use clay as a rock term and clay mineral for the relatively small number of minerals that occur as grains that are less than 2 μm in largest dimension. These are, with several important exceptions, hydrated alumino silicates with layered structures. Palygorskite and sepiolite are included with the < 2 μm layered silicates by some, but not by others because they are as much like zeolites as they are like the layered silicates. For convenience, we include them with the layered silicates. Zeolites that are < 2 μm behave much like the layered silicate clay minerals, as do the metal hydroxides, oxyhydroxides, and oxides when they are clay size. These latter two groups should be included, but we lack the expertise to do any more than mention them. Chemically and structurally clay minerals are closely related to minerals that occur in sizes > 2 μm, minerals that may be studied macroscopically. However, the small size of clay minerals, or large ratio of surface area to volume, gives clay minerals a set of unique properties, highly reactive surfaces including high cation exchange capacities, catalytic activity, and plastic behavior when moist.

Clay minerals of sedimentary rocks and soils are the primary focus of this book. Most of these are one of six types, or intergrown mixtures of two or more, of clay minerals: chlorite, illite, kaolinite, the oxides and hydrated oxides, smectite, and vermiculite. In this size range, kaolinite is the dominate mineral in the highly weathered soils that form in the wet tropical latitudes. In the most highly weathered of soils, hydrated oxides of Al, Fe, and Mn also are often present. In temperate latitudes, expandable clay minerals (smectites or vermiculites) tend to form. In the higher latitudes, where weathering is often dominantly mechanical, clay minerals (illite and chlorite) from older rocks are recycled.

Why should the minerals that are < 2 μm be dominantly clay minerals? If you examine the particle size distribution curves for quartz or feldspar in sediments, you will find that the abundance begins to drop off sharply at about 10 to 20 μm, and becomes very low in the vicinity of 4 μm. This is because the physical processes that grind or chip quartz in turbulent waters become ineffective at very small grain sizes where the inertia of the grains is small compared to the viscous forces in the water. But the clay minerals do not exist as small particles because they have been ground by natural processes. They are small because they have grown as crystals "from the ground up," so to speak. Their sizes are limited by the very slow kinetics that prevail in the low-temperature environments in which they form, and by the high density of crystal defects that would destabilize them as larger crystals. The particle size distribution curves for detrital minerals like quartz and the diagenetic clay minerals cross in the region of 2 to 4 μm, leading to relatively pure concentrations of clay minerals in the clay size fraction.

The study of clay minerals is enjoying a renaissance traceable in part to the recognition of diagenetic changes associated with the transformation of smectite to illite in samples described first from the U.S. Gulf Coast drill holes, and now found worldwide. This change is important to the oil industry because apparently it depends on temperature and time and is a potential index to the generation of petroleum and natural gas. It is also probable that this diagenetic change is an integral part of the process of petroleum generation and migration and that it is related to other processes or conditions in the accumulation of hydrocarbons, such as cementation, the preservation of primary porosity, the development of secondary porosity in shales and adjacent reservoir rocks, and the higher than normal water pressures found in some oil wells (Bethke et al., 1988). Since this transformation has become better understood, other transformations of clay-sized minerals and organic materials have been recognized for their usefulness as indicators of the degree of incipient metamorphism or maturity of hydrocarbon source rocks.

Several other recent applications of clay mineralogy have spurred research. For example, its role in the disposal of toxic and radioactive wastes (Pusch, 1992), the use of clay minerals, modified to have large and stable galleries within their crystal structures, as catalysts and molecular sieves (Schoonheydt et al., 1994), and clay minerals used to protect pesticides from breaking down from exposure to sunlight (Margulies et al., 1985). Applications such as these join the standard ones related to the study of soils, to the paper and the chemical industries, to the production of ceramic products, and to construction and engineering practices. The primary method for identifying and analyzing clay minerals is X-ray diffraction, although other instrumental methods are being used more frequently (see second paragraph in Chapter 2).

If you are interested in learning about analytical thought, you will need some perspective, some sense of what analytical thought is, and how, and by whom it has been used. In an effort to illustrate this, to provide you with examples of analytical thought, we focus on two scientists, Wilhelm Conrad Röntgen (1845-1923) and Max von Laue (1879-1960). Examining the quite different approaches of these two men, operating within essentially the same culture, illuminates both analytical thinking and our view of the different styles of contemporary scientists. Clearly, science is a human activity strongly influenced by the culture within which it is conducted. Scientists are people who eat and sleep, are timid or bold, are conservative or daring. We think that we understand, and hope that you do also, that no one uses *the* scientific method; that there are significant amounts of ambition, greed, luck, envy . . . probably even some sex, underlying the work that gets done and is called science. Nevertheless, acts of discovery have similarities despite the approach taken by the individual, the group, or the discipline.

To be gained in the bargain, as we examine these two examples, is a sense that ideas do not develop neatly. They pick up a bit from one discipline, something else from another, get a big boost from an unexpected direction, perhaps get abandoned for a time. If an idea is valuable enough to survive this tortuous path, it can become, through fortuitous combination with other ideas that have been finding their own irregular way to that moment of combination or discovery, an important part of our cultural way of viewing the world. Such combinations of ideas sometimes can change perceptions of the world.

Historically, understanding the interaction of X-rays with crystals lies at the intersection of the paths to solutions of several important puzzles: Is light wavelike or particlelike in nature? What is the architecture of the atom? What causes the sets of discrete lines in the spectra from sunlight and from electric arcs? What is the nature of crystalline solids? How can mathematics describe the regular distribution of points in space?

HISTORY

Let us look briefly at the history of ideas about X-rays and matter and begin where their paths converge. Two experiments are crucial to our discussion: the discovery of X-rays by Röntgen in 1895, and the diffraction of X-rays by von Laue and his group in 1912. These two experiments led to the primary method used today for identifying and analyzing clay minerals and all other crystalline materials: X-ray diffraction. In our discussion of these two experiments and supporting work, we identify men and women who, although they played only small roles in this drama, were important in their own right. From their names, you will recognize some of the other important, related puzzles being worked on at this time.

The Discovery of X-Rays

Let us turn, first, to Röntgen. Born in 1845 in Lennep (near Düsseldorf), Germany, he was the only child of a moderately prosperous textile merchant and his wife. They moved when Röntgen was 3 to Apeldoorn, Holland. There he attended primary and secondary school. Unable to gain admission to the best schools in Germany, he attended the Polytechnical School of Zurich, graduating in 1868 with an engineering degree, and then stayed on at the University of Zurich to do a PhD in just 1 year. At first following his major professor, August Kundt, and then on his own reputation, Röntgen moved six times in the next 18 years to arrive at the position of professor of physics and director of the new physical institute at Würzburg University in 1888. In 1894, he was elected to the highest office in the University, the post of Rector, an indication of his interest in teaching.

The scientific context of Röntgen's experiment included the work of Ampère, Faraday, Franklin, Galvani, Gilbert, Henry, Ohm, and Volta on electricity; the work of Boyle, Geissler, Hook, and Torricelli on vacuum pumps; Crookes, Goldstein, Hertz, Hittorf, and Lenard on "cathodic" rays; and Maxwell and von Helmholtz on theoretical speculations—all these preceded what Röntgen was about to do.

As for the discovery itself, details may never be known because Röntgen destroyed many documents himself and requested in his will that all notes and correspondence be destroyed unread. His executors complied (Lemmerich, 1995). We offer only a pausible reconstruction.

Having waited until the late afternoon so it would be as dark as possible in his laboratory, on Friday, November 8, 1895, Röntgen was working alone in his laboratory. He was repeating one of the experiments with cathode rays (later to be named electrons). In this experiment, he covered a Hittorf-Crookes tube, a large vacuum tube with an Al window that allowed the rays to exit the tube. He made the cover of light-tight cardboard because he knew that an electric discharge would cause the small amount of gas in the tube to glow, interfering with his observations of anticipated fluorescence caused by rays exiting through the Al window. These should cause fluorescence, but only within a few centimeters of the outside of the tube. He used a small cardboard screen painted with barium platinocyanide to detect this fluorescence-exciting radiation. As a test, in his completely darkened laboratory, he was running a high-tension discharge through the tube. It did produce fluorescence within a few centimeters of the outside of the tube. However, he also noticed a weak light on a bench about a meter away. Was the cardboard cover leaking light from the induction coil? He ran a second discharge through the tube and again the fluorescence appeared on the bench, too far from the tube to match other trials, but it didn't seem to be a light leak. Lighting a match in the dark room, he discovered that the light was coming from another of his barium platinocyanide screens. There seemed only one explanation. Something emanating from the Hittorf-Crookes tube, other than cathode rays, was producing an effect on the fluorescent screen

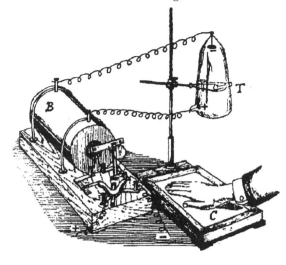

Fig. 1.1. Röntgen's experimental apparatus in 1895: B, Ruhmkorff induction coil; C, photographic plate; T, Hittorf-Crookes evacuated tube.

at a much greater distance than that in his cathode-ray experiments. Such a conclusion was contrary to general knowledge and his own experience with cathode rays. In an interview, he reported to *McClure's Magazine*:

> I was working with a Hittorf-Crookes tube which was completely enclosed in black paper. A piece of barium platinocyanide (coated) paper lay on the table next to it. I passed a current through the tube and noticed a peculiar black line running right across the paper. . . . The effect was such as could, according to our knowledge at that time, only be produced by the radiation of light. However, the possibility of light emanating from the tube was quite excluded because the paper which covered it was definitely impervious to light. . . .

The reporter: "What did you think then?" Röntgen: "I didn't think, I investigated" (Dam, 1896).

He apparently became so engrossed at the time of his discovery that he was unaware of time and surroundings. A laboratory assistant knocked, entered, looked for a piece of equipment, and left, without Röntgen noticing him. He had to be called to dinner several times. He ate in silence and immediately returned to the lab. Frau Röntgen ascribed it to a fit of bad humor. He was back in the lab early Saturday morning. There followed feverish weeks working out an explanation for these emanations, weeks in which he ate, and even slept, in the lab. Meticulous by nature, he was almost fanatically thorough in his experiments. Being shy, he preferred to work alone in his laboratory; he didn't tell colleagues, assistants, or even his wife what he was doing. One biographer (Glasser, 1958) ascribes his secrecy to self-doubts about whether his colleagues would accept his findings. He wrote to a friend, after the announcement of the discovery, that he had told his wife only that "I was doing something of which people, when they found out about it, would say: 'Röntgen seems to have gone crazy.'" He wrote further, "On the first of January I mailed the reprints, and then hell broke loose!"

His attempts to measure their wavelength, to observe reflection, and to detect interference or diffraction by passing rays through a very narrow slit were unsuccessful. He did, however, take an X-ray photograph of his wife's hand, thereby demonstrating the power of these new rays to penetrate matter. The reporter from *McClure's* asked whether the rays were light. "No," Röntgen answered. "Is it electricity?" the reporter asked. "Not in any known form." Finally, "What is it?" "I don't know," Röntgen replied. He also correctly inferred that this new kind of radiation should have all the properties of light. He demonstrated that the rays: (1) travel in straight lines; (2) are exponentially absorbed in matter with the exponent roughly proportional to the mass of the material traversed; (3) darken photographic plates; and (4) make shadows of absorbing material on barium platinocyanide-coated paper. Röntgen correctly inferred that he was observing a new kind of radiation for which he coined the term *X-rays* (Fig. 1.1). In Europe they are still referred to as *Röntgenstrahlen*, or Röntgen rays, even though Röntgen strongly objected to having them named after himself. For his discovery, he was awarded, in 1901, the first Nobel Prize in Physics, and immediately willed the 150,800 kroner award (approximately US$40,000) to the University of Würzburg.

Röntgen believed strongly that science should not be popularized (a diluting, in his view), that true science was not for the sake of gain or vanity, but should be practiced to shed light on God's truth, and that one must be absolutely certain everything one writes for a publication is absolutely verifiable—no speculation is allowed. In addition to not accepting the prize money, he also refused to patent his discovery though pressured from many directions (Nitske, 1971). And, although Bavarian Prince Regeant Luitpold offered Röntgen the Royal Order of Merit granting him the rights of nobility, including the use of the prefix *von*, he refused this sign that one belonged to the aristocracy (Glasser, 1958, p. 79). In small, informal groups he could be engaging, but to students and junior colleagues he was known as "the unapproachable." After his move to Munich, his only public address was traumatic for him. He was extremely nervous and unable to speak above a whisper. This public uneasiness was combined with an unwavering adherence to principles.

At the time of his discovery, Röntgen had already established, on the basis of about 25 to 30 published papers, an international reputation. His most important previous paper described his discovery of electrodynamic power. He found that when a dielectric, such as a glass plate, is moved between two electrically charged condenser plates, a magnetic effect is produced—a clear confirmation of part of the Faraday-Maxwell electromagnetic hypothesis. After his three papers on X-rays, he published few others. (These three are available in translation as appendices in Nitske, 1971.) His reputation, however, earned him an invitation to occupy a chair established expressly for him at the Institute for Experimental Physics, University of Munich. He accepted this position as of April 1, 1900.

An enlightening glimpse of this complex man is afforded by Max von Laue, then a junior colleague of Röntgen at the University of Munich (and really Max Laue, because he didn't use *von* until 1914, when his father was elevated to the hereditary nobility). von Laue, on a train ride to Feldafing, took the only vacant seat in a third-class compartment and found his seat mate was Röntgen. Their conversation, one of only two or three extended, private conversations he had with Röntgen, gave von Laue the impression that:

Fig. 1.2. Herr Prof. Dr. Röntgen, a caricature done soon after Röntgen's fame had spread. The original text does not explain why the head and beard are opaque. (Redrawn from an original by W.A. Wellmer.)

> the impact of his [Röntgen's] discovery was so overwhelming that he, who was 50 at the time, never recovered from it. For—as few people seem to realize—every great intellectual discovery is a heavy burden for the man who makes it. . . . The magnitude of his exploit can be recognized particularly if one considers the large number of other physicists, some of them of great renown, who experimented with the same kind of apparatus before Röntgen and did not discover X-rays. For such a breakthrough into completely unsuspected territory one must have enormous courage and in addition a self-discipline which preserves mental calm and clarity in the midst of the great joy and excitement of the first findings. Many observations had to be made and correlated to make possible [the three papers] which exhausted the subject so completely that for almost a decade nothing new could be said about it. With what ingenious care they are written! I know only few accounts of discoveries which do not contain mistakes of one kind or another. *Röntgen was correct in every detail* (von Laue *in* Ewald, 1962, p. 293).

Seldom has a scientific discovery captured the public's fancy so thoroughly and so quickly (Fig. 1.2). Contributing to this was the almost immediate recognition of the medical application of X-rays. Röntgen's preliminary communication, "Concerning a New Kind of Rays," was published on December 28, 1895, in the journal of the local Würzburg Scientific Society. As a New Year's greeting to colleagues, he sent reprints along with examples of his X-ray photos. One of the reprints, along with a photo of his wife's hand, quickly found its way to Dr. Moritz Jastrowitz, a physician, who, at a meeting of the Berlin Medical Association on January 6, 1896, told his colleagues:

The matter is obviously important to medicine. Photos of bones in the living body should at any rate be of great advantage to surgery. Fractures, dislocations, swellings, and foreign bodies will be clearly recognizable; I would draw your attention to the clear outlines of the finger joints which appear light in the photograph: it will be possible to see into the joints. It is also possible that we may be able to discern changes inside the body, in the abdominal, or other cavities, if the rays can penetrate their walls: perhaps denser tumours which are less permeable to the X-rays, as, for instance, blocked excrement in occlusion of the intestines, whereby the trouble may be located (Streller et al., 1973, p. 22).

The public was also intrigued with these strange new rays because they allowed one to see through otherwise opaque material. Rumors, perhaps facetious, of the availability of X-ray opera glasses quickly brought offers of lead underwear for sale.

Röntgen's discovery is the perfect example of serendipity. There was no basis for predicting it. The value of basic research has seldom been better illustrated. Imagine if some hospital or health ministry had made a contract with Röntgen to invent a device for inspecting broken bones. Would he have begun by building an apparatus around the Hittorf-Crookes tube and its high-voltage coils?

The Discovery of X-Ray Diffraction

The other crucial experiment took place 17 years later, in the spring of 1912, at the University of Munich. Much had been learned in the time between the two discoveries. But, as 1912 began, X-rays were still one of the enigmas of physics related to the particle-versus-wave debate. On one hand, in 1905, C.G. Barkla (1877-1944; Nobel Prize 1917) demonstrated that X-rays could be polarized and in 1907 discovered "characteristic X-rays" (see p. 35). In 1908, B. Walter and R. Pohl demonstrated the diffraction of X-rays by a tapering slit. On the side of the particle option were A. Einstein's photoelectric effect (1905) and W. H. Bragg's ionization of gases (1907). Both Einstein and Bragg considered that light might possess both properties.

At the University of Munich, we find Max Theodor Felix von Laue as the central character in this other experiment. Born October 9, 1879, near Koblenz, he was recognized as bookish and displayed an avid interest in chemical and physical experiments at an early age. Moving from city to city with his father's transfers, the family moved to Strasbourg, where he graduated from the Protestantische Gymnasium in 1898, a relatively rigorous liberal arts-type of institution. Here he developed a devotion to and respect for philosophy and the classics, was recognized for having a gift for mathematical thought and for having unreliable numerical skills. His strongest influence came from two schoolmates with similar interests and talents. In 1896, 2 years before graduation, Max and his two mates tried to duplicate Röntgen's discovery using home-made galvanometers and chemicals that burned many a hole in their clothes. After the Gymnasium, he attended the University of Göttingen, the

University of Munich, and the University of Berlin. At the latter institution, he wrote his thesis on theoretical optics under the direction of Max Planck (1858-1947; Nobel Prize 1918), which he defended in July of 1903. He returned to Göttingen for 2 years. Then, in the fall of 1905, at Planck's offer of an assistantship, he returned to the University of Berlin. In 1909, he moved to the University of Munich to work as an unpaid assistant to Arnold Sommerfeld.

In Munich, he joined an extraordinary combination of talents already assembled there, an assemblage that was crucial to what was to transpire. As von Laue (p. 292, *in* Ewald, 1962) states, there was, "an atmosphere saturated with problems concerning the specific nature of X-rays." Present, in addition to Röntgen, were Peter Debye, Peter Ewald, Walther Friedrich, Paul von Groth, Paul Knipping, and Arnold Sommerfeld. Sommerfeld (1868-1953), Director of the Institute of Theoretical Physics, was interested in a wide range of problems, including X-rays and their excitation by cathode rays. von Groth (1843-1927) was the world's leading crystallographer. Ewald (1888-1985), under the direction of Sommerfeld, was working on a doctoral thesis on theoretical aspects of the diffraction of electromagnetic radiation by resonators (or atoms) arranged in three-dimensional, ordered patterns. Friedrich (1883-1971) was an assistant to Sommerfeld and Knipping (1883-1935) and had recently completed his doctorate under Röntgen.

The trigger for the experiment was a conversation between von Laue and Ewald that the two men remember somewhat differently. Ewald recalls it as taking place in late January. He was finishing his thesis, "Optical properties of an anisotropic arrangement of resonators," and came to von Laue to discuss a surprising conclusion. They met and walked to dinner through the Englischen Garten, a large park in the center of Munich close to the University. As Ewald was explaining the results, it became clear to him that there was a point von Laue didn't understand: *von Laue had not realized that crystals were thought to have internal regularity!* He had neglected the study of mineralogy in his university studies; he worked with Sommerfeld, the person directing Ewald's thesis; and he had been an associate of von Groth's for 3 years, all reasons he should have understood. The conversation stuck there. von Laue wanted to know, "What is the distance between the resonators?" which could be atoms in a crystal. Ewald could get no further discussion on the original topic. von Laue was preoccupied and asked, "What would happen if you assumed very much shorter waves to travel in the crystal?" Probably because he had shortly before worked on a theory of diffraction from line and cross-gratings, he was considering the idea that, if X-rays really are electromagnetic radiation like visible light, but of much shorter wavelength, crystals might serve as three-dimensional diffraction gratings. Ewald showed him a formula for the general case from his thesis saying he hadn't tried solving it for very short waves. Ewald left later that evening thinking he would have to try again if he wanted von Laue's comments on his conclusion (Ewald, 1962, pp. 40-41).

von Laue remembers the conversation as taking place in February when Ewald came to ask for help with a mathematical description of the action of light waves in a lattice of polarizable atoms. He remembers saying, "If the atoms really formed a lattice this should produce interference phenomena similar to the light interferences in optical gratings" (von Laue, p. 293, *in* Ewald, 1962). He talked about it to anyone who would listen, especially at the meetings at the Café Lutz in the Hofgarten after lunch every day with members of Röntgen's and Sommerfeld's Institutes. In the informal exchange of views, diagrams and calculations were made with pencil on the white, smooth marble tops of the Café's tables—much to the dislike of the waitresses who had to clean them. "For the younger members of the group, it was most exciting to watch research in the making, and to take sides in the first tentative formulation of experiments and theory" (Ewald, 1962, pp. 33-34). Röntgen never came to these informal meetings. Sommerfeld, and perhaps Röntgen, expressed doubts about the worth of such an experiment, but the younger physicists caught von Laue's enthusiasm, and, after some diplomatic maneuvering, Friedrich and Knipping started the experiment on April 21, 1912, in Sommerfeld's laboratory (Streller et al., 1973).

Their second try yielded an X-ray photo of a copper sulfate crystal that "plunged [von Laue] into deep thought," so deep that, while walking home, "just in front of the house at Siegfriedstrasse 10, the idea for a mathematical explanation of the phenomenon came to me" (von Laue, p. 294, *in* Ewald, 1962; for the mathematics, see Klug and Alexander, 1974, pp. 126ff). His sense of the world around him was so illuminated by the moment that the street address of a stranger's house was as permanently fixed in his mind as was the theory of diffraction effects.

In this one experiment, perhaps as dramatic as any in the history of science, von Laue, Friedrich, and Knipping established three modern concepts: (1) that atomic particles within crystals are arranged in orderly, three-dimensional, repeating patterns; (2) that these regular arrangements have spacings of approximately the same dimensions as the wavelength of X-rays; and, therefore, because diffraction does take place, that (3) X-rays are wavelike in nature. Recognition was wide and quick. Perhaps the most dear comment was on a postcard from Albert Einstein, "Dear Mr. Laue, I heartily congratulate you for your wonderful success. Your experiment is among the most beautiful that physics has experienced" (Hermann, 1979). For his part in this work, von Laue was awarded the 1914 Nobel Prize in Physics.[1]

Their article describing the experiment was published in June 1912 (Friedrich et al., 1912). Probably because he knew Sir W. H. Bragg (1862-1942) was interested in X-rays, von Laue sent him a copy, which he received while on summer vacation in Yorkshire, England. His son, W. L. Bragg

---

[1] A version of the discovery of X-ray diffraction that makes the point that participants and associates are not accurate reporters is made by Forman (1969). Forman's analysis is followed by a rejoinder from one of the participants, Ewald (1969).

(1890-1971), wrote of the infectious excitement the article caused (Frisch et al., 1959, p. 147). The response was almost immediate. On November 11, 1912, the younger Bragg read a paper before the Cambridge Philosophical Society (Bragg, 1913a) corroborating and extending von Laue's work and offering what is now called Bragg's law (see Chapter 3, p. 69). In September 1913, he published the structures of NaCl, KCl, KBr, and KI, the first crystal structures ever to be determined (Bragg, 1913b). The elder Bragg, in 1912, because he was a proponent of the view that X-rays were corpuscular rather than wavelike, set up an experiment with an ionization spectrometer and demonstrated that X-rays do indeed ionize, the property of a particle, but they also are reflected from crystal planes, the property of a wave (1912). In addition, in the same experiment, he discovered the first X-ray spectrum, the L series from platinum (see Chapter 2, p. 37). Between the father's invention of the X-ray spectrometer and the son's understanding of crystal structures, *they invented X-ray crystallography* (Bragg and Bragg, 1913). The Braggs, who shared the 1915 Nobel Prize for this work, were part of a group similar to that of von Laue's in that it was made up of extraordinarily talented people pursuing closely related goals, excited about what they were doing, and eager to discuss their work. In the group were J. J. Thomson, C. T. R. Wilson (1869-1959), Ernest Rutherford, Niels Bohr (1885-1962; Nobel Prize 1922), H.G.J. Moseley, and Charles G. Darwin (1887-1962), grandson of the author of *The Origin of Species.*

This was a time of turmoil in science. Kuhn (1970) described it as a time of scientific revolution, a time between periods of normal science because the paradigms governing the way in which matter was perceived were changing. von Laue's experiment was probably the most important observational evidence contributing to this crucial shift to the way in which solid, crystalline materials are currently perceived.

After the initial investigations of von Laue and the Braggs, the first determinations of crystal structures were made, ushering in a new age in the science of materials, e.g. (an abbreviation for the Latin phrase *exempli gratia* = for example), W. L. Bragg, 1913, alkali halides; Ewald, 1914, pyrite; S. Nishikawa, 1914, spinels). Also in 1913, S. Nishikawa[2] and S. Ono published a study of diffraction photos of fibrous substances including asbestos (Nitta, 1962).

Half a world away in Japan, T. Terada in 1913, immediately following von Laue's discovery, and quite independently of W. L. Bragg, found essentially the same law of X-ray reflection that Bragg had. The approximately simultaneous discovery of X-ray reflection also was made in Russia in 1913 by G. Wulff (Shubnikov, 1962). (See Bijvoet et al., 1969, for reproductions of

---

[2]R.W.G. Wyckoff, a leading crystallographer in Europe and North America in the first half of this century, began his thesis under Nishikawa when Nishikawa was a visiting professor at Cornell in 1917, an east to west link in the science of crystals and X-ray diffraction.

most of the important early papers from Europe and Asia. There are other examples of independent, approximately simultaneous discoveries, e.g., between 1885 and 1895, the recognition of 230 space groups independently by Fedorov in Russia, Schoenflies in Germany, and Barlow in England.) We will continue to call it Bragg's law.

The crystal structures published after von Laue's discovery of diffraction strongly influenced the work of V. M. Goldschmidt (1888 1947), who, with fellow workers, used the W. H. Bragg-type X-ray spectrometer and Wasastjerna's (1923) determination of the radii of F and O to work out the principles of crystal chemistry much as we know them today. Goldschmidt published a summary of crystal chemistry in *Geochemische Verteilungsgesetze der Elemente VII* (1926) (Mason, 1992).

Nor were organic compounds neglected in the new applications of X-ray diffraction. Kathleen Lonsdale (1903-1971) published the first complete structure determination of an aromatic compound (Lonsdale, 1929). Linus Pauling (1901-1994; Nobel Prize for Chemistry 1954, Nobel Prize for Peace 1962) published the structure of micas and of chlorites, work that led to solving the structures of other macroscopic layer silicates (Pauling, 1930). Pauling then shifted his attention to organic compounds. Dorothy Hodgkins (b. 1910; Nobel Prize 1964) and her colleagues solved the structures of vitamin B12, insulin, and penicillin. James Watson (b. 1928) and Francis Crick (b. 1916) (with M. H. F. Wilkins, b. 1916) solved the structure of DNA, for which they shared the 1962 Nobel Prize in Physiology or Medicine. Max Perutz (b. 1914) and John Kendrew (b. 1917) shared the Nobel Prize, also in 1962, for establishing the structure of hemoglobin and for helping develop the techniques that made this discovery possible. Remarkably, all these people except Pauling and Wilkins were associated, directly or indirectly, with W. H. and W. L. Bragg at the Royal Institution and the Cavendish Physical Laboratory at Cambridge. The discipline of clay mineralogy also benefited directly from this sphere of influence. George Brindley (1905-1983) and S. W. Bailey (1919-1994) worked with W. L. Bragg, and will have been the last of our discipline to have done so.

History of Clay Mineralogy

While Röntgen and von Laue were making their discoveries, others were studying the chemical makeup of clay material. Several clay minerals had been assigned definite compositions, and most were identified as some sort of aluminum silicate. Millot (1970) briefly reviewed this work and cited the Europeans Fersmann, Le Chatelier, Mallard, and Vogt and the American Merrill as contributors. He ascribed to Thiebaut the first attempt (in 1925) to find a relation between the composition of an argillaceous rock and its origin and depositional environment. Grim (1902-1989) (1988) discussed the history of clay mineralogy and noted that Henrich Ries, professor of economic geology at Cornell University, publishing as early as 1906, was perhaps the first American to specialize in the study of clays. With the publication of the work of Ross and Shannon (1925), a series of important papers on clay minerals was initiated, e.g., Ross and Kerr (1930, 1931).

During the 1920s and early 1930s, not much distinction was made among fine-grained materials. Soil scientists, chemists, and geologists, all were pursuing the same questions (Bray et al., 1935): Were such materials crystalline or amorphous? Were they a mixture of oxides of Al and Si, or of specific phases or minerals? or, Was kaolinite the only essential mineral constituent or was it one of a series of minerals? Hadding (1923) in Sweden reported the results of "eine röntgenographische Methode" for white, dark gray refractory, and red clays. Rinne (1924) in Germany considered the question of whether natural materials in grain sizes smaller than what could be seen with a microscope were distinct crystalline phases or amorphous. Grim (1988) cites Hadding's and Rinne's work as the earliest use of X-ray diffraction in Europe for the study of clay-sized minerals.

X-ray diffraction settled the controversy about whether the microscopic components of soils and clays are amorphous materials of irregular composition or a limited number of distinct crystalline minerals. Although the crystalline nature of the mineral constituents in soils had been inferred from use of the petrographic microscope during the 1920s, it was not confirmed by X-ray diffraction until independent studies by Hendricks and Fry (1930) and Kelley et al. (1931) (Grim, 1988). Thus, the modern study of soil materials and the discipline of clay mineralogy began. The concept that clay-sized material was composed of specific minerals was not entirely new because Le Chatelier had stated approximately the same thing in 1887, but had no evidence to support his assertions.

J. W. Gruner (1890-1981) (1932), following Pauling's original ideas, worked out the structure for kaolinite. Hofmann et al. (1933) suggested a model for montmorillonite that featured an expanding structure. Grim et al. (1937) introduced *illite* as a general term for micalike clay minerals. (As with most new terms, illite was not accepted by everyone, and hard feelings developed, remnants of which can be detected even today.) Hendricks and Teller (1942, that's *the* Edward Teller) introduced the theoretical basis for X-ray diffraction from a stack of clay mineral layers of different kinds (mixed-layered clay minerals, p. 169).

Major roles in the growth of clay mineralogy can be viewed as having been played out in: (1) the study of soils in the agricultural industries and colleges of agriculture; (2) the study of clay as an industrial material and as raw material feedstock for the chemical and pharmaceutical industries, e.g., the study of bentonite and the development of a bentonite industry and the use of kaolinite for the coating of paper and the manufacture of porcelain; (3) the study of clay minerals as interesting mineral substances, as important parts of the earth's crust, and as analogs in material science; and (4) the study of clay as something that we must live on, build on, and make use of (from engineering properties to adobe bricks).

A major portion of soil science is concerned with the alteration of primary minerals of the earth's crust by physical and chemical weathering. The primary weathering products are clay minerals. Because soil environments change continually, there is always a stimulus to alter mineral and organic materials in the soil zone, a zone that serves to connect the inorganic part of the world with the organic. In Chapter 4, you'll learn about cation exchange capacity. This property of clay minerals allows them to hold and exchange cations in the spaces between their layers. They prefer to have K^+ rather than Na^+ between their layers. This simple fact explains why the oceans have NaCl concentrated in them rather than KCl, and why K^+ is recycled in the sediments that are buried, uplifted, and again exposed to the soil-forming processes that make K^+ available to all life forms.

As an example of a clay mineral that serves as a feedstock, let's consider bentonite. There isn't universal agreement on its definition, but, in the most commonly used sense, this material, often with peculiar, soapy properties, was named bentonite by Knight in 1898 from its occurrence in the Fort Benton unit of Cretaceous age of the northern Great Plains, United States. As early as 1917, it had been identified as a product of alteration of volcanic ash to the clay mineral smectite[3]. Two of its first uses were for making drilling mud (a dense suspension of clay minerals in a fluid circulated through the drill to cool the bit and to float rock chips to the surface) and as a bonding agent for molding sands used in foundries. The search for additional uses spawned several major companies. Filtrol Corporation marketed acid-activated bentonite for use as a decoloring agent for oils and as a catalyst for "cracking" petroleum. Government-sponsored research early in the Second World War resulted in a bentonite-organic compound that quickly hardened loose soil material and was used in building temporary landing strips. Jordan (1949) developed a group of organo-montmorillonite compounds called Bentones. These quickly found use in paints, greases, and other products for which rheological (how they flow and deform) control was important. The properties of the primary mineral in bentonite, smectite, are still the subject of research in pursuit of ways to improve its behavior as a catalytic agent (e.g., Kleinfeld and Ferguson, 1994) and for many other uses.

Intensive research also was focused on all other clay minerals for ways in which their properties could be used or modified for industrial applications. Other types of catalysts were made starting with kaolinite treated with caustic soda. This formed silica-alumina gels, which were used for hydrocarbon cracking until the mid-1960s when they were replaced by more efficient zeolites as catalysts.

---

[3]The older literature refers to this mineral as montmorillonite. Current usage places montmorillonite in the family of expandable clay minerals called smectite.

After the Second World War, workers from such disciplines as agronomy, ceramics, chemistry, engineering, geology, mineralogy, and physics began to meet, drawn together by their interest in clay minerals. By January 1947, the Clay Minerals Group of the Mineralogical Society of Great Britain was meeting regularly and had established a publication. The following year a European group meeting in London formed a committee to consider forming a clay minerals group. This led to the first meeting of the Association Internationale pour l'Étude des Argiles (AIPEA) in 1952 in Algiers. (This group's first title was Comité International pour l'Etude des Argile, or CIPEA.) The American Institute of Mining and Metallurgical Engineers met in 1951 at Washington University in St. Louis. Included on the program was a "Symposium on Problems of Clays and Laterite Genesis." It had been suggested by John J. Collins, and was implemented by the general organizer of the meeting, A. F. Fredrickson. That same year, on the basis of the enthusiastic reception of this symposium, the U.S. National Research Council was asked to establish an interdivisional committee on clay minerals, which it did. The National Research Council sponsored annual meetings on clays and clay minerals, the first of which was held the following year at the University of California, Berkeley (Grim, 1988), and an annual meeting, with an annual volume of proceedings, was held every year until 1967. The number of those involved grew large enough that sponsorship by the NRC was no longer necessary, and so the Clay Mineral Society became an independent group with its own journal in 1968.

In 1953, Ralph Early Grim's *Clay Mineralogy* was published, followed in 1962 by his *Applied Clay Mineralogy*. These books, especially the first, seem to be the vehicles upon which the discipline of clay mineralogy in North America rode to maturity. Millot's book, *Géologie des Argiles*, published in 1964, performed the same service for those in Europe. Millot, however, gave credit to J. de Lapparent, who had preceded him at the University of Strasbourg, C. W. Correns, of the University of Göttingen, and Grim, of the Illinois State Geological Survey and the University of Illinois, as the pioneers in the discipline of clay mineralogy. It is unclear under what stimulus the discipline developed in Asia. It is clear that it did, however.

During the 1950s and 1960s, descriptive information about the geographic and stratigraphic distribution of clay minerals accumulated. The question of whether clay minerals reflect important diagenetic alterations of the original detrital material or only the materials and climate of their source area was vigorously debated. For a time, it had been generally doubted that any clay minerals formed diagenetically. This debate was more or less resolved on a middle ground by John Hower and his co-workers, who were the first to present convincingly the diagenetic reactions by which smectite changes to illite and the consequences of these reactions.

THE IMPORTANCE OF CLAY MINERALOGY

Practical applications of an understanding of clay minerals abound in agriculture, engineering, and industry. In agriculture, knowledge of clay minerals is necessary to understand soils. The adsorption of water, inorganic cations and anions, and organic molecules occurs primarily on the surface of colloid-sized particles, where they are frequently catalysts for important reactions (see Box 1.1). These particles are dominated by the clay minerals, humic substances (thoroughly decomposed organic stuff), and hydrous oxides of iron, aluminum, and manganese. An understanding of the biological and physical processes (weathering) that form soils, and the nutrients exchanged or extracted from them, all begins with an understanding of the colloid-sized particle constituents of soil. For example, in many soils, micaceous or illitic materials are the principal source of potassium for plants. When illitic materials weather to vermiculite- or smectitelike minerals, they significantly increase the cation exchange capacity and affect the relative selectivity of soils for various exchangeable cations. The tilth, or workability, of soil is often a function of the kind and amount of clay mineral(s) present.

Kaolinite is the base for a number of industries: the paper industry, the sanitary ware (toilets) industry, and others (see Box 5.1, p. 145).

Box 1.1. Clay Minerals as Catalysts

In nature, clay minerals catalytically mediate reactions of organic compounds in soils (Soma and Soma, 1989) and sediments. Johns and McKallip (1989) argued that the surface charge of illites is an important catalytic agent for the formation of petroleum; Ungerer (1990) argued that clay minerals as catalysts are unnecessary for the formation of petroleum. In industry, catalysis is involved in approximately 90% of all manufactured items at some stage of production, and many catalytic materials are either clay minerals or catalysts made from clay minerals. To improve their natural catalytic activity, pillars can be emplaced in the interlayer spaces of smectites. Such modified clay minerals, known as pillared interlayered clays, combine tunable acidity, shape selectivity, regular porosity, and relatively high thermal stability (Fig. 1.3). The environmentally offensive catalysts of the past, such as sulfuric acid, are being replaced by these solid acid catalysts that are natural or treated clay minerals or natural or synthetic zeolites. Because they carry their acidic properties within crystal structural pores, they overcome a number of environmental problems. Particularly on extremely dry surfaces of aluminosilicates (clay minerals, zeolites, X-ray amorphous gels), the acidity can exceed that of concentrated sulfuric acid because of the H^+ attached to the surfaces (McBride, 1989). Figure 1.3 illustrates the steps for making a pillared clay with "superacidicity." Once the aluminum hydroxy cluster or Keggin cation has replaced the regular interlayer cation, it is heated, or calcined, changing it into aluminum oxide and releasing H^+, which attaches to the surface. Such modified natural materials can boost the yield of reactions they catalyze, sometimes manyfold (Thomas, 1992;

personal communication Tom Pinnavaia, Michigan State University, East Lansing, Michigan, 1994; Schoonheydt et al., 1994).

The shape and size of the pores are critical properties of solid catalysts. For example, a synthetic zeolite named ZSM-5 allows the rod-shaped para-xylene molecule to pass while retaining the boomerang-shaped ortho-xylene molecule. Some clay minerals can serve as catalysts with no pretreatment, others with only replacement of interlayer cations by hydrogen ions. Current research on clay minerals and catalysts focuses on forming custom-made spacings between clay mineral layers by inserting props of different sizes and composition. Some of the molecules inserted to hold the layers apart are themselves catalysts. Using some of the larger channels of these modified materials, conducting filaments have been precipitated in place by chemical reactions. This is a step toward the design of nanometer-scale electronic devices (Wu and Bein, 1994). All these aspects of clay minerals and zeolites are fertile ground for future research.

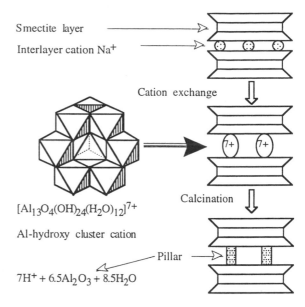

Smectite layer

Interlayer cation Na^+

Cation exchange

$[Al_{13}O_4(OH)_{24}(H_2O)_{12}]^{7+}$

Al-hydroxy cluster cation

$7H^+ + 6.5Al_2O_3 + 8.5H_2O$

Calcination

Pillar

Fig. 1.3. Schematic diagram for the preparation of an alumina pillared clay mineral (courtesy of Tom Pinnavaia).

In engineering, almost all foundation problems and slope stability studies involve a study of clay minerals. (As an entry to the subject of the use of clay mineralogy in problems of slope stability, see Haneberg and Anderson, 1995. What comes most immediately to mind is that engineering geologists and clay mineralogist must develop much closer communication.) Many large cities are built on clay-sized, unconsolidated materials. On and in these materials the engineers must build their roads, foundations, tunnels, and other structures. Engineers and clay mineralogists have not always agreed on what sort of data gathering is essential in planning for construction. Grim (1988) recalled that engineers thought they could build a subway beneath the Chicago Loop in clay that was known to lose strength when disturbed. By cutting the hole exactly the size of the liner and by lining the tunnel as it was excavated, they assumed the surrounding clay would not be disturbed. As the subway was completed, the streets above the subways gradually sank as much as 2 ft or more. One could walk along State Street in front of the Palmer House Hotel and look out over the tops of the cars. At tremendous cost, State Street was filled to bring it to its previous level. The water that had bound the clay mineral particles together liquefied and drained away, causing the decrease in volume. The engineers had tested this material exhaustively. However, without determining the kinds of clay minerals present, they could not predict how the material would perform. Perhaps more than anything else, this construction project convinced engineers that information on clay mineralogy is vital to the success of construction projects.

Another kind of failure is caused by sensitive, or quick clay in which the word *quick* is used in the same sense as in "quicksand." Although most of the rock may be made of clay-sized particles, most of them are quartz, feldspar, and rock flour, a consequence of glacial grinding. Those who work with the engineering properties of soils and clay-sized materials define sensitivity as the ratio of the shear strength (resistance to shear) of undisturbed material to that of the same material after it has been kneaded or squeezed, or remolded, as the engineers would say. Quick clays lose almost all their capacity to resist shear when they are disturbed. All known quick clays are postglacial in age, are generally found in regions associated with glaciation, and develop when clay-sized rock flour has been deposited in saline or brackish waters and then uplifted above sea level (Torrance, 1983). The clay-sized particles, most of which are tablet shaped, are in a flocculated condition, i.e., the particles are arranged in clumps—a "card house" structure with edges of particles attracted to faces in such a way as to isolate pore space. (i.e. is abbreviation for the Latin phrase *id est* = that is.) Therefore, the clay has a very high porosity, and very low permeability. The clay can then be disrupted by a shock, causing a collapse of the "card house" that, in turn, causes a change from solid to liquid—a property called *thixotropy*. The Bootlegger Cove Clay underlying the Anchorage, Alaska, region is an example. Updike et al. (1988) used the Bootlegger Cove Formation area as an example of material responsible for earthquake-induced landslides.

In other cases, slope stability seems more directly related to the type of clay minerals present. Of samples from 127 localities in the Coast Ranges of California, Borchardt (1977) found that those containing expandable clay minerals were associated with landsliding and those containing kaolinite without any expandable clay minerals tended to be stable. In some cases, something as simple as adding potassium to a weathered shale will stabilize it against landsliding (Fisher et al., 1968). (You should understand why K^+ should do this after you have thought about the material in Chapter 4.) In still other cases, problems are caused by clay minerals that swell. Some swell because water enters and exits between their layers with ambient fluctuation of humidity. Others are overconsolidated clays and mudrocks, i.e., when some clays have been deeply buried and then unloaded (e.g., after glaciation, if there have not been sufficient mineralogical changes to cause bonding), they will expand vertically. This expansion damages buildings and other structures

(Cripps and Taylor, 1987). Hays (1981) estimated that expandable clay minerals and landslides cause an annual loss of $4 billion in the United States. This number is best put into perspective by comparing it with the estimated annual loss of $3.6 billion for earthquakes, tsunamis, and floods.

In what could be viewed as an engineering, an industrial, or a societal problem, by the year 2015, commercial, medical, industrial, academic, and military procedures may have generated as much as 12 km^3 of low- and high-level radioactive wastes. Lee and Tank (1985) and Bath (1993) pointed out that clay minerals and clay-sized zeolites, because of their high sorptivity, long-term structural stability, and low permeability, are attractive candidates for isolating and retaining most natural and manmade wastes. Clay minerals are an important ingredient in clay liners for all kinds of disposal sites (Krapac et al., 1991; Benson et al., 1994). Research continues on the retention or release of organic compounds, heavy metals, and radioactive isotopes under various conditions and for different clay minerals, e.g., Pusch (1992), Yong et al. (1993), and Sawhney (1995).

Problems associated with the oil industry illustrate the value of a practical knowledge of clay minerals (Eslinger and Pevear, 1988). Understanding the diagenetic changes in clay minerals is essential to reconstructing the temperature history of potential petroleum basins. Exploration geologists know that clay minerals can be used as stratigraphic markers and environmental indicators. Petroleum geologists and engineers must understand clay minerals to develop, maintain, and manage hydrocarbon reservoirs. Clay minerals may be as little as 1% of a reservoir, but, because they coat the framework grains, they are the material in contact with any fluid in the pores of the rock. To avoid damaging a reservoir's porosity or permeability or both, we must identify the clay minerals in the reservoir rock and note their mode of occurrence, or habit. For example, clay minerals in loose aggregates can break free if too much fluid pressure is applied. A "brush pile" effect at the throats of pore spaces is then caused that reduces permeability. In other cases, smectites will swell, reducing permeability and porosity if fresh water is introduced into the reservoir, e.g., in a waterflooding operation. In yet another example, some chlorites will dissolve if the reservoir is treated with acid—a common procedure—and a precipitate of iron hydroxides and amorphous silica-aluminous gel will form that clogs pore throats. Reservoirs damaged by the introduction of improper fluids can sometimes be salvaged if the clay minerals present are understood.

It is very likely that clay minerals newly formed in the oceans, and those weathered on the land and delivered to the oceans where they are reconstituted, play an important part in the CO_2 balance among its various sinks and in the general geochemical balance for the oceans (Mackenzie and Kump, 1995; Michalopoulos and Aller, 1995).

Another topic appropriate to this section is the role of clay minerals in the origin of life. Cairns-Smith (1985), Hartman (1975, 1982), and Cairns-Smith and Hartman (1986) are major contributors to this topic. Because clay minerals grow or replicate and, in the process of replicating, new growth sometimes introduces defects, Cairns-Smith suggested that clay minerals can be treated as crystal genes. One could say there is evolution in this inorganic system. Hartman, a biochemist, has speculated about the ways in which large, complex organic molecules could be catalyzed while being held on the surfaces of clay minerals in configurations determined by patterns of charge and defects of the clay mineral substrate. As the patterns and defects change slightly from generation to generation, the results of synthesis of organic molecules would change. (Layer charges are discussed in Chapter 4. They are expressed on the surfaces of clay minerals and are an important component of what is referred to above as patterns.) This work raises intriguing questions and suggests new ways of looking at clay mineral-organic systems.

THE LITERATURE OF CLAY MINERALOGY

In any discipline, one of the first things a neophyte (and we hope you won't be offended if we call you a neophyte) must do is become acquainted with the literature of that discipline. We offer a brief list of journals and books that we think are essential for those working in clay mineralogy.

Some journals of chemistry and engineering and many journals of geology and soils have articles that include some aspects of clay mineralogy. Four journals are devoted exclusively to clay minerals and clay-sized material: *Applied Clay Mineralogy*, published by Elsevier Science Publishers, Journal Department, P.O. Box 211, 1000 AE Amsterdam, The Netherlands; *Clay Minerals*, published by the Mineralogical Society, 41 Queen's Gate, London, SW7 5HR; *Clay Science*, published by the Clay Science Society of Japan, Japan Publications Trading Co., P.O. Box 5030, Tokyo International, Tokyo, Japan; and *Clays and Clay Minerals*, published by the Clay Minerals Society, P.O. Box 4416, Boulder, Colorado, 80306. And a fifth one, started in 1993, is *Geologica Carpathica Clays*, published by the Slovak Academic Press Ltd., Posta 15, P.O. Box 57, Námestie Slobody 6, 810 05 Bratislava, Slovak Republic.

In the references that follow, we comment on the following books to guide you: Bailey (1984, 1988), Bish and Post (1989), Brindley and Brown (1980, reprinted in 1984; in the time since the publication of our first edition, this book remains as valuable as ever), Brown (1961), Dixon and Weed (1989), Drits and Tchoubar (1990), Grim (1962, 1968), Grimshaw (1971), Jackson (1969), Newman (1987), and Weaver (1989). (We use the books, Grim and Grimshaw, to remind you, and ourselves, that we ignore the past at our peril. In science, we too often discount older works and ascribe too much value to the newest data. Certainly judgment is required to sort out which facts have been replaced because of better understanding, but much of value often lies unused simply because it is in older references.) The CMS Workshop Lectures of the

Clay Mineral Society and the Reviews in Mineralogy series of the Mineralogical Society of America are quite inexpensive and loaded with first-rate information written by the leading figures on the particular topic in question. There also are books that address more specialized topics and that are, therefore, a bit more esoteric, but of great value if you need to pursue the topic they address. They are Manning et al. (1993), Odin (1988), Chamley (1989), and Velde (1985).

SUMMARY

In spite of a remarkable pace of change since the first edition of this book, there is still much we do not know about clay minerals. We are closer to general agreement on fundamental particles; we are in the midst of an intense debate about whether high-resolution transmission electron microscopy (HRTEM) or X-ray diffraction (XRD) can best represent the degree of order in clay-sized minerals; we continue to learn more about the smectite-to-illite transformation, and that perhaps there are several paths for this transformation. In one place or another, in one set of circumstances or another, it seems as if any clay mineral can be transformed into any other. We continue to struggle to decipher what these transformations are trying to tell us; to find ways to define the mineral illite; to draw boundaries among muscovite, illite, vermiculite, and smectite, or to decide if there should be boundaries. And still, many misconceptions exist. Do not despair that there will be no problems left for you to work on.

We have offered a few details from the history of the discovery of X-rays, their diffraction by crystals, and the major characters involved—reminders that every major concept and most inventions have a similar history involving interesting and chance combinations of personalities and ideas, sometimes with unexpected results. In the following pages, little historical context is apparent, but it does exist. If pursued, aspects and personalities as dramatic, as engaging, and as interesting as those presented above, will add to your understanding and appreciation of the subject matter, the people involved, and the way in which scientific activity fits into the general culture of its time.[4] (For more information on the history of crystallography and X-ray diffraction, see Burke, 1966; Bragg, 1975; Caroe, 1978; Ewald, 1962; Glasser, 1958; Nitske, 1971; and Streller et al., 1973. In addition, note that we have included citations of several of the original papers. A detailed, definitive history of clay mineralogy has yet to be written.)

The arena you are about to enter is one of continuing inquiry. When you have finished this book, you should be within sight, if not within touching distance, of the "cutting edge" of clay mineralogy. Many intriguing puzzles remain to be solved, and they will be solved, especially if you, in pursuit of

---

[4]Several VHS videotapes of people involved in the founding of the Clay Minerals Society are available from the Clay Mineral Society office in Boulder, CO.

them, can combine the practices of the careful experimental techniques of Röntgen and W. H. Bragg with the "what if" theoretical approach of von Laue and W. L. Bragg. And, it is our impression, your productivity will be greater, your insights clearer, if you forge your ideas on the eyes and ears of interested peers, and you reciprocally, offer your eyes and ears.

REFERENCES

Bailey, S. W., editor (1984) *Micas*: Vol. **13** in Reviews in Mineralogy: Mineralogical Society of America, Washington, 584 pp. Concentrates on macroscopic micas, but has excellent chapters on illite (Ch. 12) and glauconite and celadonite (Ch. 13).

Bailey, S. W., editor (1988) *Hydrous Phyllosilicates (exclusive of micas)*: Vol. **19** in Reviews in Mineralogy: Mineralogical Society of America, Washington, 725 p. The latest word on chlorites, kaolinite, serpentines, smectite, talc, vermiculite, and several other minerals and topics from several points of view.

Bath, A. H. (1993) Clays as chemical and hydraulic barriers in waste disposal: evidence from pore waters: in Manning, D.A.C., and Hughes, C.R., editors, *Geochemistry of Clay-Pore Fluid Interactions*: Chapman and Hall, London, 316-30.

Benson, C. H., Zhai, H., and Wang, X. (1994) Estimating hydraulic conductivity of compacted clay liners: *Jour. Geotechnical Engin.* **120**, 366-87.

Bethke, C. M, Harrison, W. J., Upson, C., and Altaner, S. P. (1988) Supercomputer analysis of sedimentary basins: *Science* **239**, 261-67.

Bijvoet, J. M., Burgers, W. G., and Hägg, G., editors (1969) *Early Papers on Diffraction of X-Rays by Crystals, Volume 1*: International Union of Crystallography, A. Oosthoek's Uitgeversmaatschappij N.V., Utrecht, 372 pp.

Bish, D. L., and Post, J. E., editors (1989) *Modern Powder Diffraction:* Vol. **20** in Reviews in Mineralogy: Mineralogical Society of America, Washington, D.C., 369 pp.

Borchardt, G. A. (1977) Clay mineralogy and slope stability: *Special Report 133*, Calif. Division of Mines and Geology, 15 p.

Bragg, W. H. (1912) X-Rays and Crystals: *Nature (London)* **90**, 360-61.

Bragg, W. H., and Bragg, W. L. (1913) The Reflection of X-Rays by Crystals: *Proceedings of the Royal Society of London* A**88**, 428-38.

Bragg, W. L. (1913a) The Diffraction of Short Electromagnetic Waves by a Crystal: *Proceeding of the Cambridge Philosophical Society* **17**, 43-57.

Bragg, W. L. (1913b) The structure of some crystals as indicated by their diffraction of X-rays: *Proceedings of the Royal Society of London* A**89**, 248-77.

Bragg, W. L. (1975) *The Development of X-Ray Analysis*: G. Bell & Sons, Ltd., London, 270 pp. The younger Bragg has written several books that are at a beginning level.

Bray, R. H., Grim, R. E., and Kerr, P. F. (1935) Application of clay mineral technique to Illinois clay and shale: *Bull. Geol. Soc. Amer.* **46**, 1909-26.

Brindley, G. W., and Brown, G., editors (1980) *Crystal Structures of Clay Minerals and Their X-ray Identification*: Monograph No. 5, Mineralogical Society, London, 495 pp. The most important book on clay minerals. If you are going to work with clay minerals, you must have access to a copy.

Brown, G., editor (1961) *The X-ray Identification and Crystal Structures of Clay Minerals*: Mineralogical Society, London, 544 pp. Only partly superseded by the Brindley and Brown volume. This one treats individual clay minerals as topics. You will need to be familiar with both.

Burke, J. G. (1966) *Origins of the Science of Crystals*: University of California Press, Berkeley, 198 pp.

Cairns-Smith, A. G. (1985) The first organisms: *Scientific Amer.* **252**, no. 6, 90-100.

Cairns-Smith, A. G., and Hartman, H., editors (1986) *Clay Minerals and the Origin of Life*: Cambridge University Press, Cambridge, 193 pp.

Caroe, G. M. (1978) *William Henry Bragg, 1862-1942*: Cambridge University Press, Cambridge, 212 pp.

Chamley, H. (1989) *Clay Sedimentology*: Springer-Verlag, 623 pp. Perhaps a better title would be Clay Sedimentology of the Ocean Basins.

Cripps, J. C., and Taylor, R. K. (1987) Engineering characteristics of British over-consolidated clays and mudrocks, II. Mesozoic deposits: *Engineering Geology* 23, 213-53.

Dam, H. J. W. (1896) The new marvel in photography: *McClure's Magazine* no. 5 (April), 6, 403-15.

Dixon, J. B., and Weed, S. B., editors (1989) *Minerals in the Soil Environment*, Second Edition: Soil Science Society of America, Madison, Wisconsin, 1244 pp. If you are interested in soils, with Moore and Reynolds under your belt, this is the reference you will want to have close at hand.

Drits, V. A., and Tchoubar, C. (1990) *X-Ray Diffraction by Disordered Lamellar Structures*: Springer-Verlag, Berlin, 371 pp. The cutting edge of the theoretical side of diffraction from layered structures of minerals and synthetic materials.

Eslinger, E., and Pevear, D. R. (1988) *Clay Minerals for Petroleum Geologists and Engineers*: SEPM Short Course Notes No. 22, Society of Economic Paleontologists and Mineralogists, Tulsa, Oklahoma.

Ewald, P. P., editor (1962) *Fifty Years of X-ray Diffraction*: Oosthoek's Uitgeversmaatschappij, Utrecht, 733 pp. A review of the early work and the principles and applications of X-ray diffraction plus the personal reminiscences of 35 workers associated with the development of crystallography and X-ray diffraction.

Ewald, P. P. (1969) The myth of myths; Comments on P. Forman's paper "The discovery of the diffraction of X-rays in crystals": *Archive for History of Exact Sciences*, 6, 72-81.

Fisher, S. P., Fanaff, A. S., and Picking, L. W. (1968) Landslides of southeastern Ohio: *The Ohio Jour. of Science* 68, 71-80.

Forman, P. (1969) The discovery of the diffraction of X-rays by crystals; A critique of the myths: *Archive for History of Exact Sciences*, 6, 38-71.

Friedrich, W., Knipping, P., and Laue, M. (1912) Interferenz-Erscheinungen bei Röntgenstrahlen: *Sitzungsberichte der (Kgl.) Bayerische Akademie der Wissenschaften* 303-22.

Frisch, O. R., Paneth, F. A. and Laves, F., editors. (1959) *Trends in Atomic Physics*: Interscience Publishers, Inc., New York, 285 pp.

Glasser, O. (1958) *Dr. W.C. Röntgen*:, Second Edition: Charles C. Thomas, Springfield, Illinois, 142 pp. This short biography was first published on the 100th anniversary of Röntgen's birth.

Grim, R. E. (1953) *Clay Mineralogy*: McGraw-Hill, New York, 384 pp. and (1968) *Clay Mineralogy*, Second Ed.: McGraw-Hill, New York, 596 pp. Still very useful if you are also familiar with more recent work.

Grim, R. E. (1962) *Applied Clay Mineralogy*: McGraw-Hill, New York, 422 pp. To date, the only systematic treatment for applications.

Grim, R. E. (1988) The history of the development of clay mineralogy: *Clays and Clay Minerals* 36, 97-101.

Grim, R. E., Bray, R. H., and Bradley, W. F. (1937) The mica in argillaceous sediments: *Amer. Mineral.* 22, 813-29.

Grimshaw, R. W. (1971) *The Chemistry and Physics of Clays*: Wiley-Interscience, New York, 1024 pp. Written from the point of view of the ceramics industry. Contains detailed information about pre-World War II views of clay minerals, their geology, and the processes to which they were subjected.

Gruner, J. W. (1932) The crystal structure of kaolinite: *Z. Kristallogr.* 83, 75-88.

Hadding, A. (1923) Eine röntgenographische Methode kristalline und kryptokristalline Substanzen zu Identifizieren: *Z. Kristallogr.* 58, 108-12.

Haneberg, W. C., and Anderson, S. A., editors (1995) *Clay and Shale Slope Instability*: Reviews in Engineering Geology, Vol. X. The Geological Society of America, Boulder, Colorado, 153 pp.

Hartman, H. (1975) Speculations on the origin and evolution of metabolism: *Jour. Molecular Evol.* 4, no. 4, 359-70.

Hartman, H. (1982) Life, language, and society: *Semiotica* 42, 89-106.

Hays, W. W. (1981) *Facing Geologic and Hydrologic Hazards, Earth Science Considerations*: U.S. Geological Survey Prof. Paper 1240-B, 109 pp.

Hendricks, S. B., and Fry, W. H. (1930) The results of X-ray and microscopic examination of soil colloids: *Soil Sci.* 29, 457-78.

Hendricks, S. B., and Teller, E. (1942) X-ray interference in partially ordered layer lattices: *J. Phys. Chem.* 10, 147-67.

Hermann, A. (1979) *The New Physics: The Route into the Atomic Age*: Inter Nationes, Bonn-Bad Godesberg, Heinz Moos Verlag, Munich, 175 pp.

Hofmann, U., Endell, K. and Wilm, D. (1933) Kristalstrucktur und Quellung von Montmorillonit: Z. Kristallogr. **86**, 340-48.

Jackson, M. L. (1969) *Soil Chemical Analysis, Advanced Course:* Published by the author, Dept. of Soil Science, Univ. of Wisconsin, Madison, Wisconsin, 53706, 498 pp. A catalog of laboratory procedures you will need. There are later editions, but the changes are minor.

Johns, W. D., and McKallip, T. E. (1989) Burial diagenesis and specific catalytic activity of illite-smectite clays from Vienna Basin, Austria: *AAPG Bull.* **73**, 472 82.

Jordan, J. W., Jr. (1949) Organophilic bentonites. I. Swelling in organic liquids: *J. Phys. Colloid. Chem.* **53**, 294-306.

Kelley, W. P., Dore, W. H., and Brown, S. M. (1931) The nature of the base-exchange material of bentonite, soils, and zeolites as revealed by chemical investigation and X-ray analysis: *Soil Sci.* **31**, 25-55.

Kleinfeld, E. R., and Ferguson, G. S. (1994) Stepwise formation of multilayered nano-structural films from macromolecular precursors: *Science* **265**, 370-73.

Klug, H. P., and Alexander, L. E. (1974) *X-Ray Diffraction Procedures,* 2nd ed.: Wiley-Interscience, New York, 966 pp. A very good reference for all aspects of X-ray diffraction procedures.

Krapac, I. G., Cartwright, K., Hensel, B. R., Herzog, B. L., Larson, T. H., Panno, S. V., Risatti, J. B., Su, W-J., and Rehfeldt, K. R., (1991) Construction, monitoring, and performance of two soil liners: *Environmental Geol. Note* **141**, Illinois State Geological Survey, Champaign, Illinois, 118 pp.

Kuhn, T. S. (1970) *The Structure of Scientific Revolutions,* 2nd ed.: University of Chicago Press, Chicago, 210 pp.

Lee, S. Y., and Tank, R. W. (1985) Role of clays in the disposal of nuclear waste: A review: *Applied Clay Science* **1**, 145-62.

Lemmerich, J., organizer (1995) *Exhibition Catalog, Röntgen Rays Centennial,* The Martin von Wagner Museum, University of Würzburg, Feb 13 - Nov 19, 1995: Bavarian Julius-Maximilians-Universität Würzburg, 121 pp.

Lonsdale, K. Y. (1929) X-ray evidence on the structure of the benzene nucleus: *Transactions of the Faraday Society* **25**, 352-66.

Mackenzie, F. T., and Kump, L. R. (1995) Reverse weathering, clay mineral formation, and oceanic element cycles: *Science* **270**, 586-87.

Manning, D. A. C., Hall, P. L., and Hughes, C. R., editors (1993) *Geochemistry of Clay-Pore Fluid Interactions:* Mineralogical Society Series 4, Chapman & Hall, London, 448 pp.

Margulies, L., Rosen, H., and Cohen, E. (1985) Energy transfer at the surface of clays and protection of pesticides from photodegradation: *Nature* **315**, 658-59.

Mason, B. (1992) *Victor Moritz Goldschimdt: Father of Modern Geochemistry:* The Geochemical Society, San Antonio, Texas, 184 pp.

McBride, M. B. (1989) Surface chemistry of soil minerals: in Dixon, J. B., and Weed, S. B., editors, *Minerals in the Soil Environment,* Second Edition, Soil Science Society of America, 35-88.

Michalopoulos, P., and Aller, R. C. (1995) Rapid clay mineral formation in Amazon Delta sediments: Reverse weathering and oceanic elemental cycles: *Science* **270**, 614-17.

Millot, G. (1970) *Geology of Clays:* Springer-Verlag, New York, 429 pp.

Newman, A. C. D., editor (1987) *Chemistry of Clays and Clay Minerals:* Monograph No. 6, Mineralogical Society, Wiley, New York, 480 pp. This is an important contribution comparable to and complementing Brindley and Brown (1980).

Nitske, W. R. (1971) *The Life of Wilhelm Conrad Röntgen:* Univ. of Arizona Press, Tucson, 335 pp. Provides a close look at the personality of this complex man. Draws heavily on his personal correspondence and notes. Perhaps other sources are better for details on Röntgen's science.

Nitta, I. (1962) Japan: in Ewald, P.P., ed. *Fifty Years of X-ray Diffraction:* Oosthoek's Uitgeversmaatschappij, Utrecht, 484-92.

Odin, G. S., editor (1988) *Green Marine Clays: Oolitic Ironstone Facies, Verdine Facies, Glaucony Facies and Celadonite-Bearing Facies —A Comparative Study:* Elsevier, Amsterdam, 445 pp.

Pauling, L. (1930) The structure of mica and related minerals: *Proc. Nat'l Acad. Science, U.S.A.* **16**, 123-29.

Pusch, R. (1992) Use of bentonite for isolation of radioactive waste products: *Clay Minerals* **27**, 353-61.

Rinne, F. (1924) Röntgenographische Untersuchungen an einigen feinzerteilten Mineralien, Kunsprodukten und dichten Gesteinen: *Z. Kristallogr.* **60**, 55-69.

Ross, C. S., and Kerr, P. F. (1930) The clay minerals and their identity: *J. Sediment. Petrol.* **1**, 35-65.

Ross, C. S., and Kerr, P. F. (1931) *The Kaolin Minerals*: U.S. Geol. Surv. Prof. Pap., **165F**, 151-75.

Ross, C. S., and Shannon, E. V. (1925) The chemical composition and optical properties of beidellite: *J. Wash. Acad. Sci.* **15**, 467-68.

Sawhney, B., convener (1995) Reactions of Organic Pollutants with Clays: notes for CMS Pre-Meeting Workshop, The Clay Mineral Society, Boulder, Colorado.

Schoonheydt, R. A., Leeman, H., Scorpion, A., Lenotte, I., and Grobet, P. (1994) The Al pillaring of clays. Part II. Pillaring with $[Al_{13}O_4(OH)_{24}(H_2O)_{12}]^{7+}$: *Clays and Clay Minerals* **42**, 518-25.

Shubnikov, A. V. (1962) Schools of X-ray structural analysis in the Soviet Union: in Ewald, P.P., ed. *Fifty Years of X-Ray Diffraction*: Oosthoek's Uitgeversmaatschappij, Utrecht, 493-97.

Soma, Y., and Soma, M. (1989) Chemical reactions of organic compounds on clay surfaces: *Environmental Health Perspectives* **83**, 205-14.

Streller, E., Winau, R., and Hermann, A. (1973) *Wilhelm Conrad Röntgen*: Inter Nationes, Bonn-Bad Godesberg, Heinz Moos Verlag, Munich, 71 pp. Three separate essays: one on Röntgen and the other two on the impact of X-rays on medicine and on physics, chemistry, and biology.

Thomas, J. M. (1992) Solid acid catalysts: *Scientific American*, no. 4, 112-18.

Torrance, J. K. (1983) Towards a general model of quick clay development: *Sedimentology* **30**, 547-55.

Ungerer, P. (1990) State of the art of research in kinetic modelling of oil formation and expulsion: *Org. Geochem.* **16**, 1-25.

Updike, R. G., Egan, J. A., Yoshiharu, M., Idreiss, I. M., and Moses, T. L. (1988) A model for earthquake-induced translatory landslides in Quaternary sediments: *Geol. Soc. Amer. Bulletin* **100**, 783-92.

Velde, B. (1985) *Clay Minerals: A Physico-Chemical Explanation of their Occurrence*: Elsevier, 427 pp.

Weaver, C. E. (1989) *Clays, Muds, and Shales*: Developments in Sedimentology 44, Elsevier, Amsterdam, 819 pp. An almost encyclopedic coverage of all aspects of the occurrence of clay minerals.

Wu, C-G., and Bein, T. (1994) Conducting polyaniline filaments in a mesoporous channel host: *Science* 264, 1757-59.

Yong, R. N., Galvez-Cloutier, R., and Phadungchewit, Y. (1993) Selective sequential extraction analysis of heavy-metal retention in soil: *Can. Geotech. Jour.* **30**, 834-47.

Chapter 2
Nature and Production of X-Rays

The purpose of the following material is to give you a general idea about the nature, production, and control of X-rays. You need a minimum understanding in order to use them as an analytical tool, to make judgments about the conditions under which samples should be run, to make an intelligent selection of accessories, and to select or modify equipment. It is not a coincidence that the most productive and respected workers are those who demonstrate the clearest understanding of X-rays and the equipment and programs they use. We want to help you form a picture of what is happening physically as you subject a sample to X-ray diffraction analysis. For a thorough understanding of the more abstract aspects of the nature and production of X-rays, we urge you to use Cullity (1978), James (1965), and Klug and Alexander (1974).

Box 2.1. Other Methods

X-ray diffraction (XRD) is the instrumentation most commonly used to study clay-sized minerals, and will almost certainly remain so. However, we wouldn't want you to think that it is the only method. XRD provides information about the bulk properties of whole populations, information averaged over as many as perhaps 10^{11}-10^{12} unit cells or billions of crystals. Clay minerals have formed at temperatures and pressures that, compared to the conditions under which most minerals form, are quite low. Therefore, they are often quite disordered and heterogeneous, especially in soils. Clay minerals are at, or very near, the boundary between mineral and not mineral. They are close to that world ruled by quantum effects, effects often counterintuitive to us in the macroworld. These characteristics cause us to question the meanings of terms from the macroworld like mineral, crystal, and phase. To answer questions about this nether world, instrumentation that can deal with discrete particles has been developed. Such instrumentation may well be THE revolution of the last half of the twentieth century, as viewed by historians of science in the future.

These recently available methods, along with others that are well established but only recently applied to the study of clay-sized minerals and their surfaces, can be put into four categories: (1) microscopy;

---

(2) spectroscopy and analysis; (3) diffraction and imaging; and (4) thermal methods. These methods, each of which can offer a somewhat different type of information, have been given acronyms that make an alphabet soup. We are not competent to do more than offer you a brief tour through this soup, a taste so to speak, and to suggest references that are starting points for developing an understanding of these various methods. One method doesn't fit in these categories. It is really X-ray diffraction, but the source of the X-rays is a synchrotron. This gadget can deliver a highly intense beam in a short time, 10-20 milliseconds, to an area as small as 0.2 μm.

The Clay Minerals Society and the Mineralogical Society of America publish, in their Workshop Lecture series and Reviews in Mineralogy series, respectively, some of the best values anywhere in technical books. From the Workshop Lecture series, there are Stucki et al. (1990), *Thermal Analysis in Clay Science*; Mackinnon and Mumpton (1990), *Electron-Optical Methods in Clay Science*; and Blume and Nagy (1994), *Scanning Probe Microscopy of Clay Minerals*. You should see at least two of the Reviews in Mineralogy series: Hawthrone (1988), *Spectroscopic Methods in Mineralogy and Geology*; and Buseck (1992), *Minerals and Reactions at the Atomic Scale: Transmission Electron Microscopy*. Buseck, the editor of the latter volume, has written the first chapter. It is an especially clear explanation of the transmission electron microscope made by comparing it to the petrographic microscope and an equally clear explanation of electron diffraction compared to X-ray diffraction. Two other books you need to be aware of are Buseck et al. (1988), *High-Resolution Transmission Electron Microscopy;* and Joy et al. (1986), *Principles of Analytical Electron Microscopy*. Finally, Wilson (1994) has edited a book, *Clay Mineralogy: Spectroscopic and Chemical Determinative Methods*, an extended version of his 1987 book. We refer you to these excellent texts for explanations of these methods. You may want to have copies of modern physics and physical chemistry books at your elbow as you approach these references, but they should tell you almost everything you need to know about the methods listed in Table 2.1.

We have modified a table from Mackinnon (1990), adding items from Hawthorne (1988) and other sources, to list these acronyms (Table 2.1).

Table 2.1. Acronyms for instrumental methods

Acronym	Variation	Meaning
MICROSCOPY		
AEM	STEM	Analytical electron microscopy
AFM		Atomic force microscopy
FESEM		Field-emission scanning electron microscopy
HREM	HRTEM	High-resolution electron microscopy
HVEM		High-voltage electron microscopy

SEM		Scanning electron microscopy
STEM	AEM	Scanning-transmission electron microscopy
STM		Scanning tunneling microscopy
TEM		Transmission electron microscopy

SPECTROSCOPY AND ANALYSIS

AES		Auger electron spectroscopy
ALCHEMI		Atom location by channeling-enhanced microanalysis
EDX	EDS, XES EDXS	Energy-dispersive X-ray spectroscopy, X-ray emission spectroscopy
EELS		Electron-energy-loss spectroscopy
EPMA	Probe	Electron-probe microanalysis
EPR		Electron paramagnetic resonance spectroscopy
EXAFS		Extended X-ray absorption fine-structure spectroscopy
EXELFS		Extended X-ray energy-loss fine-structure spectroscopy
FTIR		Fourier transform infrared spectroscopy
MAS NMR		Magic-angle spinning nuclear magnetic resonance spectroscopy
NA*		Mössbauer spectroscopy
NMR		Nuclear magnetic resonance spectroscopy
PEELS		Parallel electron-energy-loss spectroscopy
PIGE		Proton-induced gamma excitation spectroscopy
PIXE		Proton-induced X-ray excitation spectroscopy
WDS		Wavelength-dispersive X-ray spectroscopy
XANES		X-ray absorption near-edge structure spectroscopy
XAS		X-ray absorption spectroscopy
XPS		X-ray photoelectron spectroscopy

DIFFRACTION AND IMAGING

BEI	BSE	Backscattered electron image
CBED	CBD	Convergent-beam electron diffraction
DFI		Dark-field image
FOLZ		First-order Laue zone
HOLZ		Higher-order Laue zone
HREM	HRTEM	High-resolution electron microscopy
µDiff		Microdiffraction
mmD		Micro-microdiffraction
NA*		Neutron diffraction
OTED		Oblique-texture electron diffraction
SAED	SAD	Selected-area electron diffraction

SEI	SEM	Secondary electron image
NA*		Synchrotron radiation diffraction
ZOLZ		Zero-order Laue zone

THERMAL METHODS

DSC	Differential scanning calorimetry
DTA	Differential thermal analysis
HPDTA	High-pressure differential thermal analysis
TGA	Thermogravimetric analysis

*No acronym!!

SAFETY AND PROTECTION

The effects of short-wavelength radiation on mammals are damaging and partly cumulative, as far as is generally understood. Many of the earliest workers with X-rays lost limbs or died from overexposure. Because the effect is cumulative, it is always prudent to avoid exposure as much as possible to any form of shortwave radiation. All modern X-ray equipment has features to minimize exposure. Dangerous exposure will occur when specified routines are not followed, when an uninformed person tries to use the equipment, or when there is a failure in the safety interlocks of a machine. *Brief exposure to the direct beam can cause permanent skin injuries.* Never look into the shutter of the X-ray tube when it is on, and never look at the sample as it is being exposed to X-rays. Many early workers suffered from radiation damage. The loss of fingers and hands was relatively common.

ALWAYS block the primary beam by closing the shutter before handling the sample in the diffractometer. On some machines with an enclosed diffractometer, the shutter closes automatically, but never trust it. Listen for the click of the shutter closing and check the shutter light before you open the door. And remember, the bulbs in warning lights must burn out from time-to-time.

Box 2.2. Defining a Dose of Radiation
    You may be asked to wear a film badge when you use the X-ray machine. This is a reminder that radiation above the background that is part of our natural environment can be harmful. The units used to measure radiation are in a bit of a tangle. This tangle results from the limitations of what we understand about radiation, and this changes with time. For example, when the röntgen R was accepted as a unit in 1928, accepted theories of electricity suggested that potentials greater than 3 MeV were impossible. Its crude dose measurement was based on how long it took to redden the skin. Units have continued to change as recently as 1968, when the sievert (Sv) was adopted,

because of the realization that different kinds of radiation did more or less damage to tissue even though they were delivering the same amount of energy per unit of volume. Alpha particles do about 20 times as much damage to tissue as an X-ray photon when both are depositing the same amount of energy per unit volume. By 1985, all the units in use were supposed to be SI units. (SI = Le Systéme International d'Unités, accepted by the General Conference on Weights and Measures in 1977.) However, the older units rad and rem are still in use.

There are two kinds of units for characterizing radiation, ones that measure the amount or the flux of the radiation—these are pretty straightforward and taken from physics—and those that measure the effect of radiation on living tissue.

For the first kind, the ones from physics, the amount of radioactivity is described in terms of activity, i.e., the number of disintegrations per unit of time. Different isotopes disintegrate at different rates, which, of course, affects the total flux. Much of the older literature uses the curie (Ci) [after Marie (1867-1934) and Pierre Curie (1859-1906), who discovered polonium and radium in 1898]. Recent literature uses the Becquerel (Bq) [after Henri Bequerel (1852-1908), who discovered radioactivity in 1896]. To measure the amount of radiation, one Bequerel = one disintegration per second. To convert to curies, there are 3.7×10^{10} Bq per Ci or 1 Bq = 26.95 picocuries (1 curie = 10^{12} picocuries).

Measuring the effect the amount of radiation has on living tissue is a more complex matter. Different wavelengths have different effects; different types of tissue respond differently. The realization, made by Röntgen in his first paper on X-rays, that X-rays ionize air was used to define dosages. The röntgen, the rem (**r**öntgen **e**quivalent **m**an = rem, now, of course, a politically incorrect term), and the rad (**r**öntgen absorbed **d**ose = rad) are the non-SI units; the gray, Gy (absorded dose), and sievert, Sv (equivalent dose), are the SI units. A röntgen is the amount of radiation that ionizes 1 cm^3 or 0.001293 g of air so that it has a charge of 1 electrostatic unit (esu). Absorption in air is not quite comparable to absorption in fluid-containing tissue. Therefore, the rad had to be invented. This is the radiation that delivers 100 ergs to a gram of tissue. An absorbed dose of 1 rad is approximately equivalent to a dose equivalent of 1 rem: a subtle difference, but apparently a necessary one. The amount of energy delivered is measured in rads or grays, and the effect it has on tissue measured in rems or sieverts.

The amount of wholebody exposure to radiation per year "permissible," as recommended by the National Council on Radiation Protections and Measurements, is 5 rems. In SI units, that's 50 mSv. For perspective, annual natural radiation exposure is 1-6 mSv; it is greater if you live at high altitudes or spend much time in planes flying above 20,000 feet. (Information for this box was taken from Gollnick, 1983, and the National Council on Radiation Protection and Measurements, 1985.)

A second hazard, with more immediate and deadly effects if it is not understood and respected, is the electrical hazard. The voltage across the X-ray tube is high, usually 15 to 60 kV; across the detectors, approximately 1.5 kV. Even with the power off at the source, high-voltage capacitors can produce dangerous electrical discharges.

NEVER reach inside the X-ray generator or the electronic circuit panel unless you know exactly what you are going to do and where the potentially dangerous components are.

THE RESPONSIBILITY FOR SAFE OPERATION IS YOURS.

THE NATURE OF X-RAYS

X-rays are part of the electromagnetic spectrum and therefore have properties of both waves and particles. [This issue was resolved, to everyone's satisfaction, by Louis de Broglie (1892-1987) in 1923, for which he received the Nobel Prize in 1929.] The energy of an electromagnetic beam interacting with a medium is partly transmitted, partly refracted and scattered, and partly absorbed. The packets of energy, or photons, can "bounce" and transfer momentum, which is a property of a discrete particle. They also can be diffracted by a grating of appropriate size and have measurable wavelengths, which are properties of a wave. Recall that von Laue realized that the spacings in a crystal structure could serve as a diffraction grating. The relationship among the wavelength of the radiation λ, the angle θ between the incident beam of radiation and the parallel planes of atoms causing the diffraction, and the spacing d between these planes is called Bragg's law and is developed in Chapter 3.

This chapter discusses X-rays emitted from the target of an X-ray tube, both continuous and characteristic, and the absorption of X-rays, both general and characteristic. We also consider the electronic equipment for controlling and measuring X-rays.

Continuous or White Radiation

X-rays are usually produced by the sudden deceleration of fast-moving particles. We are concerned with those generated in an X-ray tube when electrons are emitted from a glowing tungsten filament (the cathode), pulled (accelerated) through the vacuum of the tube by voltages, usually from 15 to 60 kV, and strike a metal target (the anode). Within the target the electrons encounter crowds of other electrons, which causes a sudden deceleration. The result is X-radiation of two major varieties: one form has a broad, continuous spectrum of wavelengths, and the other has very sharp peaks of discrete wavelengths characteristic of the target material.

The variety of radiation with a broad, continuous spectrum is called white or continuous radiation, or *bremsstrahlung* (braking radiation). This broad spectrum does, however, have a short-wavelength limit that is a function of the voltage across the X-ray tube: the greater the voltage, the shorter the shortest of the X-ray wavelengths in the white radiation spectrum. Figure 2.1 shows that the radiation from the target, in this case tungsten (W), gives a shorter wavelength for each increase in voltage.

Almost all electrons will "collide" more than once with the strong electric fields near atomic nuclei in the target material. As a consequence, they lose an amount of energy ΔE with each collision. Because no energy can be lost when the whole system is considered, about 98% of it is changed into heat energy, while a small part appears as X-ray photons, each with a frequency ν (Greek lowercase letter pronounced "new") according to the equation

$$\Delta E = h\nu \tag{2.1}$$

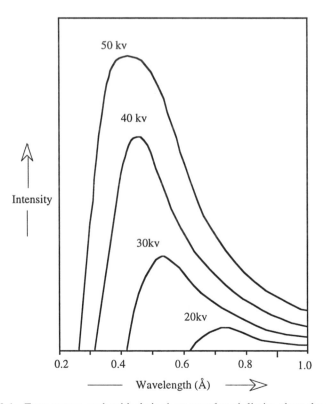

Fig. 2.1. Four curves, each with their short-wavelength limits, show the variation in intensity with wavelength for four different voltages. Such radiation is called continuous or white radiation. These curves are for radiation from a tungsten target. (Modified from Klug and Alexander, 1974, p. 82)

in which *h* is Planck's constant and ν is the frequency (c/λ) of the X-ray. Equation (2.1) is Einstein's frequency condition, which describes a quantum process that can be confirmed by experiment. The stream of billions of electrons, each losing energy in several steps, produces a continuous range of wavelengths, the white radiation. These wavelengths can be plotted as shown in Fig. 2.1, and the distribution of intensity is different for each value of applied voltage. The wavelength distribution of the white radiation is independent of the target material.

A few of the millions of electrons striking the target will lose all their kinetic energy in one collision. The photons thus released give the short-wavelength, or Duane-Hunt, limit. Only in this case does the photon released have energy equal to the initial energy of the electron. Reasoning from here, an equation has been worked out relating the short-wavelength limit to the voltage of the X-ray tube (e.g., see p. 83 in Klug and Alexander, 1974). Indeed, measurement of the wavelength of the Duane-Hunt limit is one way to estimate Planck's constant *h*.

Characteristic Radiation

Electrons orbiting close to the nucleus are more tightly bound than those farther from the nucleus. Radiation characteristic of the metal from which the target is made results when electrons accelerated across the X-ray tube strike these inner, more tightly bound electrons hard enough to knock them out of their orbital positions and away from the influence of the nucleus. This event leaves vacancies that are immediately filled by electrons dropping in toward the nucleus from orbits farther out. The energy lost in the drop, like the bang of a book hitting the floor, appears as a photon with frequency ν, again according to Eq. (2.1) and again, therefore, a quantum process. The energy difference between orbitals is a function of the number of protons in the nucleus attracting the electrons and is different for every element and, perhaps just as important, is exactly the same for all atoms of the same kind of element. It is, therefore, characteristic and periodic, as H. G. J. Moseley (1887-1915, killed in WW I) discovered in 1913. The greater the number of protons in the nucleus, the more tightly the electrons are held, and therefore the incident electrons must have greater energy to knock them out. The electron dropping from a higher orbit to fill the resulting gap is pulled more forcefully as the number of protons in the nucleus increases from element to element so that the wavelength becomes shorter as the atomic number of the target material increases. The resulting drop from a higher orbital to fill the vacancy is a larger ΔE; therefore, the X-ray photons have shorter wavelengths. [You can always remember that ΔE is reciprocally related to wavelength by recalling that frequency ν is equal to velocity divided by wavelength. For light, $\nu = c/\lambda$. You can substitute this into Eq. (2.1) to get $\Delta E = hc/\lambda$. Thus, the shorter the wavelength, the greater the energy.]

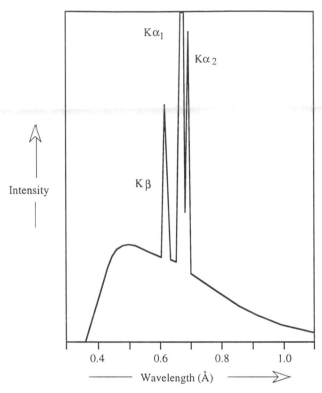

Fig. 2.2. Continuous radiation plus the characteristic radiation of molybdenum superimposed on it for a voltage of 30 kV. $K\alpha_1$ is off scale. (Modified from Klug and Alexander, 1974, p. 84)

Let us consider some diagrams to make this process clear. The potential necessary to form the characteristic radiation of molybdenum (Mo, atomic number = 42) is 20.00 kV. (The potential necessary for silver, atomic number = 47, is 25.52 kV; for copper, atomic number = 29, it is 8.98 kV.) Figure 2.2 shows that with a molybdenum (Mo) target, characteristic radiation will appear if the voltage applied to the tube is 20 kV or greater. This much, 20 kV or more, voltage is needed to drive electrons into the target with enough energy to knock bound electrons from the K-shell of Mo. The characteristic spectrum from the Mo target can be seen superimposed on the white radiation spectrum. Except for the intense peaks at 0.63 and 0.71 Å, this curve is quite similar to the curves of only white radiation shown in Fig. 2.1. (The $K\alpha_1$ peak of the $K\alpha_1$-$K\alpha_2$ doublet is off the scale in this figure.) You can see the short-wavelength limit here also.

For a physical picture of what is happening inside the atoms of the target material, Fig. 2.3 shows a nucleus with K, L, and M shells. Imagine that it has an N shell, too. Then this drawing could represent an atom of Mo, which has

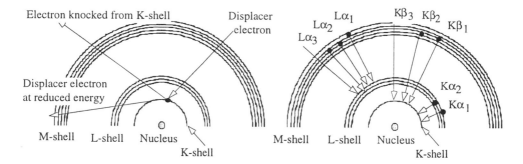

Fig. 2.3. Step 1 shows an electron being knocked from the K shell. The incident electron after impact and the displaced electron, sometimes called a photoelectron, both leave the atom at a wavelength greater than that of the displacer electron as it came into the atom. Step 2 shows many of the possibilities for replacing displaced electrons, only one of which can happen to replace the displaced electron. An X-ray photon is emitted when an electron "drops" from a higher level to refill a shell.

42 electrons in its orbitals. In Step 1, an incoming electron knocks out an electron from the K shell. To do this, the incoming electron must have been accelerated by at least 20 kV. This leaves a vacancy that can be filled by any one of the five possible "drops" shown in Step 2. When this happens, the vacancy will be filled by an electron pulled in by the charges of 42 protons. The most probable candidates are the closest electrons, those in the L shell. The sublevel in the L shell that most often contributes to the K shell is L_3. It forms radiation we call $K\alpha_1$. Then, approximately half as often, L_2 contributes. This forms $K\alpha_2$ radiation. L_1 never contributes. (For an explanation of why energy changes between some levels are not allowed, see the topics on exclusion rules or selection rules in most books on physical chemistry.)

Electrons contributed from shells M and N form Kß radiation. Electrons can also be knocked from the L shell and have their places taken by electrons "falling" from the M and N shells (Fig. 2.3, Step 2). The characteristic radiation formed when electrons fall into the K shell is called the *K series*, and that from electrons falling into the L shell, the *L series*. Because electrons are not held as firmly in the L shell (they are farther away from the protons of the nucleus), it does not take as much energy to dislodge them. Nor do the photons formed as the energy lost from falling electrons entering the L shell have wavelengths as short (i.e., as energetic) as radiation in the K series. The L series wavelengths are too long to appear in Fig. 2.2; the five lines of the L series of Mo are in the 5.5-Å-wavelength region, compared with 0.71 and 0.63 Å for Kα and Kß radiation, respectively.

The ratio of the intensities for $K\alpha_1$ to $K\alpha_2$ is always about 2:1 for all elements. The ratio Kß:Kα is less constant, usually 15-30% that of Kα. (Does

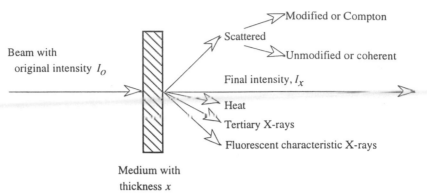

Medium with
thickness *x*

Fig. 2.4. Part of the energy of an X-ray beam is transmitted through a medium, part is transformed into heat energy, part is scattered with unchanged wavelength, part is scattered with a longer wavelength, part is reradiated as X-rays characteristic of the medium, and part is reradiated as tertiary X-rays. All the energy except that in the transmitted beam I_x is what is generally considered to be absorbed. (See Klug and Alexander, 1974, for details.)

the relation between the wavelengths and corresponding energies of $K\alpha_1$, $K\alpha_2$, and Kß and the distances they had to fall, as shown in Fig. 2.3, Step 2, make sense to you?)

General Absorption of X-Rays

The ability of X-rays to penetrate opaque material may be their most interesting property, but here we want to concentrate on their absorption. Absorption is generally considered to be the energy of the incident beam minus that of the transmitted beam [Fig. 2.4 and Eq. (2.2)]. The absorption of X-rays follows the same law as that for all other radiation passing through an imperfectly transparent medium, namely, that for a monochromatic beam the fraction of the radiation absorbed is the same for each increment of equal thickness of the absorbing material. Put another way, the intensity of the radiation decreases logarithmically as it passes through absorbing material. This law can be expressed as

$$dI/dx = -\mu I_o \quad or \quad dI/I_o = -\mu dx \quad or \quad I_x = I_o e^{-\mu x} \qquad (2.2)$$

in which I_o is the original, incident intensity, x is the absolute thickness, I_x is the intensity transmitted after passing through thickness x, and the constant μ is called the *linear absorption coefficient*. This coefficient should be thought of as a measure of the probability that radiation of a given wavelength will be absorbed per unit length of the medium through which it passes. The value of μ differs for each material, changes with the wavelength of the incident radiation, and depends on the state of the material. The relation that expresses the proportionality between absorption and the thickness of the medium traversed by the radiation is the *Bouguer-Lambert law*, or the *Lambert law*. (Incidentally, because x is the absolute thickness and voids do not count, a comparison of measured μ vs. theoretical μ gives a measure of porosity.)

Because μ varies with the density ρ of the material, it is convenient to modify μ so that absorption may be described independently of the physical and chemical states of the material. For example, diamond and graphite have different values of μ but the same value for μ/ρ. This modification is made by multiplying the exponent of Eq. (2.2) by ρ/ρ so that

$$I_x = I_o e^{-(\mu/\rho)\rho x} \qquad (2.3)$$

The quantity μ/ρ is the *mass absorption coefficient* and is the value most frequently tabulated. For mechanical mixtures, chemical compounds, or solid or liquid solutions, the mass absorption coefficient for the whole sample is simply a weighted average of the mass absorption coefficients of the constituent elements. Table 2.2 shows how to calculate the mass absorption coefficient of albite, which equals 32.6 for the wavelength of CuKα.

If the mass absorption coefficient is plotted against wavelength, absorption edges of the absorber are shown. (See Fig. 2.5 and the discussion associated with it.) Equations (2.2) and (2.3) can be used to determine the thickness of shields to protect against radiation damage, to determine the thickness of foils for making an X-ray beam essentially monochromatic (see next section), and to understand the changes caused by radiation hitting a film emulsion (Klug and Alexander, 1974, pp. 114ff). Absorption effects do slightly change diffraction peak shape and intensity, but not enough that anyone working with clay minerals will ever notice, even in the most careful quantitative measurements. (These absorption effects are called transparency; James, 1965.)

Now, using Table 2.2 as an example, can you calculate the μ/ρ for an illite with the formula

$$K_{0.75}(Al_{1.5}Fe^{3+}_{0.25}Mg_{0.30})(Si_{3.4}Al_{0.6})O_{10}(OH)_2$$

assuming CuKα and using μ/ρ for Fe = 308, K = 143, Mg = 38.6, and H = 0.435. (We ignored H. Do you see why? Why is the μ/ρ of Fe so much larger than that of Mg? The numbers you generate here should help you understand why the variation in the amount of Fe in a mineral makes such a difference in the absolute intensities of peaks in an X-ray diffraction tracing.)

Characteristic Absorption

Even as each element emits characteristic radiation that is a periodic property (i.e., related to atomic number), so each element absorbs different wavelengths of radiation in a manner characteristic of that element. We will find a useful application for this phenomenon. If we could pass a stream of X-

Table 2.2. Calculation of μ/ρ for albite, $NaAlSi_3O_8$, for $CuK\alpha$

Constituent elements	Atomic weight	Weight % in albite	μ/ρ [a]	$(\mu/\rho)(wt \%)$
Na	23	8.7	30.1	2.6
Al	27	10.3	48.6	5.0
Si $28\times3 =$	84	32.1	60.0	19.4
O $16\times8 =$	128	48.8	11.5	5.6
Totals	262	100.0	150.8	32.6

[a]Values from Klug and Alexander (1974, p. 875).

ray photons through a foil of a metallic element, and if we could control the wavelength of these photons, we would find that the degree to which the foil absorbs photons increases with increasing wavelength to a maximum at λ_A, the *absorption edge* (or *critical absorption wavelength* or *quantum wavelength*). Once past λ_A, absorption drops immediately. Absorption, a discontinuous function of the incident wavelength, is shown in Fig. 2.5. This particular figure is the plot of absorption versus wavelength for a thin foil of zirconium (Zr, atomic number 40) irradiated by X-rays from a Mo target (as in Fig. 2.2). Any elemental foil would give a similar plot; only the position of the absorption edge in relation to wavelength would be different.

To understand the phenomenon of absorption edges, we refer again to the architecture of the atom. X-ray photons striking the metal foil can excite an atom in the same way an electron can, that is, knock bound electrons out of their orbitals (Fig. 2.3, Step 1). X-ray photons with energies equal to or just greater than the energy holding the electrons in their orbits are the most likely to displace orbital-bound electrons. Photons displacing the electrons have used or transferred their energy; that is, they have been absorbed.

Obviously, photons with less energy than the bound electrons will not displace them. Such photons have wavelengths greater than λ_A (0.687 Å in Fig. 2.5) and, because they have not been absorbed, account for the precipitous drop in absorption. Not as obvious, but confirmed by observation, is the conclusion that photons with energies much greater than the bound electrons tend to go right past the electrons without displacing them. The phenomenon of the interaction of photons and bound electrons with approximately the same energy is called *resonance*.

Now for the practical use of this characteristic of every element: In the section on diffraction we will see that a monochromatic (i.e., single wavelength) X-ray beam simplifies the analysis of structural spacings in crystalline materials. However, Fig. 2.2 shows the characteristic, two-peak K spectrum of Mo—a composite or doublet $K\alpha$ peak and a $K\beta$ peak. If we could eliminate $K\beta$, we would have essentially monochromatic radiation. Imagine the plot of Fig. 2.5 superimposed on Fig. 2.2. Because absorption edges are characteristic and periodic, all we have to do is find an absorption edge that

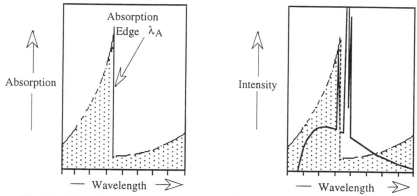

Fig. 2.5. Plot of absorption vs. wavelength for a thin zirconium foil (atomic number = 40). Note the drastic change in the ability of the foil to absorb X-rays at the absorption edge, which is at 0.687 Å. This is the characteristic absorption edge of Zr.

Fig. 2.6. This is Fig. 2.5 superimposed on Fig. 2.2. The effect is to reduce the intensity of both Kα and Kß, but Kß will be reduced to a far greater extent; so radiation from the other side of the foil will be essentially monochromatic.

lies between Kα and Kß. Both K spectra and the absorption edges are periodic, and both are functions of the energy with which electrons are bound in the K shell. You won't be too surprised to find that an element with an atomic number close to that of Mo has an absorption edge that just fits. Zr, with atomic number 40 (recall that the atomic number of Mo is 42), is just such a fit. In fact, the rule is that the element with one or two atomic numbers less than that of the target will have the right absorption edge for eliminating Kß. (Absorption edges are labeled just as the emission spectra are: There is a K absorption edge, an L, an M, etc., for each element with enough electrons.) Figure 2.6 shows Fig. 2.5 superimposed on Fig. 2.2. The Kß radiation is not entirely eliminated by absorption in the Zr foil, but its intensity is markedly decreased. Without filtering, it is 15-30% of the Kα intensity, as shown in Fig. 2.2. With filtering it is about 0.2% of the intensity of Kα. About half of the Kα intensity is also lost, but the beam is now nearly monochromatic and ready to be applied to diffraction problems. Another way to eliminate Kß is based on principles discussed in the section on single-crystal monochromators, p. 52.

Absorption edges also find practical applications in one of those analytical methods listed in Table 2.1, X-ray absorption spectroscopy (XAS). Two varieties of XAS are extended X-ray absorption fine structure (EXAFS) and X-ray absorption near-edge structure (XANES). The absorption edge in Fig. 2.5 is greatly oversimplified. If magnified and examined in detail, the curve would show discrete peaks, several clear, sharp peaks on the immediate shorter-wavelength side of the absorption edge (XANES) and the more subtle fine structure of EXAFS farther away on the shorter wavelength side (EXAFS). The energies with which electrons are bound in the K-shell (20,000 eV in the case of Mo) will vary as much as ±10 eV depending on how the element is bound to its neighbors. Therefore, XAS can yield information about the geometry, types of bonding, and oxidation state. See Brown et al. (1988) for detailed explanations of XAS.

EQUIPMENT FOR PRODUCING AND RECORDING X-RAYS

The following information should give you an idea of how to operate an older, manually controlled X-ray diffractometer. We do this because, even though many consider such a machine obsolete, machines with automated functions are not as effective for teaching. From this description based on an older machine, we hope you will be able to extrapolate to those machines that are controlled by a computer and are, therefore, relatively inaccessible for tinkering. The information we offer concentrates on adjustments that are important in maximizing responses and service from the equipment. Nothing is more valuable to your understanding the capabilities of the equipment than your own exploration of it, of varying one adjustment at a time trying to maximize a response or to eliminate an interference. More information on the operation of X-ray equipment is available in the manufacturers's manuals, in Chapter 7 in Cullity (1978), in Chapters 2 and 5 in Klug and Alexander (1974), and in recent reviews by Jenkins (1989a, b).

The organization of this section follows the path of the electric current entering the equipment from the circuit box to the voltage stabilizer, to the circuit panel, to the transformer for the X-ray generator, and then to the X-ray tube. It then follows the signal through the diffractometer, to the detector, to the amplifying circuit, and finally to the strip chart recorder. In spite of the fact that almost all new X-ray diffraction equipment is computer controlled, with results stored digitally, and displayed and manipulated on a computer screen, we refer our explanations to a Philips-Norelco X-ray diffractometer and associated gadgets, circa 1960-1970, because we think it is the best type of machine for teaching. Equipment from all other manufacturers has the same parts arranged in approximately the same manner, so comparisons are straightforward and uncomplicated, even if adjustments are not possible.

Stabilizing the Voltage

A voltage stabilizer is located in the bottom compartment of the circuit panel (Fig. 2.13). Constant voltage is the first essential step in controlling diffraction conditions because it provides an X-ray beam of constant intensity from the X-ray tube. From the wiring of the building, 220 V current runs into the voltage stabilizer. (If you have a chronically ill X-ray machine, do not assume the voltage coming into the building is 220 V.) From the voltage stabilizer, current goes to both the rest of the circuit panel and to the transformer in the X-ray generator.

Generating X-Rays

Most of the large cabinet from which the X-ray tube extends (Fig. 2.13) and upon which the diffractometer rests is filled with a transformer tank about 60 cm in diameter and 45 cm high. This unit transforms the incoming 220 V to voltages from 5,000 to 50,000 V (5 to 50 kV), which are used to accelerate electrons across the gap in the X-ray tube.

The current and the voltage for the X-ray tube are controlled by the two large knobs at the right and left of the lower part of the panel on the X-ray generator. The current, measured in milliamps (mA), heats the filament of the X-ray tube. The more the current, the greater the number of electrons that are available to be pulled across the gap to strike the target. Current and voltage must be absolutely constant when you compare the intensities of peaks within a single sample or between samples. Tubes have maximum combinations of voltage and current, or a maximum wattage, at which they may be used. Above these values, the tube will be damaged. Check the manufacturer's literature accompanying your tube to determine these maximums or you may produce an expensive flash bulb. Then experiment to find settings that optimize signal-to-noise ratios. They are almost always lower than maximum settings. A fringe benefit for lower-than-maximum settings is that the tube will last longer. The upper limits are fixed by the capacity of the tube to dissipate heat. If it is important to fit the peaks from a specific sample onto the chart paper, the voltage and current may be adjusted slightly. More about fitting peaks onto the chart paper is discussed in the section on the recorder.

To start the X-ray generator, you must start the transformer and the water circulation system, which carries away the heat generated by the X-ray tube. A switch in the X-ray generator will not close until water pressure reaches 40 psi in the circulating system. Comments about the cooling system emphasize our earlier point that about 98% of the energy from the impact of the accelerated electrons hitting the target of the tube is changed into heat energy. This is a good place to say increase the voltage slowly, and then increase the current slowly when activating the tube, in that order, *and*, just as important, turn the current down first and then the voltage, again slowly. Allow the water system to cool the tube for a few minutes before shutting it off. An X-ray tube will give faithful service for 5,000 h or more if properly cared for. Keep in mind that it costs more than $2,500 to replace one. As a tube ages, it does lose intensity.

You should keep track of tube intensity in your laboratory by designating a sample as a standard for intensity, such as a block of chert. Running it every month under exactly the same conditions and recording the intensity of one or two prominent peaks will give you a record of the tube's condition. There are at least three reasons why such a record is important. First, you may, in retrospect, wish to put an earlier diffraction run on a quantitative basis, or at least compare, e.g., the amount of quartz in a previously run sample with the quartz content of that shale that you just finished analyzing. With heavy use, a tube might have deteriorated sufficiently to reduce its intensity by one-half over a period of 1 year. Such a comparison is not feasible unless an intensity "log" has been maintained. Second, suppose your machine crashed and the serious damage required extensive service and new parts. You have paid a service engineer a small fortune to fix it, yet the nagging feeling persists that it is not as good as before the crash. Your intensity log will convince the engineer that his job is unfinished. And finally, you suspect that some knob-fixated individual has tinkered with the pulse height selection controls, or the slits, or the alignment. If running your quartz standard for the previous day (or week) shows a discontinuity in the intensity measurements, your suspicions are confirmed and you have a standard against which to readjust your machine.

The choice of tube is not usually open to students, but Brown and Brindley (1980, pp. 312ff) discuss choices. Copper tubes are the most common in X-ray diffractometry, but as they point out, the cobalt tube is being chosen more often, primarily because its wavelength, 1.79 Å, is longer than that of copper, 1.54 Å. This difference shifts low-angle peaks to slightly higher angles. These low-angle peaks are crucially important in the study of clay minerals, as you will see later. You will come to appreciate wavelengths and the angular values of diffracted beams after we have considered Bragg's law.

The Diffractometer

You need to understand the diffractometer thoroughly because its alignment and operation are most important for generating reliable data. For clay minerals, we need to obtain maximum intensity even if some resolution must be sacrificed.

It is crucial that the surface of the sample is held at and rotated about the axis, G, of the diffractometer (Fig. 2.7) because any misalignment here shifts the positions of peaks. (The terms *diffractometer* and *goniometer* are interchangeable for our purposes, although diffractometer is more specific than goniometer, which is any device for measuring an angle.) The beam leaves the X-ray tube through a window of beryllium (Why Be? Why not Pb?), then passes through the shutter opening into a Soller slit and a divergence slit. These slits collimate, limit, and direct the beam. The beam strikes the sample, and the reflected part passes through the receiving slit, another Soller slit, and an anti-scatter slit before it reaches the detector. In some machines the second Soller slit precedes the receiving slit; in others the second one has been removed to increase intensity. The antiscatter slit may be placed at one of several places in the beam path. When a single-crystal monochromator is used, the second Soller slit and the antiscatter slit are

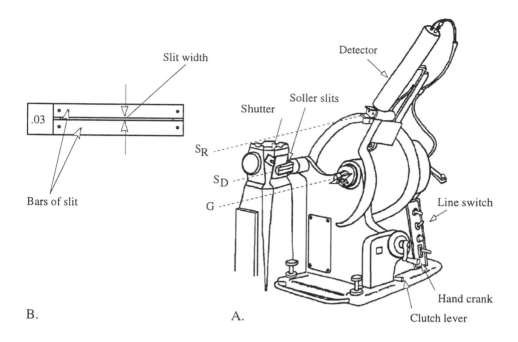

Fig. 2.7. A. Philips-Norelco diffractometer showing the axis of the diffractometer G, the axes of the divergent slit S_D and the receiving slit S_R. B. Blown-up view of a slit.

eliminated, and the receiving slit is mounted just in front of the single-crystal monochromator. (See section on single-crystal monochromator, p. 52.)

A Soller slit consists of closely spaced, parallel, highly absorbing metal plates, usually of molybdenum or tantalum. The plates are normal to the aperture of the receiving slit. A Soller slit allows only X-ray beams to pass that are traveling in a plane perpendicular to the axis G (Fig. 2.7). The divergence slit S_D is mounted on the Soller slit immediately in front of the exit port for the tube. It controls the area of the sample exposed to the X-ray beam. The separation of the two bars, or the width of the slit (given in degrees in Fig, 2.7B), controls the length of the sample that is exposed perpendicular to the goniometer axis G, and the length of the opening of the slit controls the length of the sample exposed parallel to G. For a given slit width, the length of sample exposed, perpendicular to G, will decrease as the incident angle increases (see Fig. 9.1, p. 303). The receiving slit S_R (Figs. 2.7, 2.8) is on the diffractometer circle and in front of the detector (or the single-crystal monochromator). The reflected rays focus approximately on the diffractometer circle, and the receiving slit eliminates rays that do not. The axis G and lines parallel to the long dimensions of the divergent and the receiving slits, which also may be called axes of the slits, must be parallel to

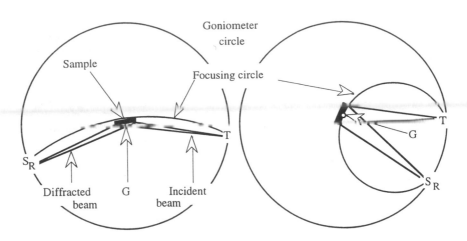

Fig. 2.8. Focusing geometry for flat specimens showing change in size of the focusing circle. G = diffractometer axis, S_R = receiving slit, T = tube or source.

one another, as shown in Fig. 2.7. The detector and the receiving slit move around the circumference of the diffractometer or goniometer circle (Fig. 2.8).

The source of the X-rays, the target of the tube, is also on the diffractometer circle; so the distance from the target to the sample equals that from the sample to the receiving slit. Both distances are radii of the diffractometer circle. As noted in the next chapter, the incident beam, the normal to the plane causing diffraction, and the diffracted beam all must be in the same plane (Fig. 3.8). This plane must be parallel to the plates of the Soller slits and perpendicular to the axes G, S_R, and S_D. Can you picture this geometry without referring to Fig. 2.7?

The diffractometer circle is fixed in size. Another circle, called the focusing circle (Fig. 2.8), changes in radius as the sample rotates through degrees θ and the receiving slit-detector assembly rotates through degrees 2θ. (There are some machines in which the sample does not move and the tube and the detector both rotate through the same angular value θ, one clockwise, the other counterclockwise.) Ideally, the sample should be curved so that it is everywhere tangential to the focusing circle. Even flat, however, it serves to focus the diverging incident beam into a beam that converges at the receiving slit (Fig. 2.8). If we did not have this focusing effect, measured diffraction intensities would be too weak to be of much use. Recall that the size of the divergence slit governs the length and breadth of the sample surface that will be struck by the incident beam. Ideally, the same amount of surface should be exposed at all angles. Do you see a problem here? Incidentally, do you see why the detector rotates 2° for every degree the sample rotates?

It is important to understand the relation among slits, peak shape, and peak intensity. The receiving slit (also called the detector or aperture slit) is the most important of the slits that control peak sharpness and resolution. Some clay mineralogists retain the slit used by the field engineer who set up the instrument. That is a mistake in most cases. We are a small portion of the customers who purchase diffractometers, and invariably the instrument is tuned to achieve the maximum 2θ resolution required by metallurgists and materials scientists. Clay minerals have *intrinsically* broad peaks that cannot be improved upon by highly refined optics. Fine slits diminish intensity—a patently undesirable factor for recording reflections from clay minerals that are very weak compared to those from many other kinds of crystalline materials. To paraphrase this discussion, the use of a fine receiving slit will not improve your diffraction patterns in terms of peak resolution, but it will surely degrade your patterns in terms of distinguishing peak from background.

To a fair approximation, the breadth B of an experimentally observed diffraction peak is described by the equation

$$B^2 = b^2 + \beta^2 \qquad\qquad (2.4)$$

where β is the pure, undistorted, diffraction peak width, in degrees 2θ, of the peak at half its height above background (sometimes called full-width at half-maximum FWHM), and b is the same quantity for a peak from a perfect crystalline substance that should produce an infinitely sharp reflection. The variable b represents the effects of the instrumental optics that broaden such an infinitely sharp peak into the shape that you would observe if you recorded one from a material such as fluorite or quartz or LaB_6 (a standard for peak shape available from the National Institute of Standards and Technology, formerly called the National Bureau of Standards). To a first approximation, that breadth is equal to the width of the receiving slit expressed in degrees. Now, suppose that a typical clay mineral reflection has particle-size broadening that gives it a breadth (β) of 0.5° 2θ, and further suppose that you select a receiving slit with a 0.05° aperture ($b = 0.05°$). The observed peak breadth B calculated from Eq.(2.4) is equal to 0.502. So the 0.05° slit added 0.002° 2θ to the width of the peak. An instrumental broadening of 0.002° 2θ is undetectable. Now let us try a 0.1° detector slit. The result of the calculation for B is 0.51. This slit has broadened the clay mineral peak by only 2%— barely detectable. How about a 0.2° slit? The results from this option produce a peak breadth of 0.54° 2θ, detectably different from the pure breadth of 0.5. We had better stop now because further increases in slit width will begin to deteriorate the sharpness of our diffraction pattern. If the slit were available, it might even be a good idea to back off a bit and select a 0.15° slit, which turns out to be a good compromise between maximum intensity and acceptable 2θ resolution for clay mineral work.

What has happened to intensity as we've changed the slit width? Diffraction intensity is proportional to receiving-slit width. If you elect to use the 0.05° slit instead of the 0.15° option, your peak sharpness will increase by an undetectably small amount, but you will have to live with a threefold loss of intensity. As we will see later, this intensity loss requires that you increase your instrument scanning time by threefold in order to achieve the signal-to-noise ratio provided by the 0.15° slit.

This analysis of instrument broadening is only approximate and should not be used for quantitative corrections of line breadth. It does, however, provide good ground rules for slit selection. Of course, if you are working with nonclay minerals such as quartz, calcite, and feldspars, then you must select a finer receiving slit and live with the intensity loss, for these minerals have intrinsically sharp lines, and their identification and quantification might require conditions of maximum peak 2θ resolution. A final note: Increasing the aperture of the receiving slit to very large values will not affect line breadth beyond some limiting value if your instrument incorporates a diffracted beam monochromator. The reason is that the monochromator focuses the diffracted beam and serves in much the same optical capacity as a large-aperture receiving slit.

There are three important factors in making a diffraction tracing that faithfully represents the variation in intensity with degrees 2θ; one is electronic and two are mechanical. The electronic one is the time constant, and the mechanical ones are the receiving slit and the scanning rate. This point is particularly important because much of the art of interpreting diffraction tracings is visual, i.e., based on the appearance of the tracings. The relation among time constant, receiving slit, and rate at which a sample is scanned is discussed later in the section on signal-processing circuitry.

As you first approach the diffractometer, it will probably be in alignment—but never assume this. Mechanical alignment is critical to obtaining precise and accurate tracings. Anyone using a given machine is responsible for monitoring alignment. Aligning a diffractometer is not difficult, but it is tedious. The manufacturer supplies detailed directions and often includes special alignment tools and materials with the equipment. There are standard materials for checking the operational alignment. For example, quartz is a particularly handy material because its structural spacings are relatively constant and because it is a common constituent of most samples. A diffraction tracing of pure quartz can be made and compared to a table, e.g., the one in Frondel (1962, pp. 25ff). The quartz peaks do not, however, test alignment in the low-angle regions so important to clay mineralogy. Although we are not aware of a muscovite that has been standardized with respect to its $d(001)$ value, one from a simple pegmatite usually gives a $d(001)$ value of 10.00 to 10.05 Å. The National Institute of Standards and Technology has offered a synthetic fluorophlogopite, No. 675, with $d(001) = 9.98104 \pm 0.00007$ Å.

For peaks representing even larger spacings, Brindley (1981) recommended keeping a slide of tetradecanol ($C_{14}H_{29}OH$) on hand. (After reading the section Interaction with Organic Compounds in Chapter 4, you should be able to tell what kind of organic compound this is.) It can be prepared by melting a few milligrams on a glass slide, placing a warmed glass slide over it, and allowing it to cool and crystallize (some cool it in liquid N). Then break the two slides apart. The $d(001)$ for tetradecanol is 39.68 Å; so the first-order peak (from Bragg's law, p. 69), if CuKα radiation is being used, is at 2.2° 2θ, the second is at 4.45° 2θ, the third at 6.68° 2θ, etc. Brindley recorded peaks up to the twenty-first order. The equipment he used was a 1960 vintage Philips-Norelco instrument. Parrish and Lowitzsch (1959) have written a detailed paper on the crucial topic of the geometry, alignment, and angular calibration of diffractometers if you ever have to start from scratch.

The receiving slit and detector assembly can be moved around the diffractometer circle by using the hand crank (Fig. 2.7). During the recording of a diffraction tracing, this assembly is driven by a motor that is engaged and disengaged with the clutch lever (Fig. 2.7). One revolution of the dial at the base of the crank equals 1° 2θ. The dial has 100 divisions. In either case, hand crank or motor, the 2θ position is shown in the small window next to the hand crank. The rate at which the assembly is driven is governed by the gears the operator selects and places on the motor. These gears are not shown in Fig. 2.7, but are located just to the right of the hand crank.

Step-Scanning with Automated Diffractometers

Let us step away from the older machines for a moment to consider one of the advantages of newer, more automated diffractometers. The operating procedures that have been discussed so far should give you a good idea of the basic physics and geometry behind the diffraction process. It is our opinion that some of the new machines are still operated as if they were the old-fashioned strip-chart recording types. In fact, we are dismayed at the poor quality of many currently published diffractograms when the remedy is so easily at hand. There are many advantages in automation, but the most overwhelming of these is the ability to produce almost any signal-to-noise ratio that you require for any sample.

Look at a diffraction pattern (Fig. 2.9A, lower trace) and you will see diffraction maxima on which are superimposed high-frequency random noise or "grass." Plotting such a pattern at an enlarged vertical scale gives empty magnification because the increase in peak height is proportional to the increase of the amplitude of the noise.

Consider now a pattern produced by a step-scanning procedure in which steps in 2θ of a certain size, say 0.05°, are counted for a specific time like 1 second. The wavelength of the grass will now be 0.05° 2θ, but its amplitude will depend on the count time. You can easily see that if you double the count time, the heights and areas of the peaks will double. The amplitude of the grass will increase only by a factor of the square root of 2 because the "noise" is random with respect to time, but the peak height is not. The standard deviation of the count for a peak is equal to ± the square root of the total count ($\pm\sqrt{TC}$). The precision of counting depends only on the number of counts

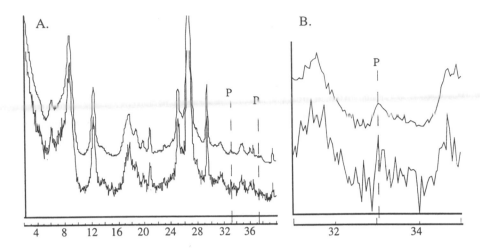

Fig. 2.9. A. Paleozoic shale, < 2 μm, ethylene glycol treated. Lower trace, 1 sec/step, 15 mA; upper trace 10 sec/step, 40 mA. P = pyrite. B. Detailed portion of A.

accumulated. For example, the standard deviation for an accumulation of 2,500 counts is ±50 or 2.0%; that of 10,000 counts is ±100 or 1%; that of 10^6 counts is ±1000 or 0.1%; etc. Klug and Alexander (1974, pp. 360-65) give a thorough explanation of counting statistics.

Suppose the peak-to-noise ratio is unity—a pretty feeble diffraction signal. Now increase the count time tenfold. The peak-to-noise ratio becomes $10/\sqrt{10}$ or 3.16. If the count time is increased 100-fold, the peak-to-noise amplitude ratio is 10. The law of diminishing returns catches up with you, but if you are willing to spend enough time on the experiment, you can obtain very well formed peaks from samples that may show no peak at all if analyzed with "quick-and-dirty" methods. The power of the step scan is made possible by the remarkably stable electronics that control beam intensity in modern X-ray machines.

We will illustrate what can be gained by long count times. Figure 2.9A shows two diffraction patterns from the < 2 μm fraction of a Paleozoic black shale solvated with ethylene glycol and prepared as an oriented aggregate. The lower trace was obtained with a count time of 1 sec per 0.05° step and the tube set at 40 kV and 15 mA. These conditions roughly simulate what we obtained from the older strip-chart machines operating at 2° 2θ per minute and a current setting appropriate for the older tubes. The upper trace was recorded at 40 kV, 40 mA, and 10 sec per step. Most of the essential features of the diffraction pattern can be seen in the lower trace, but the upper one certainly inspires more confidence in any detailed interpretations. For example, the slight low-angle bulge at about 8° looks real on the upper trace, but you would not have the courage to claim it if you only had the lower pattern.

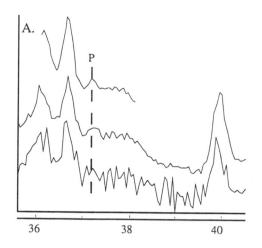

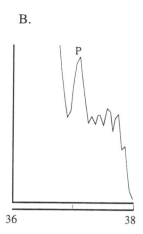

Fig. 2.10. A. The higher angle pyrite peak (P). The lower two traces are from the patterns of Fig. 2.9A. The uppermost trace was run at 100 sec/step, 40 mA. B. Vertical magnification of the upper trace of A.

But let us look deeper into the mineralogical composition of this shale. It is black; so we suspect the presence of pyrite. The two dashed vertical lines on Fig. 2.9A mark the positions of two strong pyrite lines. Both traces indicate a peak for pyrite at about 33° 2θ, though only the upper trace is definitive. Figure 2.9B shows a portion of these same two diffraction patterns that have been expanded vertically. The upper trace, obtained with a count time of 10 sec per step, shows a pyrite reflection that is good enough for quantitative analysis, but the lower pattern is simply an example of empty magnification.

How about the higher-angle pyrite reflection that is not evident on either of the patterns of Fig. 2.9A? Figure 2.10A shows the two patterns of Fig. 2.9A that have been magnified and, in addition, a trace (upper) of a portion of the diffraction range obtained with a count time of 100 sec per step. The latter clearly shows a well-formed pyrite reflection. Yes, pyrite is present, and if a randomly oriented sample were prepared, it would be possible to analyze quantitatively for it.

We have not reached the limit of practical magnification for this pattern, as evidenced by the absence of grass. The five tiny peaks to the right of the pyrite peak on the upper trace of Fig. 2.10A are real and can be assigned to minerals present in the sample. Replotting it with increased vertical magnification makes it possible to do something with those five peaks, if we so desire (Fig. 2.10B). The truncated tops of some of the peaks on Fig. 2.10B result from the relatively large 2θ steps used (0.05° 2θ), which could have been minimized by smaller steps, though that would have further lengthened the counting procedure.

So the question comes up—what step size and count time should be used? We suggest steps of 0.1° for clay mineral samples and 0.05° for bulk rock powders, if the application is routine qualitative analysis. The accuracy of integrated peak areas does not depend much on step size, as long as it is at least 1/5 of the peak width at half-height. For accurate *d*-values from nonclay minerals, a step size of 0.02 is adequate, though 0.01 if you are determining lattice parameters. A rule of thumb used in our labs is to set the count time so that the heights of the strongest important reflections are about 10,000 counts above the local background. If you cannot practically approach that value, then try another sample preparation.

Time constraints can be mitigated by running only selected portions of the pattern. Note the example given above for pyrite. Long count times lower detection limits dramatically. Dr. David Bish (Los Alamos National Laboratories, personal communication, 1995) suggested that a practical limit for quartz, for example, lies in the range of 5 to 10 ppm.

The Single-Crystal Monochromator

Returning to our treatment of components of older or teachable machines, a single-crystal monochromator is essentially a small spectrometer using a curved graphite crystal, usually placed just in front of the detector, which is positioned so that a prominent set of spacings of the graphite crystal diffracts the Kα-part of the X-ray beam into the detector. However, because Kα is diffracted at a slightly different angle than Kß, only Kα enters the detector. Therefore, an absorbing filter in the X-ray beam is no longer needed, which is an advantage because, if you recall, not only is the Kß absorbed by a metal foil, but part of the Kα radiation also is absorbed. This is another application of Bragg's law (see p. 69).

The Detector

The detector (Fig. 2.11), sometimes called a quantum counter, catches and reacts to the diffracted beam or any other kind of short-wavelength, ionizing radiation. There are five types of detectors: film, proportional, Geiger, scintillation, and semiconductor (solid state). Proportional and Geiger detectors are gas-filled tubes. Photons of an X-ray beam interacting with the gas produce ion pairs, positively charged gas ions and electrons. Depending on the initial energy of the X-ray photon and the kind of gas in the detector, a single photon may create hundreds of ion pairs. The gas ions migrate to the negatively charged cathode, or external housing of the detector, and the electrons migrate to the anode, or thin central wire, of the detector (Fig. 2.11). An electron contacting the central wire causes a voltage pulse to travel out of the detector and to be amplified and processed by the electronics in the circuit panel. There must be a voltage difference, or potential, between the anode and the cathode of the detector for it to work. This potential is supplied from the circuit panel. The detector responds properly only within a range of voltages characteristic of the individual detector, whether gas-filled detector or solid state. This range can be determined by setting the detector at an angular position and adjusting voltage and current so that the detector receives a diffracted beam of moderate strength, say, 200 to 1,000 counts per second (cps). With the detector at this position, beginning from essentially no voltage, increase the voltage across the detector in steps of 20 to 40 V and record the counting rate at each step. Your plot of counting rate versus voltage should look like Fig. 2.12. There will be no counts until a threshold voltage V_T is reached. Then the counting rate will increase rapidly until the front edge of the voltage plateau V_{PF} is reached. The count rate will increase slightly with increase in voltage until a second point at the back of the plateau is reached, V_{PB}. If you increase the voltage past V_{PB}, the count rate again will begin to increase sharply, but don't go any further. The detector can go into a state of continuous discharge and be damaged. The voltage you want across the detector is the voltage in the middle of the plateau, an optimal voltage V_O. Adjustments of voltage across the detector for Philips-Norelco equipment are made on the circuit panel (Fig. 2.13).

All detectors have dead time. We can illustrate this by considering the Geiger tube (Fig. 2.11). Compared to other types of detectors, the Geiger tube takes a long time (about 4×10^{-4} sec) to process one hit or one incident photon. During this time, it cannot process any other hits. This time is called *dead*, a time when the detector is said to be saturated with ions. Ideally, we would like a detector that gave a straight line count; i.e., for each hit, one pulse would be sent to the counting circuits. For Geiger detectors, the response becomes nonlinear at rates above 100 to 300 cps, not a good condition if you are trying to establish relative intensities of peaks for making quantitative estimates of the amounts of a mineral. One variety of the proportional counter, which is quite similar to a Geiger tube (Fig. 2.11), flushes gas through the detector during operation. This greatly reduces the possibility that the gas will be saturated with ions and therefore reduces the occurrence of dead time. The Geiger detector is seldom used these days, although some of the starred cards in the ASTM Powder Diffraction File (a star on a card indicates that the data on it are highly reliable) have values of relative intensity that were measured using a Geiger tube. This means, for these cards, the strongest peaks probably have been assigned intensity values

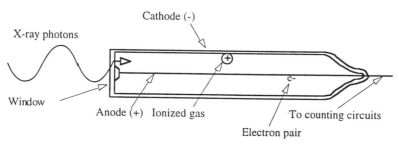

Fig. 2.11. Diagram of a Geiger detector tube.

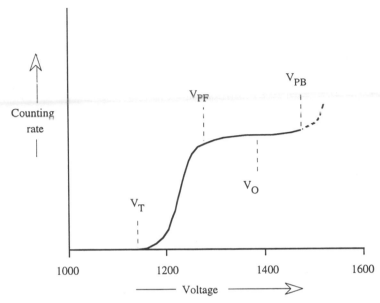

Fig. 2.12. Typical response of a Geiger tube detector showing the voltage plateau. (Other counters would have a similarly shaped curve.)

that are too low. The old quartz card (5-0490, which has been replaced with 33-1161) is a good example. See Klug and Alexander (1974, pp. 313ff) for a comprehensive review of all types of detectors.

Signal Processing Circuitry

There is much signal processing circuitry you cannot see nor do much about. Although it is distinctly to your advantage to know what is going on in there, we will bypass most of the circuitry and refer you to Klug and Alexander (1974, Chapter 5) and the manufacturer's manuals of the X-ray equipment you are using. The part you must see is the signal display. The signal has been amplified at several stages. Thus for each pulse, at the time it is displayed or recorded, the voltage is 10^3 to 10^5 times larger than the initial signal. The first amplification is usually in the detector itself, e.g., the single photon producing hundreds of ion pairs. For most detectors, the voltage pulse sent to the counting circuitry is proportional to the energy of the incident photon. Pulses can be sorted according to their voltage, and most circuitry will allow pulses higher or lower than a certain voltage to be rejected. These high and low limits can be adjusted. This is called *pulse height discrimination*. With some detectors, rather than using an adsorption edge to decrease the intensity of the Kβ line (as in Fig. 2.6), it can be screened out by excluding it because its pulse has a higher voltage than that of Kα radiation. See Cullity (1978, p. 214) or Klug and Alexander (1974, p. 344) for more information.

The Strip-Chart Recorder

Diffraction tracings from the type of machines we have been discussing are made on paper strips fed through a recording pen potentiometer that continuously records the output signal from the ratemeter circuit as the diffractometer scans the sample. This results in a graph of diffracted X-ray intensity vs. the angle, in degrees 2θ, of reflection or diffraction of the X-ray beam. Frequently, the simplest part of the recorder causes the diffractionist the most persistent and nagging problem: the pen does not work. Your ingenuity and patience will be tested. (This is certainly one of the reasons that most modern machines record diffraction data in digital form and display it on the screen of a terminal.) Other than this problem, you should be familiar with the manual provided by the manufacturer. There are usually adjustments for dampening the response of the pen and for changing chart speeds.

Weak signals on these strip-chart-type machines can be enhanced by using a combination of time constant, rate of chart speed, and rate of diffractometer scan speed. For example, two tracings covering exactly the same angular span and covering exactly the same amount of chart paper can be made with the settings given in Table 2.2. Note that the time constant was increased by a factor of four, while the chart speed and diffractometer speed were reduced by a factor of four. These changes allow four times as many counts to accumulate per increment of scan. You can then increase the magnification (the scale factor) four times so that you can resolve very weak reflections from the background. As in the discussion above for automated step scanning, if you only increase the magnification for the fast scan, it is empty magnification.

We recommend the values in Table 2.2 for clay minerals because their peaks are intrinsically broad and not much affected by time-constant distortion. It is a better procedure for nonclay minerals to use a time constant of 0.4 sec for the 2° 2θ/min condition and a value near 2 sec for the 0.5 to 2° 2θ/min scans.

Adjustments for fitting peaks onto the chart paper are made on the circuit panel. You will find controls allowing you to set the number of counts per second needed for a peak to reach the top of the chart, i.e., allow you to adjust the sensitivity. The best way to do this before making a diffraction tracing is to guess where the most intense peak will be, set the detector at this angle, and then fit the peak onto the chart paper. For example, most samples contain

Table 2.2. Comparison of settings for enhancing weak signals

Time constant	Chart speed	Diffractometer speed
1.0 sec	2 cm/min	2° 2θ/min
4.0 sec	0.5 cm/min	0.5° 2θ/min

quartz, which has its strongest peak at 26.6° 2θ for CuKα radiation. So use the hand crank to move manually the diffractometer to this position and then maximize the signal before choosing a value for full-scale limit.

An Example of a Checklist for Operating XRD Equipment

These operating instructions are offered as a case study for a particular machine. Every lab should have such a set of instructions written specifically for that lab (e.g., circuit breaker for the X-ray generator is on the north wall behind the X-ray generator). As we have emphasized elsewhere, we think the less automated type of machine is ideal for teaching and learning purposes.

Before using the machines in your lab, assuming you are the student, you will need the instruction and permission of a qualified person. Most institutions also have some sort of a health physics functionary who will want to show you pictures of people who have burned their fingers or arms or eyes through accidental use of X-rays. The instructions we offer will serve as a checksheet for you after you have been checked out on the equipment. Instructions are written as if you are facing the X-ray generator (Fig. 2.13).

1. Enter in the logbook the date, purpose, name of user, hours on the generator (find the rectangular "odometer" on the panel, which records the accumulated time), and any other information requested. (The logbook is also a good place to record safety checks, intensity checks, and repairs or problems.)
2. Switch on the water-cooling and circulating equipment. The machine won't run without this. Remember, all that heat has to be carried away.
3. Switch on the line voltage to the X-ray generator.
4. Put on your film badge.
5. Make sure all beam ports and the diffractometer shutter are closed.
6. Turn on the power switch in the middle of the voltage regulator, which is behind and to the left of the X-ray generator. There will be a slight delay as the unit warms up. (In many laboratories the voltage regulator is left on unless the machine is not expected to be used for several days.)
7. Turn on the power to circuit panel. The switch is in the lower part of the leftmost NIM module.
8. Make sure the key is in the on position in the control panel of the generator and that there is plenty of time on the timer or that it is set at infinite.
9. Check the knobs for kV and mA on the generator. They should be turned as far down as they will go, or counterclockwise (if they are not at this position, the X-ray tube may be shocked when unit comes on).
10. Check the mA stabilizer (upper left-hand corner on the right side of the generator cabinet) by looking through the hole about 1.5 cm in diameter. It should be on medium for diffraction; medium or high for fluorescence.

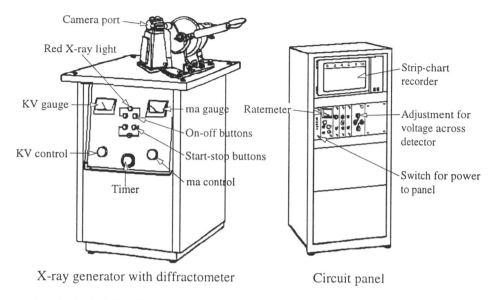

X-ray generator with diffractometer   Circuit panel

Fig. 2.13. A Philips-Norelco X-ray generator with diffractometer and a circuit panel, both manufactured about 1970.

11.   Push the black **START** button on X-ray generator.
12.   Wait about 30 sec; then push black **ON** button. It will take a minute or two before the red X-ray light comes on. There is a small audible pop or snap when the X-rays come on. At this time the kV needle will jump to about 10 and the mA needle to about 3.
13.   Turn kV up *slowly* to proper setting. If mA is turned up first, both X-ray tube and milliamp potentiometer could be damaged.
14.   Turn mA up *slowly* to proper setting. This will cause kV setting to "fall off" a bit, so readjust kV setting as required. (Maximum values for diffraction are 40 kV and 20 mA for a Cu target tube for most older machines. Loadings are considerably higher for some modern, high-intensity tubes.)
15.   Turn on the power to the strip-chart recorder (not the drive).
16.   Double check that the shutter for the diffractometer is closed. Then carefully remove the X-ray shield around the sample holder. [The shield slides on with one aligning pin on the sample holder fitting into the widest slot on the shield. Be very gentle with the shield because the diffractometer only rests (it is not fastened) on the top of the X-ray generator and too vigorous pushing and pulling could put the diffractometer out of line.]
17.   Place the sample in sample holder of the diffractometer and replace the shield.

18. Make sure the CLUTCH lever (directly below the crank that when turned moves the diffractometer) is disengaged and then turn on the LINE switch on the right-hand side of the diffractometer .

19. Using the crank, turn the diffractometer to the approximate position of what you anticipate will be the strongest peak, perhaps the quartz peak at 26.6° 2θ. Open the shutter and, again using the crank or by gripping the perimeter of the dial at the base of the crank, find the position of maximum intensity. At this position, one usually would like to have this most intense peak at full scale on the chart paper. If this is not the case, then the RANGE setting on the amplifier module in the circuit panel needs to be adjusted.

20. Synchronize the diffractometer and the strip-chart recorder. For clay minerals, we usually start at 2° 2θ. Synchronizing is accomplished by allowing the chart to run toward a predetermined starting point, e.g., 2° 2θ and, when it reaches this point, pulling the clutch so that the diffractometer begins to scan. The diffractometer has been preset at 2° 2θ also. If manual synchronization is not good enough, you may need to add an internal standard to your sample.

21. You should know beforehand how far you want to scan. For preliminary identifications of clay minerals, 33° 2θ is usually enough. To look at quartz and feldspar carefully, or to check the 060 peak of a clay mineral, you might want to go past 60° 2θ. *NEVER* allow the diffractometer to run past approximately 100° 2θ. If the diffractometer runs into the X-ray tube mount, it will be damaged. There are switches to reverse the direction of the diffractometer drive, but if they have been moved or are loose, the diffractometer will be damaged. Do not trust a $20 switch to protect a $20,000 goniometer. In addition, *NEVER* set the goniometer to 0° 2θ, or the intense primary beam will damage the detector.

22. When you are ready to quit, turn off chart drive switch and the instrument switch on the recorder. Disengage the clutch and close the shutter on the beam port for the diffractometer. Turn first the mA and then the kV knobs SLOWLY all the way counterclockwise. Give the water a few minutes to cool the tube and then push the red OFF button and red STOP button. And finally, enter the hours reading in the logbook. Now you are ready to identify and analyze the diffraction tracing you have made.

A final word on safety. If something drastic goes wrong with the equipment (smoke, steam, etc.) do not touch anything but the wall emergency switch, and shut that off. If the machine is shorted to ground, you will be of no use to the machine or yourself if you are in parallel with it.

SUMMARY

The main point of this chapter is that you need to know your equipment and how it works in order to (1) optimize its response and (2) maximize the precision and accuracy of its response. Another reason for being familiar with the equipment is that all electronic equipment has the capacity to embarrass its operator. When a machine is not functioning properly, 80% of the time it is something very simple: a switch that hasn't been flipped, a timer that has reached zero time, a fuse that has blown, humidity or temperature that is too high. Don't expect to learn all the idiosyncrasies in one term or one year or one decade. When the equipment isn't working properly is when it's important to have a checklist (as in the preceding section) . . . and a cool head.

It's fair to assume that the equipment you use as you learn about X-ray diffraction, or any other laboratory technique, will become obsolete, if it isn't already. Having learned one system, however, you will find it relatively easy to understand other systems. The understanding of the physics of X-rays almost certainly will not change in your lifetime.

Two other factors to consider in order to have complete command of X-ray diffraction analysis are the phenomenon of diffraction and the preparation of samples. These subjects are covered in Chapters 3 and 6, respectively.

REFERENCES

Blum, A., and Nagy, K., editors (1994) *Scanning Probe Microscopy of Clay Minerals*: Vol. 7, CMS Workshop Lectures, Clay Minerals Society, Boulder, Colorado, 239 pp.

Brindley, G. W. (1981) Long-spacing organics for calibrating long spacings of interstratified clay minerals: *Clays and Clay Minerals* **29**, 67-68.

Brown, G., and Brindley, G. W. (1980) X-ray diffraction procedures for clay mineral identification: in Brindley, G.W., and Brown, G., editors, *Crystal Structures of Clay Minerals and Their X-Ray Identification:* Mineralogical Society, London, 305-59.

Brown, G. E., Jr., Calas, G., Waychunas, G. A., and Petiau, J. (1988) X-ray absorption spectroscopy and its applications in mineralogy and geochemistry: in Hawthorne, F. C., editor, *Spectroscopic Methods in Mineralogy and Geology*: Reviews in Mineralogy, Vol. **18**, Mineralogical Society of America, Washington, D.C., 431-512.

Buseck, P. R., editor (1992) *Minerals and Reactions at the Atomic Scale: Transmission Electron Microscopy*: Vol. 27 in Reviews in Mineralogy: Mineralogical Society of America, Washington, 508 pp.

Buseck, P. R., Cowley, J. M., and Eyring, L., editors (1988) *High-Resolution Transmission Electron Microscopy*: Oxford University Press, 645 pp.

Cullity, B. D. (1978) *Elements of X-Ray Diffraction*, 2nd ed.: Addison-Wesley, 555 pp. Clearly presented theory and practice.

Gollnick, D. A. (1983) *Basic Radiation Protection Technology*: Pacific Radiation Press, Temple City, California, 436 pp.

Hawthrone, F. C., editor (1988) *Spectroscopic Methods in Mineralogy and Geology*: Vol. **18**, Reviews in Mineralogy: Mineralogical Society of America, Washington, 698 p.

Frondel, C. (1962) *The System of Mineralogy*, Seventh Edition: *Volume III, Silica Minerals*: John Wiley and Sons, Inc., New York, 334 pp.

James, R. W. (1965) *The Optical Principles of the Diffraction of X-Rays:* Vol. II of *The Crystalline State*: Series edited by Sir Lawrence Bragg, Cornell University Press, Ithaca, 664 pp. A thoroughly readable account of almost every theoretical aspect of the physics of X-ray diffraction. An old book, but the principles are ageless.

Jenkins, R. (1989a) Instrumentation: in Bish, D. L., and Post, J. E., editors, *Modern Powder Diffraction*, Reviews in Mineralogy, Mineralogical Society of America, Washington, D.C., Vol. **20**, 19-45.

Jenkins, R. (1989b) Experimental Procedures: in Bish, D. L., and Post, J. E., editors, *Modern Powder Diffraction*, Reviews in Mineralogy, Mineralogical Society of America, Washington, D.C., Vol. **20**, 46-71.

Joy, D. C., Romig, A. D., Jr., and Goldstein, J. I., editors (1986) *Principles of Analytical Electron Microscopy*: Plenum Press, New York, 448 pp.

Klug, H. P., and Alexander, L. E. (1974) *X-Ray Diffraction Procedures*, 2nd ed.: Wiley, New York, 966 pp. The most complete book on methodology available. Very useful.

Mackinnon, I. D. R. (1990) Introduction: in *Electron-Optical Methods in Clay Science*: Vol. **2**, CMS Workshop Lectures, Clay Minerals Society, Boulder, Colorado, 1-13.

Mackinnon, I. D. R., and Mumpton, F. A., editors (1990) *Electron-Optical Methods in Clay Science*: Vol. **2**, CMS Workshop Lectures, Clay Minerals Society, Boulder, Colorado, 159 pp.

National Council on Radiation Protection and Measurements (1985) *SI Units in Radiation Protection and Measurements*: NCRP Report 82, National Council on Radiation Protection and Measurements, Bethesda, Maryland, 64 pp.

Parrish, W., and Lowitzsch, K. (1959) Geometry, alignment and calibration of X-ray diffractometers: *Amer. Mineral.* **44**, 765-87.

Stucki, J. W., Bish, D.L., and Mumpton, F. A., editors (1990) *Thermal Analysis in Clay Science*: Vol. **3**, CMS Workshop Lectures, Clay Minerals Society, Boulder, Colorado, 192 pp.

Wilson, M. J., editor (1994) *Clay Mineralogy: Spectroscopic and Chemical Determinative Methods*: Chapman and Hall, London, 367 pp.

Chapter 3
X-Ray Diffraction

Diffraction is a phenomenon with which we are all familiar, if only as the cause of the rainbow on the surface of a compact disc, or on a puddle of water with a film of oil on top, or from the wing of a butterfly. These effects are dependent on the properties of light: wavelength, amplitude, and phase. For light in the optical range, we experience a change in wavelength as a change in color; a change in amplitude, we experience as a change in brightness; we cannot, however, directly experience a change in phase. We need to examine these phenomena as they apply to the interaction of X-rays and crystals. *The essential feature of the diffraction of waves of any wavelength is that the distance between scattering centers be about the same as the wavelength of the waves being scattered.* The dimensions of X-rays and the spacings between atoms in crystals meet these conditions, as von Laue and his group showed in 1912. The wavelength of X-rays and the structural spacings of crystals both have dimensions about 10^{-8} cm (10^{-8} cm = 1 Å). In brief, this chapter develops the concept that X-rays are scattered by the electrons around the nuclei of the atoms composing a unit cell. This scattering is modified in three ways: (1) by the way in which the electrons of a particular atom are distributed within the field of influence of that atom; (2) by the thermal vibrations that tend to blur the atoms as scattering centers as temperature increases; and (3) by the way atoms are arranged within the unit cell. Each scattering center contributes a diffracted beam that can be represented by a vector. Atoms, unit cells, or crystallites (very small crystals) can be treated as

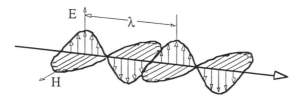

Fig. 3.1. A single, polarized ray of the X-radiation part of the electromagnetic spectrum. E is a vector representing the electrical field, and H is a vector representing the magnetic field. Wavelengths in this region of the spectrum range from 0.5 to 10 Å.

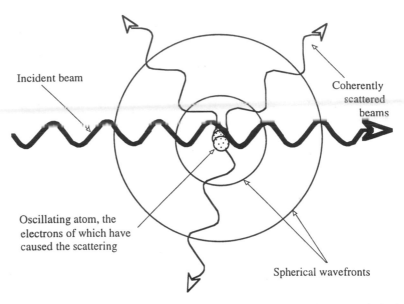

Fig. 3.2. An atom oscillating about its lattice point in the crystal structure behaving like a point source for radiation scattered spherically (like an isometric optical indicatrix).

scattering centers. All the vectors from scattering centers within an optically coherent domain must be summed into a single vector, and then the vectors from each domain must be summed. This will bring us to a resultant, diffracted beam with a specific amplitude and phase, and we will see that we can calculate the intensity of this beam to compare it to our experimental results.

SCATTERING

Like all electromagnetic radiation, we can characterize X-rays in terms of a single, polarized ray with an electric vector **E** vibrating perpendicular to the direction of propagation in one plane and a magnetic vector **H** vibrating perpendicular to both the direction of propagation and to that of the electric vector (Fig. 3.1). **H** will be of little interest to us, but **E** of an incident photon is the stimulator of scattering. As an incident ray encounters an electron, the electron vibrates in resonance with **E** (i.e., it vibrates with the same frequency). Because the electron is a charged particle, and a vibrating charged particle is an emitter of electromagnetic radiation, the vibrating electron becomes a beacon absorbing a small amount of energy from the incident beam and reradiating it in all directions at the same wavelength as the incident beam (Fig. 3.2). This phenomenon is called *coherent scattering*. (*Incoherent scattering* is scattered radiation with a wavelength longer than that of the incident or stimulating radiation. The term *scattering* in this text refers to coherent scattering. For more about incoherent scattering, see Fig. 2.4, p. 36 and the topic Compton scattering in Cullity, 1978 or Klug and Alexander, 1974.)

The electrons doing the scattering that interests us are held in orbits around nuclei. Because each electron scatters, the scattering power f of an atom increases with the number of electrons bound to that atom. The scattering power of an atom is somewhat less than the scattering power of a single electron multiplied by the number of electrons in the atom. The reason is that outgoing waves scattered by electrons in different parts of the same atom will be slightly out of phase with each other and some destructive interference results. These effects increase with scattering angle; so scattering efficiency of an atom declines as the Bragg angle θ increases. However, for our purposes we can consider each atom in a unit cell as a point source of scattered radiation and ignore most of these other details. We will discuss phase and Bragg angle shortly.

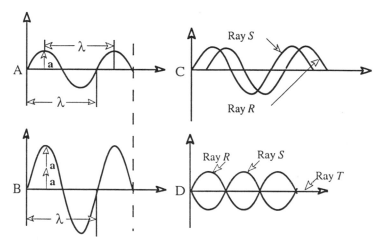

Fig. 3.3. Constructive and destructive interference. A. A sine wave representing an X-ray and showing amplitude **a**, a vector, and wavelength λ. B. Two rays exactly the same as the ray in A added together by constructive interference. Note that the same wavelength is maintained but the amplitude is doubled. The dashed line perpendicular to the rays in A and B is a wavefront, so called because it marks the same place in the progress of both rays. See the wavefronts in Fig. 3.4. C. Rays R and S are not in phase as were the two rays that added to make the ray in B. The resultant ray has an amplitude less than that in B. As the two rays approach being exactly out of phase the resultant amplitude will decrease to zero. D. Here rays R and S are exactly out of phase and the result is ray T or essentially no ray at all.

INTERFERENCE

Next, you need to understand what is meant by *constructive* and *destructive* *interference* of electromagnetic radiation. In short, there is constructive interference when two or more rays are in phase, and destructive interference when the rays are out of phase. In Fig. 3.3, A shows the wavelength λ, and that wavelength is the same from point to similar point. A line drawn perpendicular to the direction of propagation touches both rays in A and B at exactly the same state: at the point of zero amplitude. (Such a perpendicular line is called a *wavefront.*) Think of the amplitude of each ray at each increment of propagation as a vector (Fig. 3.3). In all cases, whether interference is constructive or destructive, the amplitude portion of the vectors add, like the **a** vectors in B, which represents the adding, or constructive interference, of two waves equal to the wave in A. C shows two rays, *R* and *S*, that are not in phase. Try putting a line perpendicular to their direction of propagation. Does it meet the criterion given for that shown in A and B? The amplitude vectors will add here also, but in some cases, one vector will be positive and the other negative, thus partially or completely canceling one another out reducing the amplitude to something less than that shown in B. Additional rays, also slightly out of phase with *R* and *S* would again reduce the amplitude and change the phase of the resultant ray, as would each additional ray that was out of phase. Now, in Fig. 3.3D, imagine that the amplitude vectors of rays *S* and *R* are added. Do you see that each would exactly cancel out the other resulting in ray *T*? This is completely destructive interference.

When two or more rays constructively interfere, a diffracted beam is produced, but to be useful there must be many mutually constructive rays. *A diffracted beam that we can observe and measure is a beam composed of an enormous number of constructively interfering rays, rays that are mutually reinforcing one another.* Now, what sort of condition will lead to such an observable beam?

Scattering from a Row of Atoms

Figure 3.4 shows how a diffracted beam can be formed as rays move through a row of regularly spaced atoms. Figure 3.4A shows wavefronts of the incident beam advancing parallel to a row of eight atoms, each of which, stimulated by the passing wavefront, becomes a beacon sending out waves of the same wavelength as the exciting radiation. Just like the single atom in Fig. 3.2, the scattering from the eight atoms is spherical and is represented here by concentric circles with radii of one, two, and three wavelengths. Any lining up of spherical wavefronts makes a mutually reinforced wave moving in the direction perpendicular to a tangent common to the spheres. The most obvious case of this is shown in Fig. 3.4B where wavefront *A* is the wavefront of a diffracted beam traveling straight ahead and parallel to the incident exciting rays. This wavefront is a line tangent to the sphere one wavelength from each atom. Though this shows the simplest case, this particular diffracted beam is overlaid by the incident beam. Therefore, it cannot be observed. We can find other diffracted beams, however.

Figure 3.4C shows wavefronts for three diffracted beams. Imagine that the spherical wavefront from atom 8 has just begun to scatter so that it is a point rather than a sphere. A line from this point tangent to the spherical wavefront that has traveled one wavelength from atom 7 is tangential to the spherical wavefront that has traveled two wavelengths from atom 6, that has traveled three wavelengths from atom 5, and so on. This line makes wavefront *B*. Diffracted beam *B'* is traveling perpendicular to wavefront *B*. Do you see the patterns for the other diffracted beams? For example, notice that wavefront *C* is tangent to wavefronts from atoms 2, 4, 6, and 8 or every second atom. What is the pattern for wavefront *D* and its associated diffracted beam? Do you see that if the atoms were not evenly spaced or periodic there could be no constructive interference and, therefore, no diffraction?

The three wavefronts in Fig. 3.4C also show that diffraction occurs at specific angles, with no diffraction in the angular regions between diffracted beams; i.e., there is no diffraction effect between *B'* and *C'*. We will demonstrate later that the angles of *B'* and *C'* to the line of atoms are functions of the wavelength of the incident radiation and the spacings of the atoms (Bragg's law, p. 69). Perhaps you can see that from Fig. 3.4C.

Another way to look at Fig. 3.4A to see where diffracted beams are formed is to realize that anywhere two spherical wavefronts intersect, rays coincide and their amplitudes add; in other words, there is constructive interference. For example, at points *a*, *b*, and *c* there is mutual reinforcing by spherical wavefronts from atoms 4 and 5. In addition, the third wavefronts from atoms 3 and 6 add to the amplitude at point *b*. These reinforcements make a diffracted beam, shown by the line drawn through *a*, *b*, and *c*, traveling in the same direction as the incident beam.

Where the first wavefront from atom 5 (Fig. 3.4A) and the second wavefront from atom 4 intersect (point *d*) and the second wavefront from 5 and the third from 4 intersect (point *e*), a second diffracted beam is created. Whether we picture the diffracted beams as shown in Fig. 3.4A or as defined by the tangents as shown in Fig. 3.4C, there is a small, finite number of beams at discrete angular spacings. There is *not* an infinite number of diffracted beams.

Now, before we go on to visualize diffracted beams created by scattering from atoms in three dimensions, let us consider a row of atoms such as that shown in Fig. 3.4A, scattering not in two dimensions as shown, but in three dimensions. Picture the intersection above the plane of the paper of the first spherical wavefronts around atoms 4 and 5. Do you see that their intersection would describe a circle on a plane common to the beam defined by points *a*, *b*, and *c* and this plane would be perpendicular to the plane of the paper? Likewise, the intersection of the second and third spherical wavefronts from

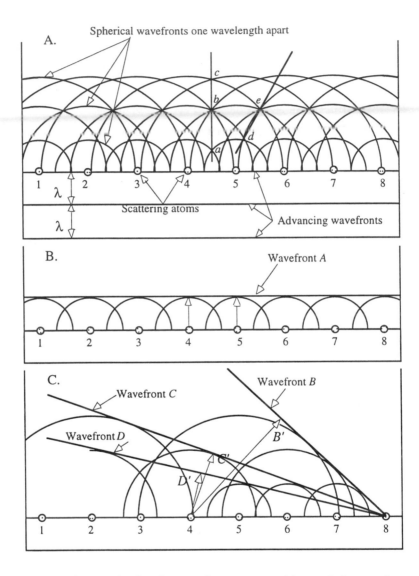

Fig. 3.4. Constructive interference of waves scattered by regularly spaced row of atoms. A. Wavefronts from each atom with reinforcing or constructive interference at points *a*, *b*, *c*, *d* and *e*, e.g., wavefronts from atoms 4, 5, 6, and 7 contribute to the constructive interference at point e. B. A single wavefront of scattered, reinforcing waves. C. Reinforced scattering at other than the direction of the primary beam. Ray *B'*, for example, is perpendicular to wavefront *B*.

atoms 4 and 5 would trace circles on the same plane perpendicular to the paper. You should be able to picture a planar disk perpendicular to the paper that contains points *a*, *b*, and *c*, and these points would be on concentric circles representing the intersecting spherical wavefronts of atoms 4 and 5.

Now that you have visualized this you hardly need Fig. 3.5, right?

If the sum of the diffracted beams represented by the line *abc* in two dimensions takes the shape of a disk in three dimensions, then what shape will be made by the diffracted beams represented by the line *de*? Try to make it swing out of the plane of the paper from Fig. 3.4A toward you. Points *d* and *e* will each trace the intersections of spherical wavefronts. Concentrating on point *e* and the intersection of the second wavefront from atom 5 and the third from atom 4, can you see that as point *e* traces the intersection of these two spherical wavefronts it will trace a circle in a plane perpendicular to the paper? Point *d* also will trace a circle in a plane perpendicular to the paper, but it will be smaller. The sum of all diffracted beams connecting or defined by the two circles traced by *d* and *e* forms a cone, the axis of which is the row of atoms. This is represented in Fig. 3.5 and a cross-section of this figure containing lines *abc* and *de* would look like Fig. 3.4A.

From each pair of atoms in the row, a similar set of cones of diffracted beams will radiate. A cross-section of such a row would show whole sets of lines parallel to lines *abc* and *de*. For the circumstances shown in Fig. 3.5, a set of lines or rays would be in phase and therefore would mutually reinforce one another, resulting in a diffracted beam.

Scattering from a Three-Dimensional Array of Atoms

We are now ready to consider the manner in which a diffracted beam is formed as a result of scattering by the atoms of a three-dimensional crystal structure. To imagine a three-dimensional array of atoms, picture three rows of atoms like the one in Fig. 3.5. Figure 3.6 shows rows of atoms *X*, *Y*, and *Z* intersecting at right angles (as if the rows were a set of orthorhombic reference axes). In response to an incident beam, scattering from a single atom at the point of intersection is combined with scattering from an adjacent atom on each of the axes to form the three cones. The angle at which the cones open is related to the angle at which the incident beam strikes. A, B, and C of Fig. 3.6 show the cones opening at larger and larger angles as the incident beam approaches a Bragg angle (an angle at which constructive interference will occur). The cones represent diffracted beams. Where the cones intersect, the diffracted beams undergo additional constructive interference. In Fig. 3.6A the cones do not intersect. In B they just touch, so there would be three minor diffracted rays coming out of the quadrant enclosed by rows *X*, *Y*, and *Z*. In C they penetrate so that all three cones have one common line of intersection. The diffracted beams shown in Figs. 3.4 and 3.5 would have significantly less intensity than the diffracted beam of Fig. 3.6C. In addition, there will be a set of cones associated with each pair of atoms throughout the unit cell. All of this will result in whole families of parallel diffracted beams, the sum of which will produce a diffracted beam with intensity enough for us to observe. Do you see that this sort of diffracted beam will have greater intensities than those of a beam like *de* in Fig. 3.5

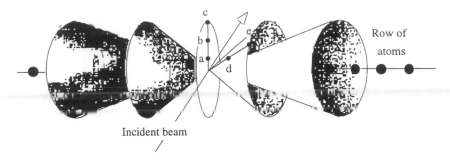

Fig. 3.5. Diffraction cones from the interaction of wavefronts from two adjacent atoms in a row of atoms. Points *a, b, c, d* and *e* are from Fig. 3.4.

because there are more atoms contributing to the scattering? The diffracted beam in Fig. 3.6C. can be described in terms of direction cosines. For this and for a more rigorous development of these steps, see Buerger (1942, pp. 29-43).

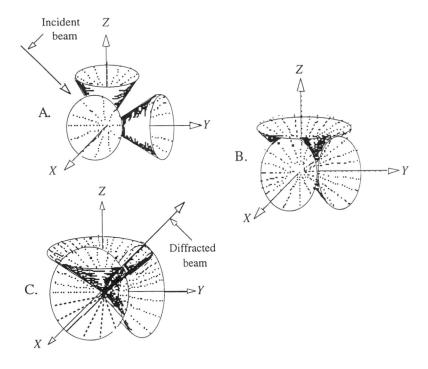

Fig. 3.6. Change in angle of cones of diffraction with change in the direction of incident beam until at a Bragg angle a reinforced diffracted beam is formed as in C.

Bragg's Law

Using a diffractometer to study clay minerals, we seldom deal with the case where an incident beam passes through the crystal scattering X-rays on the other side. Far more commonly we need to visualize what is referred to as reflection of X-rays.

The situation for a single plane of atoms (represented by a row in two dimensions) reflecting an X-ray beam incident at relatively low angles is portrayed in Fig. 3.7. Rays 1 and 2 are moving toward the row of atoms in phase. *X-X'* and *Y-Y'* are wavefronts. Atoms *A* and *B* are stimulated by the incident rays to scatter in all directions. The incident beam, the normal to the reflecting plane, and the diffracted beam are all in the same plane.

How do we determine in what direction or directions the scattered beam or beams will constructively interfere and form a diffracted beam? First, we must remember that for constructive interference, reflected rays 1' and 2' must be in phase. To be in phase, they must be at exactly the same place in their sinusoidal cycling, which is the same as differing by one or more whole wavelengths (see Fig. 3.3) because if one gets ahead or behind the other, there will be destructive interference. Notice that *AC* and *BD* are the same length if $\theta = \theta'$, and we have given that as a condition, or

$$AC - BD = AB \sin\theta - AB \sin\theta = 0$$

Both *AC* and *BD* are related to *AB* in the same way, all of which says there is zero wavelength difference between rays 1-1' and 2-2' so we can expect a diffracted beam. Each atom in row (plane) *R* could scatter rays, and the same condition would hold: They would constructively interfere only when $\theta' = \theta$,

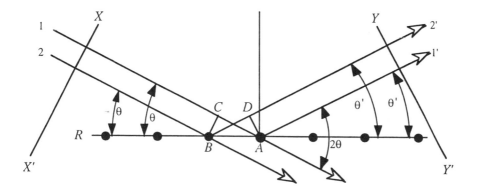

Fig. 3.7. Diffraction from a single row of atoms. 1 and 2 are incident rays; 1' and 2' are diffracted rays.

so there would be no phase difference. (This situation is very similar to that shown in Fig. 3.5; ray 2' is similar to line *de*, a line on a cone of diffracted rays, except in Fig. 3.7 the rays are shown originating at the atom instead of between pairs. This is a convenience for the graphics. The diffracted beams shown in Figs. 3.4 and 3.5 would have significantly less intensity than the diffracted beam of Fig. 3.6C.) For a single row, diffraction can take place at any angle with the condition that $\theta' = \theta$. This is the way visible light is reflected from a mirror. However, no crystals are this thin, so we must consider additional planes of atoms.

Figure 3.8 adds planes (again, represented in two dimensions by rows) of atoms *S* and *T* to the plane *R* shown in Fig. 3.7. Again, in order to have a diffracted beam, the rays at the wavefront *Y-Y'* must all be exactly in phase. From Fig. 3.8 it is obvious that some rays have traveled farther than others to reach *Y-Y'*. If rays 1', 2', 3', and 4' are to arrive at *Y-Y'* in phase with one another, rays 3', and 4', must have path lengths exactly one, two, or three (or some whole integer) wavelengths longer than rays 1', and 2',. If the rays satisfy this condition, they will be in phase with one another. You won't be too surprised that there is a way to express this geometrically. You can see that ray 3-3' has traveled farther than ray 1-1'. Do you see that the additional distance is *EG* + *GF*? This distance must equal some whole number of wavelengths. Now notice that the two triangles *EGA* and *FGA* have a common side *AG* that is also the spacing *d* between rows of atoms. Also notice that the angle *EAG* equals θ. (Briefly, the proof is that *EA*, one side of angle *EAG*, is perpendicular to ray 1; the other side of the angle, *AG*, is perpendicular to row *R*. Then, because row *R* and ray 1 make the angle θ, and because *EAG* represents a 90° rotation of this angle, the two angles are equal.)

To arrive at a simple equation that will relate angle of incidence, spacing between rows, and wavelength difference in terms of the wavelength of the incident X-rays, we need to express *AG* in terms of sinθ, and *EG* and *GF* in terms of an integral number of wavelengths. Therefore, see that sinθ = *EG/AG* (because *AG* is the hypotenuse of the triangle and *EG* is the side opposite the angle θ) and sinθ' = *GF/AG*, so *AG*sinθ = *EG* and *AG*sinθ' = *GF*. Then,

---

Box 3.1. Diffraction and Reflection

Though the X-ray beam is really *diffracted*, the term *reflected* is in such common use that we will continue to use it here with the understanding that it is not reflection in the sense of visible light reflected by a mirror. The angle of incidence equals the angle of reflection, or diffraction, for X-rays and visible light. Reflection of X-rays differs from reflection of visible light in that visible light is reflected at all angles, whereas X-rays are reflected at only a very few angles rationally related to one another. And one more point: In general optics the angles of incidence and reflection are measured from the normal, whereas in X-ray diffraction the angles of incidence and reflection, or diffraction, are measured from the reflecting or diffracting plane.

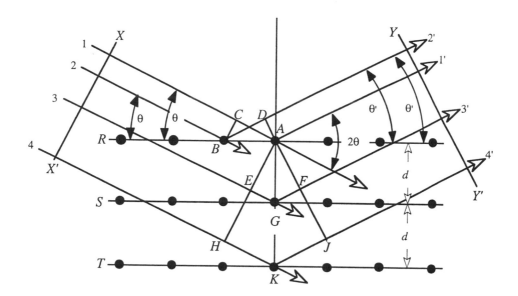

Fig. 3.8. Diffraction from more than one row of atoms illustrating Bragg's law. 1 is one ray of the incident beam; 1' is one ray of the diffracted beam; *KGA* is the normal to the planes causing diffraction; and the axis of the diffractometer is perpendicular to the page at *A*. There are two unit "cells" in this sketch, each marked by separation of equivalent planes by *d*.

because $\theta = \theta'$, we can put $2AG\sin\theta = EG + GF$; substituting $d = AG$, we have $2d\sin\theta = EG + GF$. Then the wavelength difference that must equal some integral number of wavelengths $n\lambda$ is equal to $EG + GF$, so

$$2d\sin\theta = n\lambda \qquad (3.1)$$

(What is the wavelength difference between ray 1-1' and ray 4-4'? Work it out assuming the difference of ray 1-1' and ray 3-3' is $n = 1$. You should be able to draw Fig. 3.8 and derive this relation without referring to the text. If you can do that, you understand it.)

This equation (3.1) is Bragg's law or the Bragg equation, first worked out by W. L. Bragg in 1912 (Bragg, 1913). You may safely infer from the detailed manner in which we've developed this equation that it is important. You will see it frequently, sometimes in different forms, and you will find it very useful. It is *the* most important relation for the use and understanding of X-ray diffraction. [In W.L. Bragg's original paper (1913), he used the relation

$$a\alpha = h_1\lambda, \quad a\beta = h_2\lambda, \quad a(1 - \gamma) = h_3\lambda$$

where a is the side of the cube (unit cell), α, β, and γ are the cosines of incident radiation, and h_1, h_2, and h_3 are integers. Are these direction cosines? Can you relate these equations to the development shown in Fig. 3.6?]

THE ARITHMETIC OF SCATTERING

The Summation of Scattering Amplitudes

Now we need to return to scattering and discuss scattering from a unit cell. To briefly review, we considered scattering by the electrons around an atom and concluded that we could assume that each atom was a point source sending out spherical wavefronts. Upon considering atoms in rows (two-dimensional array, Fig. 3.4), we find that there are only a few directions, not all possible directions, in which diffracted beams are formed by constructive interference of these spherical wavefronts. In three-dimensional arrays the diffracted beams represent much more constructive interference—more beams composed of more rays interacting, causing much more intensity (Fig. 3.6). You were asked to picture a family of diffracted beams emanating from each atom. In what follows we will get a bit closer to what really happens. For an incident beam at a Bragg angle, you will be asked to picture a single resultant diffracted beam emanating from each unit cell. Each of these beams will constructively interfere with its counterpart from every other unit cell. To describe the entire diffracted beam, we need to describe a beam from just one unit cell.

Just as the scattering efficiency f of an atom is somewhat less than the sum of the scattering from individual electrons because the beams from individual electrons are slightly out of phase with one another, so the scattering efficiency F of a unit cell is always less than the sum of the scattering of the individual atoms in the unit cell because the beams that the individual atoms scatter are usually at least partially out of phase with one another unless the diffraction angle is zero. F, then, is a measure of the intensity of the diffracted beam we observe. Because its value depends on the way atoms are arranged in the unit cell (the physical distance between them determining how much their scattered rays will be out of phase), F is called the *structure factor*. To find F, we need to sum the amplitude f of the beams scattered from each atom in the unit cell adjusted for the amount of interference caused by phase differences, $\emptyset$. (And phase differences will be directly related to how the atoms are placed in the unit cell. See Fig. 3.21 and Table 3.1 at the end of this chapter.) Phase differences are measured relative to one atom designated as the origin of the unit cell. In order to be compatible with our calculations, we need to express phase differences for interacting beams in terms of radians. We need to do all these things in order to select equations that experience has shown are useful. With the proper equations we can model the diffraction of X-rays; i.e., we can calculate X-ray diffraction tracings.

You may remember that when representing harmonic or wave motion, the wave or ray is represented by a sine wave, and a point on the wave can be described in terms of its *phase angle*, Ø (Fig. 3.9B or C). The phase angle can be expressed in terms of a fraction of a circle, which is 360°, or 2π radians, as indicated by the sine wave in Fig. 3.9A. At point a in Fig. 3.9A the phase angle of the sine wave is said to be 0°, 60° at b, and 180° at c. Using radians instead of degrees, point a would be at 0 rad, b at $(60/360)(2\pi)$ = $(0.167)(6.28)$ = 1.047 rad, and c at 3.14 rad.

We also need to make use of vectors. Recall that vectors can be plotted on an *X-Y* coordinate system. The amplitude of a wave can be represented by the length of a vector, and the phase angle is represented by the angle between the *X* axis and the vector (see vectors A_1, A_2, and A_3 and phase angles $Ø_1$, $Ø_2$, and $Ø_3$ in Fig. 3.9B). For two waves, the resultant vector A_3 representing the resultant amplitude and the resultant phase angle $Ø_3$ can be found by simply adding the initial vectors A_1 and A_2 by the parallelogram law.

You may or may not recall that physicists and mathematicians find it convenient for problems of this sort to refer to the *X* axis as the axis of real numbers and the *Y* axis as the axis of imaginary numbers, and that any point in this *X-Y* space is called a complex number. The unit of measure along the axis of imaginary numbers is *i*, which is defined as the square root of negative one, $i = \sqrt{-1}$. (For further information see, e.g., Vance, 1963, pp. 417ff.) The reason for introducing complex numbers is that complex exponential functions may be used to describe both phase and amplitude in the same equation.

In the development that follows, the primary goal is to help you understand how the diffracted beam is a consequence of both the kinds of atoms present and their positions within the unit cell. To that end, there is an arithmetic problem at the end of this chapter that will allow you to see how to simulate, or model, an X-ray diffraction tracing. There are several equivalent

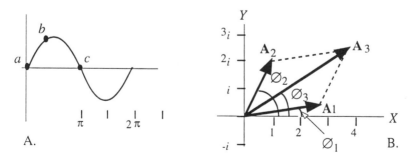

Fig. 3.9 A and B. Diagrams to aid in understanding the sine wave, radians, vectors, and complex numbers. See text for discussion.

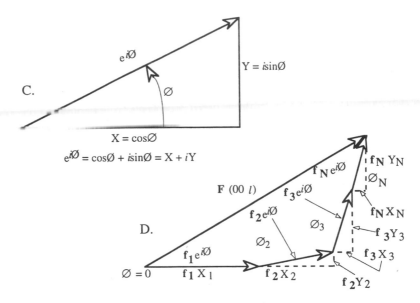

Fig. 3.9 C and D. Diagrams to aid in understanding the sine wave, radians, vectors, and complex numbers. See text for discussion.

mathematical paths to solving this problem. They all do the same thing—they add amplitudes in accordance with phases of the diffracted beams. A complex exponential function is presented in what follows, but most of the development is carried out using trigonometric relations. Our development is not comprehensive. There are several jumps in the logic, and some steps are taken by analogy rather than by rigorous mathematical reasoning. The goal is to offer you enough so you may gain insight into the process and a degree of faith in the development. See Chapter 2 in James (1965) or Klug and Alexander (pp. 152ff, 1974) for a more complete, rigorous development.

To begin our development, recall that there are very particular geometric constraints that must be satisfied before diffraction can take place. The incident and reflected beams must be in a plane perpendicular to the planes of atoms responsible for the scattering. In Fig. 3.10, the (00l) planes are perpendicular to the page and parallel to the lines of print. Now, assume that diffraction takes place from these (00l) planes, that A, B, and C are lattice positions occupied by atoms in a unit cell, and that for a beam incident at angle θ, the Bragg law is satisfied. We know then that $DC + CE$ must make the path of ray 2, which is scattered from the atom at position C, some whole number of wavelengths longer than the path of ray 1 scattered from the atom at position A. Another way to say this is, because the Bragg law is satisfied, $DC + CE$ must be $n\lambda$, and we will assume $n = 1$ for this argument. Then, from Bragg's law, the wavelength difference between rays 1' and 2', $\Delta_{1'2'}$ is

$$\Delta_{1'2'} = DC + CE = \lambda = 2d(001)\sin\theta$$

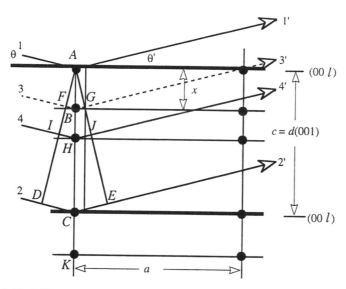

Fig. 3.10. Diffraction from a two-dimensional unit cell, assuming an orthogonal cell.

Now we ask how scattering from the atom at position B will affect scattering from atoms A and C. Certainly $FB + BG$ must be less than 1λ, if $n = 1$, and therefore the ray scattered from atom B will be out of phase with rays from atoms A and C. (Because A and C are equivalent positions in this unit cell, they must be occupied by the same kind of atom, and likewise, atoms B and K must be the same but not necessarily the same as A and C. The distinction between positions and the atom at that position will now be dropped with the understanding that this distinction is valid but inconvenient.) How much out of phase? Do you see that because the rays scattered by atoms A and C are exactly one wavelength apart and therefore in between rays 1' and 3', $\Delta_{1'3'}$ is

$$\Delta_{1'3'} = (AB/AC)\lambda = (FB + BG)/(DC + CE)\lambda$$

In anticipation of where we are going with this development, we substitute x for AB and, from the definition for Miller indices, $AC = c/l$ (see Fig. 3.10). Then

$$\Delta_{1'3'} = [x/(c/l)]\lambda$$

Now we need to express this phase difference in terms of angular measure or radians by multiplying the phase difference by $2\pi/\lambda$ (there are 2π radians per wavelength or per complete cycle on a sine wave, as in Fig. 3.9A). Now we don't need to know the wavelength to express phase difference (see below that λ cancels out).

$$\emptyset = (\Delta_{1'3'})2\pi/\lambda$$

Do you see that if Δ, the difference in distance the two waves travel, is equal to λ, then $\emptyset$ equals 2π, which equals $360°$, and the waves would be perfectly in phase? From above

$$\emptyset = (\Delta_{1'3'})2\pi/\lambda = [x/(c/l)\lambda](2\pi)(1/\lambda) = 2\pi l\,(x/c)$$

Recall that fractional coordinates in a unit cell are expressed as (*uvw*). Then because x/c is a fractional coordinate, $x/c = w$, and

$$\emptyset = 2\pi lw \tag{3.2}$$

If we consider this relation for three dimensions (*hkl*) instead of the one dimension, we have

$$\emptyset = 2\pi(hu + kv + lw)$$

This general relation is applicable to a unit cell of any shape. But because we often work with clay minerals oriented so that diffraction will take place only from the 00*l* spacings (at least that's what we try to do), the development is easier to follow, and the following development will be made in terms of the 00*l* diffraction series.

The development above is OK for orthogonal cells, but, as you're aware, there always are complications. For monoclinic or triclinic cells, c is still the unit cell dimension along the monoclinic axis, but is no longer perpendicular to $d(001)$, but is now equal to $c\sin\beta$, or, if you'll look ahead to the section on reciprocal lattice, $c*$ is the reciprocal unit cell dimension along $Z*$. Invariably, in the figures and in our thinking, we visualize the atom positions in the cell as arrayed along the normal to $d(001)$ as in Fig. 3.20, and this dimension is $Z*$ or $c\sin\beta$.

The Structure Factor F

We have now developed an expression to describe phase differences for the beams scattered by atoms A and B in Fig. 3.10 [Eq. (3.2) and its general form]. This equation is applicable to atoms in all planes causing diffraction, but there are also differences in the amplitude A of each beam. We need to return to the atomic scattering factor f (p. 72), which is a ratio of the amplitude from an atom to that from an electron

$$f = \frac{\text{Amplitude of a wave scattered by the atom in question}}{\text{Amplitude of a wave scattered by an electron}}$$

We need an expression that will give us the phase and amplitude of the single diffracted beam that results from combining the phases and amplitudes from the waves scattered by each of the atoms making up a unit cell. One way to see this is from Fig. 3.10, from which we can imagine summing the atomic scattering factors f and the phase angles Ø for each of the atoms of the unit cell, f_a, f_b,...,f_n and $Ø_a$, $Ø_b$,...,$Ø_n$

$$F(00l) = \sum_n (f_n, Ø_n) \tag{3.3}$$

F is the structure factor defined on p. 72.

Equation (3.3) is too general to be of much use. Therefore, we need to develop it into some form that will allow us to calculate the intensity of a diffracted beam. Referring again to Fig. 3.9B, note that the length of the resultant vector and its angle with the X axis can be described in trigonometric terms but, because the Y axis is imaginary, the imaginary number i is introduced. The equation $e^{iØ} = \cos Ø + i\sin Ø = X + Yi$ (Fig. 3.9C) is the well-known Euler identity. Figure 3.9D is called an Argand diagram and is directly applicable to our problem. It is a more involved form of Fig. 3.9C. $F(00l)$ is a vector representing the phase and magnitude of the amplitude of a resultant wave. It is formed from the summing of a set of vectors representing the contributing waves, each with its scattering factor f_n and its phase angle $Ø_n$. The same operation can take the form of an exponential sum

$$F(00l) = f_1 e^{iØ_1} + f_2 e^{iØ_2} + f_3 e^{iØ_3} + ... + f_n e^{iØ_n} \text{ or}$$

$$= \sum_n f_n e^{iØ_n} \tag{3.4}$$

This equation (3.4) is more useful than Eq. (3.3) because it has established solutions. Then, using the identity given in Fig. 3.9C, $e^{iØ} = \cos Ø + i(\sin Ø)$, we have

$$F(00l) = \sum_n f_n \cos Ø_n + i\sum_n f_n \sin Ø_n \tag{3.5}$$

When a center of symmetry is present in the unit cell and the origin for the calculation is set on that center, the sine series equals zero and the sine function is eliminated along with the imaginary number i. Because there is a center of symmetry on projection to Z for most clay minerals, the sine term drops out and we have the useful result (lucky us)

$$F(00l) = \sum_n f_n \cos Ø_n = \sum_n P_n f_n \cos(2\pi l z_n /c) \tag{3.6}$$

The second summation of Eq. (3.6) shows the expanded form of Ø [from Ø $= 2\pi l\,(z/c)$]. P_n is the number of atoms of type P per atomic layer, f_n is the scattering power of each of them, l is the order of the reflection, z_n is the displacement of the atomic layer, in Ångstroms, from the center of symmetry measured along a line normal to the (00l) plane, and c is the unit cell length. The intensity, which is what the detector measures, is (neglecting Lorentz-polarization effects) equal to $|F(00l)|^2$. Squaring F eliminates its sign, so we don't know whether F is positive or negative; we can only measure its magnitude.

Some structures have no center of symmetry on projection to Z. Then the intensity I is given by

$$I(00l) = |F(00l)|^2 = [\sum_n P_n f_n \cos(2\pi l z_n/c)]^2 + [\sum_n P_n f_n \sin(2\pi l z_n/c)]^2 \quad (3.7)$$

[This result is obtained by multiplying Eq. (3.5) by its complex conjugate,[1] thus eliminating i.] In this case, we can't talk about F as simply positive or negative, because it is a complex number. Now we turn to a consideration of the uses of the information gained from intensity.

Information from Intensity

Not all reflections are of the same intensity, not even those from the same set of diffracting planes, i.e., from the (00l) = (001), (002), (003), . . . , (00n). In fact, there are some useful systematic absences (and variations) in the intensity from different members of a set of planes. These tell us something about the positions of atoms in the unit cell. We deal with essentially one dimension of the unit cell in the case of diffraction from clay minerals when we deliberately try to orient them on our sample mounts. If our orientation is perfect (and that never happens) only the (00l) spacings contribute to the diffraction pattern. By using the simplified situation in Fig. 3.10, you can see how the odd and even members of the (00l) series of spacing can account for the presence, absence, or variation of diffracted intensity. This same reasoning applies, of course, when interpreting X-ray tracings from random powder mounts.

Return to Fig. 3.10 with its five rows with atoms, A, B, C, H, and K. First, picture the diagram without the row of atoms H so that d is from the plane of atoms A to that of atom C. Next, imagine that ray 2-2' is exactly in phase with ray 1-1' because DC + CE = 1λ so that there will be diffracted intensity. This is the first case. Now, if we place the row with atoms H exactly halfway between rows with atoms A and C (i.e., d is still the same), do you see that IH + HJ will equal 1/2, and that, therefore, ray 4' will be perfectly out of phase

---

[1]The complex conjugate of, e.g., $A + iB$ is given by $A - iB$.

with rays 1' and 2'. If we imagine that atom H is exactly the same as atoms A and C there will be no diffracted intensity? (If the cell dimension c is really as shown in Fig. 3.10, why is it not possible for H to be the same as A and C?

Remember that a spacing can cause diffraction at more than one angle. [When working with glycol-saturated Na-smectite, for example, its 17 Å (001) spacing will produce diffraction at 12 to 14 angles, i.e., give reflections for (00l) spacings as high as (00,12) to (00,14). A few of the less intense (00l) peaks are undetectable.] Again in reference to Fig. 3.10, the Bragg angle for rays 1 and 2 is about 30°. Leaving the row with atoms H in place, picture rays 1 and 2 incident at about 60° or high enough so that ray 2-2' would travel exactly 2λ farther than ray 1-1'. Then ray 4-4' would travel 1λ farther than ray 1-1' instead of λ/2 as in the first case. (Try drawing this to see if it makes sense.) In this case, ray 4' would constructively interfere with rays 1' and 2', and diffraction would be observed, whereas when the diffracted ray 4' traveled only λ/2, it would destructively interfere with ray 1' and 2' and no diffraction would be recorded. In the first case, i.e., Bragg angle = 30°, we refer to the spacing from which there could be diffraction if row H were not present as (001), and in the case in which ray 4-4' traveled 1λ, we call the spacing and the peak it forms on the diffraction tracing (002). From all this we generalize for the case with H present that we have no diffraction when l is an odd number, (00l) = (001), and we do have diffraction when l is an even number, e.g., 002, 004.

In our example we assumed that atoms A, C, and H were all the same. If, however, atom H is different, and had different scattering efficiencies f than atoms of A and C, then there could be observable peaks for the odd-ordered reflections and the intensity would be a function of how different the atoms were in atomic number and hence scattering factor. For example, if atoms A and C were Si and H were Al, there would be little difference in their scattering efficiencies because they have 14 and 13 electrons, respectively, or 10 and 10 as ions. There would still be no observable peaks for odd-numbered (00l)'s. If H were Fe, and A and C were Si, there would be a large difference in their scattering powers, and even after the amplitudes of the waves from the Si atoms were subtracted from those of the Fe, there would be sufficient intensity to show peaks for odd-ordered reflections. The situation in which this probably will be of most use to you is in distinguishing Mg and Al-rich chlorites from Fe-rich chlorites (see Chapter 7, pp. 234ff and Fig. 7.5). (Do you see that we could have used Fig. 3.8 as a basis for this discussion?)

The same reasoning applied to three-dimensional unit cells leads to similar generalizations that are helpful in determining lattice type in the study of crystal structures. For example, in the isometric system, structures with the simple Bravais lattice may have reflections from all possible sums of the (hkl) values. Those with face-centered lattices may have reflections for h and k or for k and l that are both even or both odd, but reflections are not possible when h, k, and l are mixed odd and even. For body-centered lattices, reflections might be present when the indices add to an even number, but no reflections are possible when the indices add to an odd number. Cullity (1978, pp. 115-26) offers a detailed discussion of these relations. He also carries through, as a specific example, the calculation for the F of NaCl, a face-centered cubic structure. You might like to try following his exercise.

Deducing the kind of atoms and their positions from intensity data constitutes an analysis of structure. In classical X-ray crystallography such an analysis was all that was desired. For the clay minerals and some other substances, however, we need and can obtain other types of information on the size and perfection of the crystallites. In addition, for any X-ray diffraction study, there are certain instrumental and geometric factors that "get in the way," so to speak, and we must take these into account. After discussing other factors that influence the amount of intensity we can observe, we will offer you some specific calculations to illustrate the effect of some of these factors.

The Reciprocal Lattice

We are sure that no one reading this has ever tried to impress anyone with their erudition by dropping words like post-modern deconstruction, Kierkegaard's existentialism, . . . etc. Unfortunately, reciprocal lattice has joined this company of things we have designated as unapproachably abstract, something difficult to master, whereas it is actually relatively simple. It was invented in the summer of 1912 by Ewald (1921; 1962, p. 42) immediately after von Laue's discovery, and helps us understand the distribution of diffracted intensity in space. Granted that it is an abstraction of an abstraction, it is quite simply derived from the lattices, the first ab-straction, that we use to describe the arrangement of atoms in the structures of crystals. We will present it in two steps. The first follows immediately, and the second will be in Chapter 10 in the discussion of diffracted intensity in three dimensions.

Not many mineralogists use film to record the diffracted beams of X-rays anymore, but you will see electron diffraction recorded this way. We're sure you can picture a film with spots arranged in a symmetrical pattern. Each spot represents a beam of X-rays diffracted from a specific set of *hkl* planes in the crystal. Therefore, the spot also represents the geometrical position of a family of *hkl* planes in the crystal and the distance d between those planes, or an integer fraction of d. (And the density of the spot is a measure of the intensity.) That's a lot of information in one spot.

To label and analyze all of the spots, there is need of an organizational framework. Start by picturing a simple unit cell, let's say a monoclinic one. The cell is defined by edge lengths a, b, and c along axes X, Y, and Z (Fig. 3.11A). To define the coordinates of reciprocal space, image an axis X^* (we say X star) perpendicular to the (100) plane[2], an axis Y^* perpendicular to the

---

[2]If you need to refresh yourself on crystallographic notation, see **Box 4.1 Nomenclature**, in Chapter 4, p.133.

(010) plane, and an axis $Z*$ perpendicular to the (001) plane. These three axes form a reference framework that is the reciprocal of the one defined by X, Y, and Z. The unit cell in reciprocal space is marked off by

$$a* = 1/d(100), \quad b* = 1/d(010), \text{ and } \quad c* = 1/d(001), \quad (1/d) = d*$$

where d is the same as in Bragg's law, the spacing between planes (Fig. 3.11B). Angles in real space are represented in reciprocal space by their supplements. Thus if the monoclinic angle β is 100°, then $\beta*$ (in reciprocal space) is 180 - 100 = 80°. The feature of clay minerals that we work with the most is, of course, their 00l spacings. Let's use the 10 Å d(001) of illite as an illustration. If d(001) = 10 Å, then $c*$[3] must equal 1/10 Å. So, along the axis normal to the (001) of an illite crystallite, we mark off points at $n 1/d$(001) along $Z*$ or 1/10 Å, 1/5 Å, 1/3.33 Å, ... n/10 Å. (Likewise, if we wished to work in the other two dimensions, units of $a*$ can be marked off along $X*$, and units of $b*$ marked off along $Y*$.) Crystallographers refer to these points as nodes. The scale of the reciprocal lattice is impossibly small for the first few values on n; so we ignore scale and pretend it is just about the same as the scale we use to show the crystal structure. We can, of course, convert the other two principal directions in the crystal into their reciprocal equivalents.

Figure 3.11 is too busy, but should prove helpful in understanding some of this stuff. Let us take you through it step by step. Figure 3.11A is the two-dimensional part or XZ plane of the lattice for an ideal monoclinic crystal structure and shows two members of the 302 family of planes. Figure 3.11B shows the reciprocal lattice for the regular lattice shown in Fig. 3.11A. Now the 302 family of planes is represented by a spot and $O*P$, a vector the length of which is $1/d$(302) or $d*$(302), is perpendicular to the (302) family of planes. In addition, it can be shown (but what we mean is, will you take this on faith from us?) that a line drawn from $O*$ (Fig. 3.11B and C) to any hkl node in the reciprocal lattice is perpendicular to the family of (hkl) planes in the real crystal. Furthermore, the distance from $O*$ to the hkl node = $d* = 1/d$. In other words, to relate this to something with which you are familiar, picture the illustrations used to explain the stereographic projections you learned to use in mineralogy. For these, you pictured a crystal at the center of a sphere with lines perpendicular to each crystal face extending to the surface of the sphere where the crystal face was represented by a point. The reciprocal lattice system is quite similar.

Because our application differs from former uses of the reciprocal lattice, we need to describe two ways in which it can be visualized. The conventional way to use this graph (Fig. 3.11C) is to ask you to imagine $O*$ as the origin of the lattice, C as the point at which diffraction occurs in the real crystal, $CO*$

---

[3]Note that $c*$ is a dimension along $Z*$ and not a direction. Therefore we shouldn't use the $c*$ when we mean a direction perpendicular to d(001), and $Z*$ is perpendicular to d(001) by definition.

parallel to, and as an extension of, the incident beam, and the direction of the diffracted beam represented by *CP*. If we have a randomly oriented powder sample, there should be at least one crystal oriented so that its reciprocal lattice is in the position shown in Fig. 3.11C. Then, any time a crystallite is in an orientation that puts a node of its reciprocal lattice on the circle (or sphere in three dimensions, this is called the Ewald sphere of reflection) that has a radius of the reciprocal of the wavelength of the incident ray $1/\lambda$, Bragg's law is satisfied and diffracted intensity will be obtained. Ewald's method of finding diffraction intensity was to rotate the reciprocal lattice to see which nodes intersected the surface of the sphere for each increment of θ. [Later in this chapter we will point out that to get the right amount of intensity at a node, the intensity intercepted will have to be multiplied by the structure factor for this node, i.e., $F(hkl)^2$.]

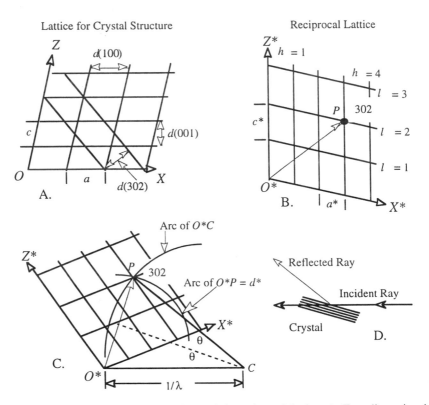

Fig. 3.11. Diagrams to help understand the reciprocal lattice. A. Two-dimensional view of a monoclinic lattice. B. The reciprocal lattice derived from A showing the node for 302. The line $O*P$ equals $1/d(302)$, and is perpendicular to the 302 planes of A. Notice that the h and l values are marked on B. Do they read correctly for the 302 node? C. The geometric condition for satisfying Bragg's law for $d(302)$ and to show the arcs made by the two radii, $O*C$ and $O*P$. Bragg's law is developed below. D. Sketch of the crystal responsible for the reflected ray CP in C.

A method suggested by Brindley and Méring (1951) for identifying the distribution of diffracted energy uses this same construction (Fig. 3.11C) but sweeps through a sphere of radius $d*$ originating at $O*$. In this variation, the reciprocal lattice does not move and diffracted intensity is sought with the sweep of the $d*$ radius for each increment of θ. Arcs of both spheres are shown in Fig. 3.11C. To calculate all the diffraction intensity that could occur within a diffraction scan, the integrated intensity within each increment of 2θ is plotted against 2θ (see Chapter 10). Figure 3.11D shows you the parallel relation of the incident ray to $CO*$ and the reflected ray to CP in Fig. 3.11C.

To derive Bragg's Law (Eq. 3.1) from Fig. 3.11C, note that

$$O*P = (2) (O*C)(\sin\theta)$$

and because

$$O*P = d*, \quad d* = (2) (1/\lambda) (\sin\theta)$$
$$\text{then } 1/d = (2) (1/\lambda) (\sin\theta)$$

and that is the same as

$$\lambda = 2d\sin\theta$$

which is, of course, Bragg's law (Eq. 3.1).

So, any time a crystallite is in an orientation that puts a node of its reciprocal lattice on the circle that has a radius of $1/\lambda$ or $1/d$, Bragg's law is satisfied and diffracted intensity will be recorded. This approach is necessary to visualize diffraction from a randomly oriented powder instead of diffraction from a single crystal.

Now, although these arguments may not seem fully intuitive, and you may not yet be fully comfortable with them, we need to move on. We can use your understanding of the machinations related to Fig. 3.11C to deal with the peak shapes, missing or weak peaks of semidisordered clay minerals, the band shapes of the turbostratic and close to turbostratic clay minerals, and with the diffraction patterns produced by polytypes, all of which we will discuss in subsequent chapters. We will see this topic again in Chapter 10.

REAL VERSUS IDEALIZED PEAKS ON XRD TRACINGS

We made simplifying assumptions to derive the Bragg equation (Eq. [3.1] and Fig. 3.8), as is always the case when we begin to model nature: (1) that the incident beam is perfectly monochromatic; (2) that the rays are perfectly parallel; (3) that we have only three rows of atoms and they are a perfectly ordered part of a perfect and infinite crystal; and (4) that the crystal is perfectly oriented for diffraction to occur. As we apply the Bragg equation to diffraction tracings, we need to consider how real circumstances differ from the idealized ones. Figure 3.12 shows a real diffraction tracing, and Fig. 3.13

will help us understand some of the characteristics of the peaks shown on this tracing.

The first thing we consider is why diffraction peaks are so narrow or, said in another way, how destructive interference is produced in all directions except those of the diffracted beam. Recall that the condition for constructive or reinforcing interference is that the rays must be in phase and differ by some whole number of wavelengths (includes 0). In Fig. 3.3C, the two waves R and S interfere destructively; they are out of phase but do not completely destroy one another. Their amplitudes simply add to form a beam of smaller amplitude compared to those of Fig. 3.3B, which are in phase. If another wave, out of phase with R as much as S is out of phase, interacts with the combined amplitude of R and S, the amplitude is reduced more, and so on until there is no amplitude at all.

To understand the effect of destructive interference on the peak shapes we see in Fig. 3.12, we need to consider the contributions from planes deeper into the crystal. Let us refer to Fig. 3.13. The ray 1-1' scattered by the first plane

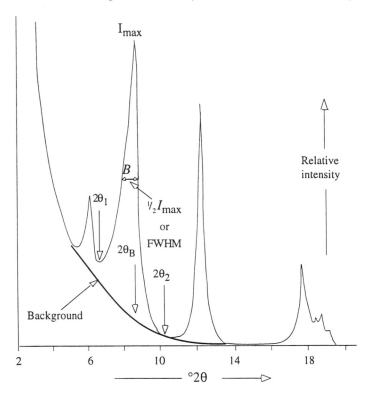

Fig. 3.12. X-ray diffraction tracing of the < 2 μm fraction of the Purington Shale from an oriented aggregate mounted on a glass slide, Cu radiation. $2\theta_B$ is the Bragg angle for the most intense peak, $2\theta_1$ and $2\theta_2$ are limiting angles as in Fig. 3.13, and FWHM = full width at half-maximum peak height above background.

below the surface has a path exactly one wavelength longer than ray 0-0'. Therefore, all rays reflected from planes in the crystal at an angle so that this condition is met unite to form a diffracted beam. This angle is known as a Bragg angle θ_B. As the angular relation between the incident beam and the planes of the crystal shift to either slightly greater or lesser angles, angles that are close but not quite Bragg angles, rays reflected from all planes 0 to N will destructively interfere.

Consider rays A-A' and C-C'. Let's say ray C-C'. is one and one-tenth wavelengths longer than A-A' The amplitudes of A-A' and C-C' will add, but because they are out of phase by one-tenth of a wavelength, the resultant amplitude will be less than double the amplitude. The ray reflected from row 2 will be two-tenths of a wavelength out of phase with ray A-A', and so on, until we reach row 5. The ray reflected here will be five-tenths, or one-half wavelength out of phase. The sum of the amplitudes of rays from rows 0 and 5 will be exactly zero; from 1 and 6, exactly zero; from 2 and 7, exactly zero; etc. If A-A' and C-C' were one-one hundredth out of phase, do you see that the sum of amplitudes from rows 0 and 50 would be exactly zero?

To generalize from this specific example, once there is the slightest deviation from the Bragg angle θ_B then at some depth in the crystal, there will be a row N that reflects a ray that is exactly one-half wavelength out of phase

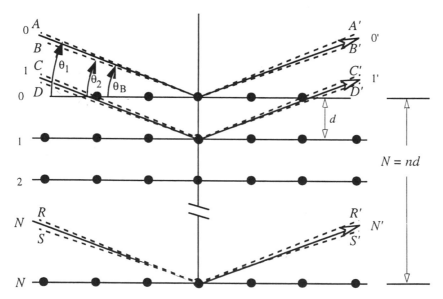

Fig. 3.13. Diffraction at a Bragg angle with two limiting angles at which there is destructive interference. The figure shows beams 0, 1, and N, and planes of atoms 0, 1, 2, etc., and N.

with the ray reflected from the top row 0. Further, row $N + 1$ will be exactly one-half wavelength out of phase with row 1, and so on. You can see that the same argument applies to ray $B\text{-}B'$ at an angle θ_1, just less than θ_B, so we have two limiting angles on either side of θ_B, θ_1 and θ_2, when ideally the intensity should drop to zero. We can now see that peaks on diffraction tracings tend to be so narrow and sharp because *destructive interference is just as much a consequence of diffraction from a periodically repeating array of atoms as is constructive interference.*

Suppose now that the crystal represented in Fig. 3.13 contained only rows 1, 2, and 3. Then slight departures from θ_B would not result in complete cancellation. There would be measurable intensity at these angles. In other words the peak would be broadened to include the beams at angles θ_1 and θ_2, although they would have less intensity than that part of the beam at θ_B. From this example you can see the reason that peaks get sharper as crystals get thicker, or contain more rows like N.

Based on this discussion, the question should arise, "Why is there any finite width to the diffraction peak at all if it is produced by a thick crystal?" There are several reasons, some of which will help you understand the real, as opposed to the ideal, nature of analytical procedures and minerals. First, although we assumed perfectly parallel incident rays and perfectly monochromatic radiation, neither is technically possible and both of these deviations from ideal add something to peak width. For Cu radiation, the $K\alpha$ peak has a finite range of wavelength values of about 0.001 Å. For the peak at 8.85° 2θ (representing a spacing of about 10 Å (does that work out right with the Bragg equation, p. 71?), the finite width of $CuK\alpha$ radiation adds about 0.002° 2θ to the peak width. Other instrumental factors are more weighty and guarantee that finite peak breadth will be observed for diffraction from any crystal studied by powder diffractometry. The diffractometer is an optical device, and, like all optical devices, it necessarily introduces distortion and defocusing into any image that it records. The effects can be minimized by careful alignment, proper slit selection, and perfectly flat samples that are perfectly aligned. The best compromise between resolution and intensity that you can achieve with modern diffractometers is a peak breadth at half height of about 0.08° 2θ, and even that shows additional broadening at high values of 2θ due to separation of the $CuK\alpha_1$-$K\alpha_2$ components of the incident radiation. The latter can be eliminated by the use of certain monochromators which are rarely used in mineral analysis because they cause prohibitive intensity losses. Graphite monochromators are another matter, and these are almost standard today, but their resolution is insufficient to eliminate $CuK\alpha_2$ radiation. For most clay mineral studies, you don't need this kind of resolution anyway because the minerals themselves produce broad peaks whose sharpness cannot be improved by any optical refinement.

A characteristic of all minerals is imperfection in crystal order, especially for clay minerals, because of their small crystal size. Almost all real crystals have an imperfection known as mosaic structure (diamond and calcite can be exceptions). In real crystals, a row of atoms does not extend unbroken across the whole width of the crystal; only within small volumes is there perfect order. These small volumes join other small volumes close to, but not quite in, perfect alignment. The size of these mosaic pieces or subgrains is about 1000 Å or less, and the disorientation between adjacent subgrains is a matter of seconds to minutes of arc, seldom as much as 1°. Many clay minerals are often 1000 Å in the largest dimension and may have subunits within that distance. These small crystals, or crystallites for very small crystals, cause noticeable line broadening.

Particle-size broadening or, in our case, thickness of crystal, is conveniently inferred from the width of the peak at peak half-height (see Fig. 3.12) for the reasons discussed related to Fig. 3.13. It is used to estimate particle size and, under ideal conditions, the distribution of particle sizes. As we have seen, the width of the peak becomes broader for smaller particles. This is expressed quantitatively by the Scherrer equation

$$L = \frac{\lambda \, K}{\beta \cos \theta} \tag{3.8}$$

where L is the mean crystallite dimension in Ångstroms along a line normal to the reflecting plane, K is a constant near unity, and β is the width of a peak at half-height expressed in radians of 2θ (i.e., measure 2θ in degrees and then multiply by $\pi/180$).

Another source of line broadening on which we merely comment is called *strain broadening*. This results from random displacements of unit cells or groups of unit cells from their ideal positions of perfect registry. It can be mathematically separated from the effects of particle-size broadening. The treatment is too complicated for the scope of this book, especially since it has yet to be demonstrated for clay minerals. Line broadening is treated briefly by Cullity (1978, pp.101-02) and in more detail in Klug and Alexander (1974, pp. 299-306 and Chapter 9).

The effect of "particle" size on the peak width for clay minerals has been called into question by work published by Nadeau et al. (1984), in which they show high-resolution transmission electron microscope (HRTEM) photographs of individual particles of smectite that are as small as 10 Å thick. The same smectite sample when deposited as an oriented aggregate gives diffraction peaks whose broadness indicates that there are on the average nine layers per scattering domain, or 90 Å. To resolve the discrepancy between the XRD and the HRTEM data, Nadeau et al. proposed that in preparing samples for X-ray diffraction, the 10 Å-thick particles stack together to make an array from which coherent scattering can be obtained. They call this effect *interparticle diffraction*.

A final cause of broad peaks is mixed layering, or interstratification, of different kinds of clay minerals. Several clay minerals, usually with different *c* dimensions, are frequently intergrown, and a resulting diffraction peak represents an averaging of the two spacings. We will look at this in more detail in the sections on characteristics of clay minerals (Chapter 5) and identification of mixed-layered clay minerals (Chapter 7).

The Interference Function Φ: Diffraction from a Crystal Whose Unit Cell Has a Unitary Scattering Factor

For those working with large crystals, essentially infinite in thickness and more ordered than those of the clay minerals, the shape of the diffraction peaks carries little information. For larger crystals, only *F* is needed because almost all the scattering is concentrated at the Bragg angles of diffraction. However, for small, thin crystals such as those of the clay minerals, there is additional scattering adjacent to the *hkl* diffraction peak positions, as shown in the discussion of Fig. 3.13, and even in small amounts between these positions. (To date, most analyses of clay minerals have been based on peak position and intensity. Peak shape and breadth may become as important as peak position in the future.) Characteristics of diffraction peak shapes that result from this extra scattering, and their correlation with the nature of the crystalline aggregates of unit cells, can be separated from the structure of the unit cell itself, which is described by *F*. We simply set the amplitude of scattering (*F*) from each unit cell equal to unity. Calculation of the diffraction pattern from any assemblage of such cells, using this additional scattering, is called the *interference function*, and it is given the label Φ. Don't worry, though. When we want to return to the real world, we simply multiply the interference function by the appropriate value of F^2 for each point on the 2θ-continuous profile of Φ. But there is a problem here: *F* is defined by *hkl*. We need a formulation for *F* that yields the amplitude of scattering as a *continuous* function of θ, not simply a value of *F* at a single 2θ value. The necessary alteration of *F* is greatly simplified by the one-dimensional case. The alteration we will use is essentially the same as *F* except that it is continuous with respect to θ. It is the layer structure factor, or layer scattering factor *G*.

In addition, Eq. (3.6) must be converted to the units of θ. To accomplish this, we make use of the Bragg relation $n\lambda = 2d\sin\theta$ where $d = d(001)$. The analysis here is one-dimensional, so we can replace the order *n* in the Bragg law, with the index *l*. Rearranging gives $l = 2d\sin\theta/\lambda$. The *d* in this equation is equal to $d(001)$ (which is different than the unit cell dimension *c*, if the symmetry is monoclinic or triclinic) and substituting $2d\sin\theta/\lambda$ for *l* in Eq. (3.6) leads to [symbols as in Eq. (3.6)]

$$G(\theta) = \sum_n P_n f_n \cos(4\pi z_n \sin\theta / \lambda) \tag{3.9}$$

where z_n is the atomic coordinate times $\sin\beta$. Now we have an equation in terms of θ instead of the index l and a scattering factor that is continuous (G) instead of discontinuous (F). F is reserved for the structure factor defined only at hkl. Crystallographers are tidy people who don't like ambiguities.

The interference function Φ is also a continuous function and is given by

$$\Phi(\theta) = \frac{\sin^2(2\pi\, ND\sin\theta\, /\, \lambda\,)}{\sin^2(2\pi\, D\sin\theta\, /\, \lambda\,)} \tag{3.10}$$

where N is the number of unit cells stacked in coherent scattering array along the Z axis and $D = d(001)$. You can compute and plot diffraction profiles from this equation and verify that the breadths of the peaks at half-height do indeed square with the predictions of the Scherrer equation (3.8). The value of L in the Scherrer equation is equivalent to N times D in Eq. (3.10).

If we multiply G by Φ for each increment of 2θ, we are pretty close to a final calculation of a complete one-dimensional X-ray diffraction pattern. However, two additional factors must be considered. Both factors vary with θ. One is called the *polarization factor*, the other the *Lorentz factor*. Their combined effect is seen in Fig. 3.12, where the background and peak intensities at angles less than 12° 2θ increase rapidly. A brief qualitative statement about both is given here. Detailed explanations for both factors can be found in Cullity (1978, pp.107-39) and Klug and Alexander (1974, pp.142-44).

The Lorentz-Polarization Factors

The polarization factor accounts for increases in peak and background scattering from a maximum at 0° 2θ to a minimum at 90° 2θ. The maximum effect on intensity is a factor of two. The basis for this change is that as radiation comes from the tube it is unpolarized, but the process of scattering causes a degree of polarization related to the angle of the incident and diffracted beams. The X-rays are polarized as they are diffracted from a plane of atoms for the same reason a beam of light is polarized by reflecting off the hood of a car, and polarized in the same way; i.e., the vibrations surviving the reflection are those parallel to the hood of the car or the plane of the atoms. As amplitude in directions other than that of the plane of polarization is lost, the intensity decreases. The total energy of the scattered beam is proportional to the polarization factor $(1 + \cos^2 2\theta)/2$, where θ is the angle between the incident beam and the reflecting plane.

The Lorentz factor is a combination of two geometrical factors that we will deal with trigonometrically. The first factor is a formulation for the volume of the crystal that is exposed to primary irradiation. The second one relates the number of crystals favorably oriented for diffraction at any Bragg angle θ_B.

The Lorentz factor is different for random powders and single crystals.

Reynolds (1976) discussed application of the Lorentz factor for basal reflections from clay minerals in oriented aggregates. He concluded that there are two cases. In the first, for diffractometers that have two Soller slits, the smaller the slit openings, the closer the random powder Lorentz factor is approached. In the second case, the more nearly perfect the orientation of the clay minerals on their substrate, the closer the single crystal Lorentz factor is approached.

The two portions of the Lorentz factor are combined with the polarization factor $(1 + \cos^2 2\theta)/2$ and given in the following forms (neglecting overall factors of 2):

The Lorentz-polarization (Lp) factor for random powders (integrated intensity)[4]

$$Lp = (1 + \cos^2 2\theta)/(\sin\theta\sin 2\theta) \qquad (3.11)$$

The Lorentz-polarization factor for single crystals

$$Lp = (1 + \cos^2 2\theta)/\sin 2\theta \qquad (3.12)$$

If you plot these functions and examine them, you will see that they are very different at low values of 2θ, so the question arises: "What is the best form to apply to oriented aggregates of clay minerals?" There is no exact answer to this question because usually the preferred orientation is unknown, and the correct Lp equation is somewhere between the two unique cases as given. Experience with clay mineral samples prepared by either the glass slide or the Millipore® transfer process indicates that a realistic average degree of preferred orientation is $\sigma^* = 12°$, where σ^* is approximately the rocking angle of the sample about the goniometer axis; i.e., all of the intensity of a given 00l peak is collected through the 12° arc indicating that the diffracting crystals are perfectly oriented ±6° (see Reynolds, 1986). The solution of the Lorentz factor, using this value, $\sigma^* = 12°$, suggests that the random powder form is appropriate for results from diffractometers that use two Soller slits. For instruments with one Soller slit, Eq. (3.13) provides a rough approximation for integrated intensities.

$$Lp = (1 + \cos^2 2\theta)/(\sin 2\theta \sin^{0.8}\theta) \qquad (3.13)$$

It is not very good (±15%), but it is much better than using either the random powder or the single-crystal versions. Attempts to improve on this approach are unwarranted unless the preferred orientation (σ^*) is measured and the correct Lorentz factor calculated using the equation given by Reynolds (1986).

---

[4]For point-by-point corrections on 2θ continuous diffraction profiles, the correct Lp powder factor is $1/\sin^2\theta$.

You have probably been asking yourself, "Where is all of this leading?" The answer is that computer programs have become available (e.g., Reynolds and Hower, 1970; Reynolds, 1980; Reynolds, 1985; Reynolds, 1994; and see Chapter 10 and the Appendix) that allow the simulation, or modeling, of X-ray diffractometer tracings. NEWMOD©, for example, is a powerful, interactive program for deciphering real tracings, because to model a tracing you have to think about which atoms are in what positions in the unit cell under consideration. Modeling is particularly helpful in analyzing diffraction patterns from mixed-layered clay minerals [i.e., the stacking of two or more kinds of layers with the stacking direction along a line perpendicular to (001)]. The modeling done here [except for that for identifying polytypes, measuring the (060), and the treatment in Chapter 10] assumes that oriented aggregates are used so that all diffraction is due to the (00l) spacings. Then the scattering is reduced to a one-dimensional problem, which greatly simplifies the arithmetic. You will see the application of this program, NEWMOD©, in Chapters 7, 8, and 9, and a description of how it was written in the Appendix. The exercise in the section following the next one will illustrate how calculations are made.

Putting It All Together—Building an 00l Diffraction Pattern

We have discussed the three essential ingredients for modeling a diffraction tracing, the Lorentz-polarization factor, the interference function, and the scattering amplitude of the unit cell. All three change with the diffraction angle θ, and the final diffraction pattern is simply the product of these three quantities. Let's now generate some 2θ-continuous, one-dimensional patterns for illite.

The intensity is calculated by taking the products of the interference function Φ, the layer scattering intensity G^2, and the Lorentz-polarization factor Lp, at each increment of θ

$$I(\theta) = [G^2(\theta)][\Phi(\theta)][Lp(\theta)] \qquad (3.14)$$

We start with an examination of Φ. Figure 3.14 shows the calculated results for illite that consists of crystals that contain five unit cells each, or $N = 5$. Note the broad peaks that represent the 00l diffraction series. We expect such broad reflections because of the very small crystallite thickness assumed here. The peaks are sharpened a great deal by increasing the crystallite thickness to 15 unit cells (Fig. 3.15). The following observations are in order: (1) The interference function produces peaks with equal areas for all 00l diffraction positions (including the 000) and (2) The peak breadth is inversely related to N.

The background contains N - 2 weak, evenly spaced ripples. They result from the unrealistic assumption that diffraction occurs only from crystals made up of exactly 5 (or 15) unit cells. We can do something about this and obtain more realistic patterns. The values calculated for Φ so far describe a single crystal that is N unit cells thick, but if we multiply Eq. (3.10), the equation for the interference function, by $1/N$, the result applies to an infinitely thick aggregate of such crystals.

We can eliminate or minimize the spurious ripples in the interference function if we assume the array causing diffraction is composed of a series of crystallites with different values of N—a much more physically realistic proposition. Each of these crystallites will produce ripples at different positions; so the sum will become a smooth background if we add a sufficient number of thicknesses. Let $q(N)$ be the proportion of crystallites of thickness N normalized so that the sum of all values of $q(N)$ is 1. Equation (3.15) shows the final result.

$$\Phi(\theta) = \sum_{N=n_1}^{N=n_2} q(N) \; \frac{\sin^2(2\pi\,ND\sin\theta/\lambda)}{N\sin^2(2\pi\,D\sin\theta/\lambda)} \tag{3.15}$$

Figure 3.16 shows the results from a sum of equal proportions of crystallites for which N varies from 2 to 15. The ripples have been almost eliminated, and the shapes of the peaks have changed. They now are more peaked; the tails are broader for a given width at half-height. Their shape illustrates the dependence of the shape of a diffraction line on the distribution pattern of crystallite sizes that make up the hypothetical crystallite aggregate.

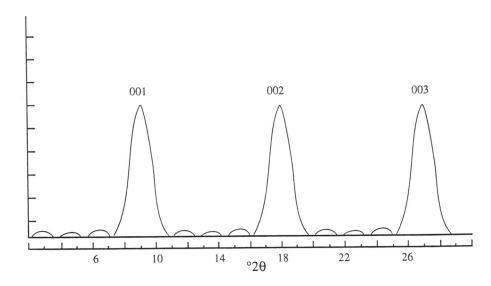

Fig. 3.14. Interference function (Φ) for $N=5$, $d(001) = 10$ Å.

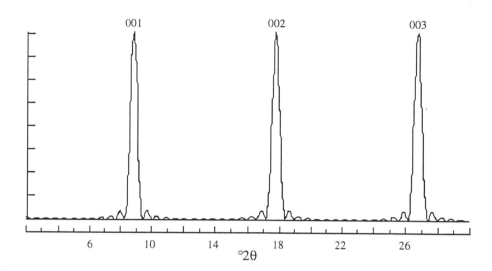

Fig. 3.15. Interference function (Φ) for N=15, $d(001)$ = 10 Å.

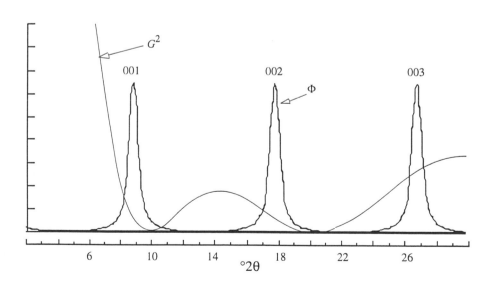

Fig. 3.16. Interference function (Φ) for N = 2-15, $d(001)$ = 10 Å, and $G^2(\theta)$ for an illite unit cell.

At this point we have a fairly realistic peak shape, so let us go on and build the rest of the diffraction pattern. Figure 3.16 also shows plots of G^2. Remember, Φ refers to the intensity of scattering from a crystallite with layers that scatter with unit intensity. A real crystal contains layers that do not have unit scattering power, but scatter with different intensities at different values of θ. We are a big step closer to the diffraction pattern of a real crystal if we multiply G and Φ for each value of θ. The results of such a point-by-point multiplication are given in Fig. 3.17, which also shows a curve for an appropriate Lorentz-polarization factor, the factor for taking into account the geometric effects of the diffraction system. The profile is not very impressive because of unrealistic relative intensities, but multiplication by the Lorentz-polarization factor produces, finally, a fairly representative illite 00l diffraction pattern (Fig. 3.18).

Ergun (1970) demonstrated that large crystals that are broken into small diffracting domains by stacking defects behave like an aggregate of crystallites whose abundances diminish exponentially with increasing thickness. Many clay minerals produce diffraction patterns whose line shapes are consistent with this model. An example has been computed for the range of $N = 2$ to 40. The term $q(N)$ in Eq. (3.15) was replaced by $\exp[(2-N)/\delta]$, a weighting coefficient, in which δ signifies the mean defect-free distance, six unit cells for this example (Fig. 3.19).

If you closely compare Figs. 3.18 and 3.19 with the X-ray tracing of natural illite, you can see that Ergun's model has produced more realistic line shapes. In addition, in this model, the troublesome background ripples have been eliminated. More work needs to be done in applying the defect-broadening model to clay minerals, but it invariably produces the most

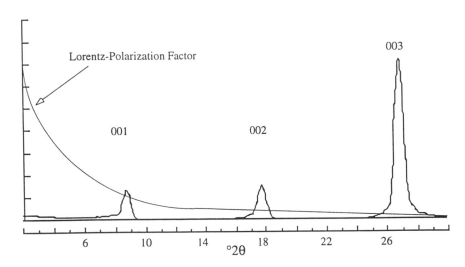

Fig. 3.17 The Lorentz-polarization factor and the product of G^2 and Φ.

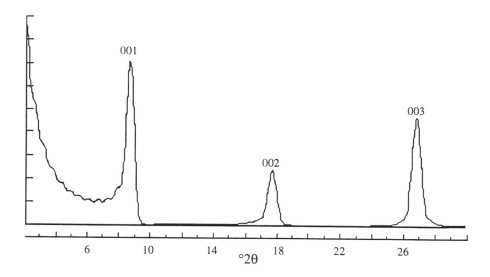

Fig. 3.18. Calculated diffraction patterns for illite, *N*= 2-15.

realistic simulated diffraction patterns. Before leaving this subject, we need to point out that if you attempt to duplicate these calculations, your pattern will not be as realistic on the low-angle side of the 001 peak as that of Fig. 3.19. This is because NEWMOD© assumes that the surfaces of the terminal 2:1 layers of a crystallite are naked, i.e., there is no K there. This is probably realistic, but the equations developed here do not allow for this and position $^{1}/_{2}$K on the outside of each terminal 2:1 layer.

We have discussed the effects of mean *N* and crystallite particle-size distributions on line profile, but a complete treatment of 00*l* line shape requires a consideration of the effects of the interplay of *G* and Φ on peak symmetry. The Lorentz-polarization factor is an important component of peak intensity, but it has little or no effect on peak shape except at low diffraction angles, and we will not consider it further at present. Look at the shape of the G^2 function for the illite 001 reflection (Fig. 3.16). It has low values near the high-2θ side of the illite 001 reflection and rises abruptly with diminishing 2θ. When G^2 is multiplied by Φ, the resulting peak shape is weakened on the high-2θ side and extended considerably on the low-2θ side, thus producing a marked asymmetry. By contrast, G^2 is a good deal flatter through the 002 and 003 peak positions, with the result that their shapes are essentially unmodified from the form of the peak in Φ at the 002 and 003 positions. The degree of asymmetry for a given 00*l* reflection increases with the slope of G^2 across that peak, and asymmetry increases as *N* diminishes because the broadest peaks are the most affected by a given slope in G^2.

For extreme cases like the illite 001 reflection, the asymmetry introduced by G^2 is so great that the top of the peak is shifted toward lower 2θ values. A calculation of the profiles of various 00*l* reflections from an illite consisting

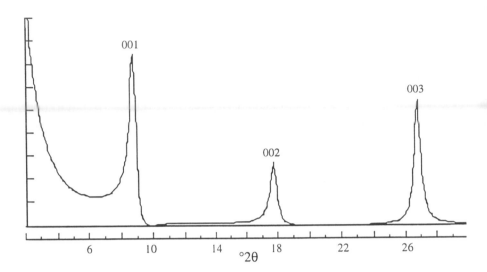

Fig. 3.19. Diffraction pattern for illite. Defect-free distance (δ) = 6 unit cells.

of thin crystals produces peak positions that depart from Bragg's law, and such a situation can be easily misinterpreted as evidence for interstratification with another clay mineral species.

Small crystallite or coherent domain size is often equated with poor "crystallinity." It is ironic that more information can be obtained about the structure of the unit cell from a poor crystal than from that of a good one. Extra information can be obtained because we can examine G over a small but finite range on either side of a normal Bragg peak, as we discussed in connection with Fig. 3.13. This is something that we cannot do with a large crystal because its peaks are infinitely sharp and have measurable intensities only at one point in the 2θ continuum. To be sure, the experimental patterns for large crystallites show some diffracted intensity on either side of the peak position, but remember that is caused by instrument distortion and tells us nothing about the diffraction intensity from the unit cell at angles that deviate somewhat from the ideal ones. Small crystals give information not only on the magnitude of G^2 but also on the direction and magnitude of the slope of G^2 in the vicinity of a peak in the interference function. W. F. Bradley (1908-1973) perceptively remarked that the interference function Φ provides windows through which we can examine the layer transform G.

Exercise: Calculation of the Intensity from d(001) for Illite

To show you how the computer program simulated the diffraction tracings in the preceding section, let's calculate intensity values for diffracted beams scattered from a simplified and stylized illite unit cell. Figure 3.20 shows one unit cell layer of illite with interlayer ions of K and a fragment of the overlying unit cell. Each layer of illite is made up of an octahedral (o) sheet sandwiched by two tetrahedral (t) sheets. The sheets are made up of planes of atoms. We have chosen to use the plane of Al and Mg in the center of the octahedral sheet as the origin for measuring distances to neighboring planes. We have listed the separation in Ångstroms starting at this plane. The number and kind of atoms in five of the planes of the illite unit cell are noted at the far right of Fig. 3.20. We will use this figure to explain the calculations associated with Table 3.1.

See that it is 10 Å from the center of the K atom at the bottom of Fig. 3.20 to the center of the K atom at the top. It is also 10 Å from the base of the lowest tetrahedral sheet to the base of the uppermost one. Is it understood that the distance from any atom in the lower unit cell to its equivalent in the upper unit cell is 10 Å? (An important way to see the geometry of this situation is to realize that there is a series of 10 Å spacings interleaved, and each will diffract at exactly the same time when the incident beam is at a Bragg angle.)

What kind of intensities will this arrangement of atoms yield? Intensity is proportional to the layer scattering factor G squared. We developed an expression for G in a preceding section (Eq. 3.9); so to calculate intensity at a chosen value of θ we use Eq. (3.14).

In more careful work, other factors would be needed, but we need only the general idea. We will assume, for this exercise, that the Lp factor is close enough to constant for the values of θ we use so that we may ignore it. Recall that crystallographers like to use F, the structure factor, only for specific *hkl* values, and therefore, for a continuous function of θ we use G, the layer structure factor, or layer scattering factor. We will ignore a factor to allow for effects of temperature, one that allows for some of the incident radiation being absorbed, and one called the multiplicity factor that allows for differences in the relative number of a spacing of one size compared to the number of spacings of other sizes. The general power of Eq. (3.14) can be clearly demonstrated without including any of these factors. Recall how we got from Eq. (3.6), p. 78

$$F(00l) = \sum_n f_n \cos\emptyset_n = \sum_n P_n f_n \cos(2\pi l z_n/c) \qquad (3.6)$$

to Eqs. 3.9 and 3.14:

$$G(\theta) = \sum_n P_n f_n \cos(4\pi z_n \sin\theta/\lambda) \qquad (3.9)$$

and

$$I(\theta) = [G^2(\theta)][\Phi(\theta)][\mathrm{Lp}(\theta)] \qquad (3.14)$$

From these equations, we have

$$G(\theta) = \sum_n f_n(\cos\emptyset_n) \quad \text{and} \quad |G(\theta)|^2 = [\sum_n f_n(\cos\emptyset_n)]^2 \qquad (3.16)$$

Table 3.1 shows how the parts of Eq. (3.16) are evaluated. Recall from the discussion of Fig. 3.9 that the phase angle $\emptyset$ is found by assuming there is a 360° difference [for the (001) reflection], or one full wavelength, between the wave scattered from the lower plane of Al and Mg atoms and that scattered from the equivalent upper plane of Al and Mg atoms, i.e., the phase angle of the lower plane is 0° and that of the upper is 360°. (This is true only when the diffractometer is set at a Bragg angle, i.e., when constructive interference takes place.) Then the phase angles for waves scattered from other planes are the proportional distance of each plane times 360°, as we did for the phase difference for the ray scattered from atom B in Fig. 3.10. (For example, see Fig. 3.20 and Table 3.1. The plane labeled 3.4 Si's and 0.6 Al's is 2.68 Å above the plane with 3.6 Al's and 0.4 Mg's, the plane that has been chosen as the starting point, or $z = 0$. Then the phase angle for the plane with 3.4 Si's and 0.6 Al's is 2.68/10 x 360° = 96.5°, i.e., it is 96.5° out of phase with the wave scattered from the Al-Mg plane. Or, another way to visualize this is to

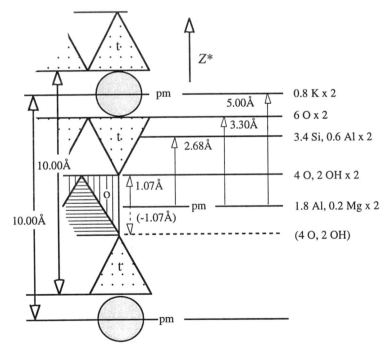

Fig. 3.20. A diagram of an edge of the illite structure showing the separation in Ångstroms of each plane of atoms above the mirror plane marked pm (because it is a pseudomirror). The contribution of each plane to the amplitude of the beam diffracted from the unit cell will be related to its fractional position in relation to the 10 Å spacing. t marks a tetrahedral sheet and o marks an octahedral sheet.

recall that at the Bragg angle for diffraction from a 10 Å spacing, 360° represents the full cycle of one sine wave. Then the 2.68 Å above the starting point is 2.68/10 of a full sine wave.)

Focus on Fig. 3.20 for a moment. See that there are three mirrors labeled pm. (These are mirrors, or pseudomirrors pm, we have invented for the sake of this description. They are not mirrors in the strict crystallographic sense.) We have given the distances between planes of atoms starting from the central atom plane in the octahedral sheet as zero. We have stopped at the mirror plane that cuts the K atom in the interlayer space of this illite crystallite. How can we do this if the unit cell dimension is 10 Å and the center of the K atom is only halfway through the 360° cycle? Do you see that the plane above the origin with 4 O's and 2 OH's is duplicated at an equal distance below a mirror plane? Correspondingly, all the planes for which we have given the atom populations are duplicated below this mirror. By the happy consequence of this mirror, the arithmetic is reduced. If you review your trigonometry, you will rediscover that $\cos(x) = \cos(-x)$. Starting at the origin, $z = 0$, the atom plane above the mirror has the coordinate 1.07 Å. There is an identical plane at the coordinate -1.07 Å. Likewise, planes of atoms are present at 2.7 and -2.7 Å, at 3.3 and -3.3 Å, and finally at 5 and -5 Å. Thus we need only consider five planes (counting the zero plane) in the cosine summation. The cosine terms for planes 2, 3, and 4 must be multiplied by 2 to account for the identical contributions of the coordinates at z and $-z$. You don't double the contributions of the Al and Mg plane and the K plane because they do not have mirror images above and below the center of symmetry in one unit cell. You might say that they have half an image on each side.

The column headed $f(\theta)$ is the scattering power of the neutral atom for the Bragg angle for which the problem is being solved. Because f is a function of the Bragg angle, as well as the number of electrons of an atom, we must specify the angle for which we want the scattering power. In this case, for CuKα radiation, we know that the (001) spacing of 10 Å causes diffraction at 8.8° 2θ or 4.4° θ; i.e., 4.4° θ is a Bragg angle for this spacing, so we will use the scattering values for these atoms at 4.4°. Values of f are listed in tables (see Klug and Alexander, 1974, pp. 880ff). They also can be calculated by means of a simple polynomial published by Wright (1973). For $nf(\theta)$, simply multiply the number of atoms in the particular plane by the scattering factor, e.g., for 3.6 Al and 0.4 Mg, $(12.1 \times 3.6) + (11.25 \times 0.4) = 48.06$, which we've rounded off to 48.1. Notice that for the planes containing O and OH, H is ignored and only the O atoms are counted (how could this be?) The last column is then summed to find the total scattering efficiency of the unit cell (G) of $K_{1.6}(Al_{3.6}Mg_{0.4})(Si_{6.8}Al_{1.2})O_{20}(OH)_4$ arranged as shown in Fig. 3.20. In this case it is 35.3 electron scattering units.

G calculated from Eq. (3.16) will not simulate an experimental diffraction tracing. G plotted against 2θ would give the curve shown in Fig. 3.21. To get a simulation of a diffraction tracing, we must multiply G^2 by the interference

Table 3.1. Evaluation of terms for Fig. 3.20 and Eq. (3.16) for a 10 Å spacing using CuKα radiation.

Atoms, P_n	Ø (deg)	$f(\theta)$	$r*f(\theta)$	cosØ	$nf(\theta)cosØ$
1.6 K	(5.00/10)(360) = 180.0	17.9	28.6	-1.000	28.6x1 = -28.6
6 O	(3.00/10)(360) = 118.8	7.6	45.6	-0.481	-21.9x2 = -43.8
0.4 Si, 0.6 Al	(2.68/10)(360) = 96.5	13.2; 12.1	52.1	-0.113	-5.9x2 = -11.8
4 O; 2 OH	(1.07/10)(360) = 38.5	7.6	45.6	0.783	35.7x2 = 71.4
3.6 Al; 0.4 Mg	(0.00/10)(360) = 0.0	12.1; 11.25	48.1	1.000	48.1x1 = 48.1
				$G(001) = \Sigma =$	35.3

*r = number of atoms in plane

function Φ, as defined in Eq. (3.10). For our sample calculation, assume the illite crystallite is 20 layers thick, each layer is 10 Å thick, and the wavelength of the CuKα radiation is 1.54 Å. We put these values in Eq. (3.10) as $N = 20$, $d = 10$ Å, $\lambda = 1.54$ Å, and $\theta = 4.4°$, as stated earlier

$$\Phi = \frac{\sin^2[(2\pi)(20)(10\text{ Å})(\sin 4.4°)/1.54\text{ Å}]}{\sin^2[(2\pi)(10\text{ Å})(\sin 4.4°)/1.54\text{ Å}]}$$

$$= \frac{\sin^2 62.57}{\sin^2 3.13}$$

[We need to introduce a qualification here and advise you that this function is very sensitive to errors when it is solved right at the Bragg angle. In fact, if you had θ, λ, and d exactly correct, the value of Φ would be equal to $\sin^2(n\pi)$ over $\sin^2(\pi)$, where n is integral and that would be indeterminate. But if you are very close, but not precisely on the Bragg position, the value of Φ is simply equal to N^2. What allows a computer to arrive at a useful answer is that π is an irrational number, i.e., it is without an exact value. Play with these numbers and see for yourself. We will ignore this qualification and go on from here.]

The numbers 62.57 and 3.13 are in radians, and we know that there are 2π radians in 360°, or one cycle, or one wavelength. We are interested only in how much a wave or cycle is out of phase; so we ignore all whole cycles because whole cycles will not tell us how close two waves are to being in phase. By simple arithmetic, you can find that 6.28 (i.e., 2 x 3.14, the number of radians in one cycle) times 9 = 56.52, and 62.57 - 56.52 = 6.05. Or, to say the same thing a different way, there are nine complete cycles and 6.05/6.28 of another in 62.57 radians. We don't need to do anything to 3.13 because it is less than one cycle. We now have a ratio of two cycles, 6.05/3.13, or a measure of phase difference. To convert these to degrees, we multiply 360° by the following ratios

$$(6.05/6.28)(360°) = 346° \ 49'$$
$$(3.13/6.28)(360°) = 179° \ 26'$$

The sine of 346° 49' = -0.2281 and that of 179° 26' = 0.01. Now we must square these values

$$\frac{(-0.2281)^2}{(0.01)^2} = \frac{0.052}{0.0001} = 520 \text{ for } \Phi \text{ at } 4.4°\theta$$

Then

$$I(001) = |G(001)|^2 \ [\Phi(001)][Lp(001)] = |35.3|^2 \ (520)[Lp(001)]$$

$$= 647,967 \ [Lp(001)]$$

or an intensity of 647,967 (in some arbitrary unit) times the Lp factor. [The 35.3 is the sum of $G(001)$ taken from Table 3.1. Notice that about the same value can be read from Fig. 3.21.]

After you have followed this argument to its conclusion, solve Φ for 4.2° and 4.6° θ, just 0.2° on each side of this intense, diffracted beam (these could

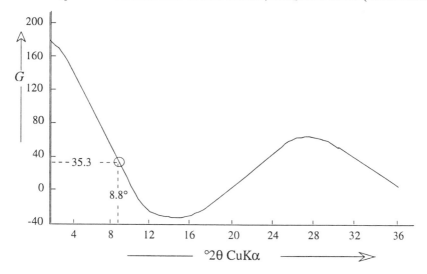

Fig. 3.21. Plot of the layer structure factor for dioctahedral illite *vs.* ° 2θ for CuKα. This does not model a diffraction tracing. We need one more step. Note that the value read from this graph is approximately the same as the *G* of Table 3.1.

be rays A' and B' of Fig. 3.13). Next to the main diffracted beam is precisely where G provides useful information for small crystals. By definition F doesn't exist away from the 001 peak position. Φ, which has a simple solution at 4.4°, still gives finite intensities 0.2° on each side of the 001 peak position. Assume that G [as in (Eq. 3.15), p. 92] stays constant over the range of the peak. Then, if you would like further insight, or just like arithmetic, you may wish to figure the intensity for the second- and third-order reflections, (002) and (003).

Are you impressed that you can calculate X-ray intensity? Do you see that this could be done for an entire diffraction tracing? This is what a computer program can do for you.

We took advantage of planes of pseudosymmetry in the illite unit cell to simplify the arithmetic. There are two reasons why this simplified approach won't work for noncentrosymmetric structures. See if you can figure them out. A note of caution. It is easy to make mistakes in assigning the correct number of atoms to the origin and to the last plane in the calculation. Look at the quantities of all the atoms you have assigned to the different atomic planes. They had better give the correct formula! (A more detailed discussion of these calculations can be found in Reynolds, 1980, pp. 255ff.)

POINTS TO REMEMBER

To summarize:
1. The 00*l* diffraction pattern, as a continuous function of θ, is calculated from the expression $I(\theta) = [G^2(\theta)][\Phi(\theta)][Lp(\theta)]$. Lp is the Lorentz-polarization factor whose values are controlled by the geometry of the instrument and the preferred orientation of crystallites within the crystal aggregate, G is the amplitude of scattering of the unit cell in the direction θ, and Φ is the interference function that describes the intensity of scattering of a crystallite or mixture of crystallites, each of which contains unit layers that scatter like single electrons.
2. The breadths of 00*l* reflections at half-height are inversely proportional to N or mean N, where N is the number of unit cells in a coherent diffracting array along the direction normal to $d(001)$.
3. The widths of 00*l* peaks, in regions of 2θ where G^2 is constant or nearly so, are determined by particle-thickness frequency distributions, i.e., by the relative proportions of crystallites that have different thicknesses N.
4. Peak asymmetry is caused by steep slopes of the function G^2 where this function crosses a peak in the interference function Φ. The asymmetrically extended tail of the illite 001 reflection lies in the up-slope direction of G^2, and is a good example of this. Any reflection may be displaced in the direction of increasing G^2 if the slope is steep enough, or if N is small enough to produce broad peaks in the interference function, or both.

REFERENCES

Bragg, W. L. (1913) The diffraction of short electromagnetic waves by a crystal: *Proceeding of the Cambridge Philosophical Society* **17**, 43-57.

Brindley, G. W., and Méring, J. (1951) Diffraction des rayons X par les structures en couches desordennees: *Acta Crystallographica* **4**, 441-47.

Buerger, M. J. (1942) *X-Ray Crystallography*: Wiley, New York, 531 pp.

Cullity, B. D. (1978) *Elements of X-Ray Diffraction*, 2nd ed.: Addison-Wesley, Reading, Mass., 555 pp.

Ergun, S. (1970) X-ray scattering by very defective lattices: *Phys. Rev. B* **131**, 3371-80.

Ewald, P. P. (1921) Das 'reziproke Gitter' in der Strukturtheorie: *Zeitschrift für Kristallographie* **56**, 129-56.

James, R. W. (1965) *The Optical Principles of the Diffraction of X-Rays*: Vol. II, *The Crystalline State*: Series edited by Sir Lawrence Bragg, Cornell University Press, Ithaca, 664 pp.

Klug, H. P., and Alexander, L. E. (1974) *X-Ray Diffraction Procedures*, 2nd ed.: Wiley, New York, 966 pp.

Nadeau, P. H., Tait, J. M., McHardy, W. J., and Wilson, M. J. (1984) Interstratified X-ray diffraction characteristics of physical mixtures of elementary clay particles: *Clay Minerals* **19**, 67-76.

Phillips, F. C. (1971) *An Introduction to Crystallography*, Fourth Edition: John Wiley & Sons, New York, 351 pp.

Reynolds, R. C., Jr. (1976) The Lorentz factor for basal reflections from micaceous minerals in oriented powder aggregates: *Amer. Minerl*. **61**, 484-91.

Reynolds, R. C., Jr. (1980) Interstratified clay minerals: in Brindley, G. W., and Brown, G., editors, *Crystal Structures of Clay Minerals and Their X-ray Identification*: Monograph No. 5, Mineralogical Society, London, 249-303.

Reynolds, R. C., Jr. (1985) NEWMOD©, *a Computer Program for the Calculation of Basal X-Ray Diffraction Intensities of Mixed-Layered Clays*. R. C. Reynolds, Hanover, N.H. 03755.

Reynolds, R. C., Jr. (1986) The Lorentz-polarization factor and preferred orientation in oriented clay aggregates: *Clays and Clay Minerals* **34**, 359-67.

Reynolds, R. C., Jr. (1994) *WILDFIRE©, A Computer Program for the Calculation of Three-Dimensional Powder X-Ray Diffraction Patterns for Mica Polytypes and their Disordered Variations*: R. C. Reynolds, 8 Brook Rd., Hanover, N.H. 03755.

Reynolds, R. C., Jr., and Hower, J. (1970) The nature of interlayering in mixed-layer illite-montmorillonite: *Clays and Clay Minerals* **18**, 25-36.

Vance, E. P. (1963) *An Introduction to Modern Mathematics*: Addison-Wesley, Reading, MA, 534 pp.

Wright, A. C. (1973) A compact representation for atomic scattering factors: *Clays and Clay Minerals* **21**, 489-90.

Clay minerals are hydrous aluminum silicates and are classified as phyllosilicates, or layer silicates. There is considerable variation in chemical and physical properties within this family of minerals, but most have in common platy morphology and perfect (001) cleavage, a consequence of their layered atomic structures. Historically, useful information was collected by using staining techniques and the petrographic microscope, but much of what we understand, to this point, about the structural and chemical details of these minerals has been *extrapolated from X-ray diffraction studies of their macroscopic counterparts* because clay minerals are usually < 2 μm and, therefore, too small for study by optical or single crystal X-ray methods. Surprising amounts of information and insight into the nature of clay minerals has been gained in the last two decades by an interplay between modeling of the XRD tracings of clay minerals and improved XRD equipment and accessories. Modeling requires that we know just what happens as X-rays interact with the atoms of the clay minerals that serve as scattering centers.

In this chapter, first let's consider the structural features common to the five most common clay minerals. Then we'll deal with some of the properties that are consequences of structural features such as layer charge, cation-exchange capacity, and interaction with water and other compounds, and finish with general nomenclature and classification for clay minerals.

GENERAL STRUCTURAL FEATURES

All layer silicates can be imagined as constructed from two modular units: A sheet of corner-linked tetrahedra and a sheet of edge-linked octahedra. (At the end of Chapter 5 you will find patterns for making paper octahedra and tetrahedra and directions to guide you in the construction of models of clay minerals.) Our discussion of structural features is quite dependent on figures. You probably will need to flip back and forth to relate one figure to another to follow our descriptions.

Tetrahedral Sheets

In the tetrahedral sheet (Fig. 4.1A), the dominant cation T is Si^{4+}, but Al^{3+} substitutes for it frequently and Fe^{3+} occasionally. The T/O (oxygen) ratio for layer silicates is T_2O_5. Picture this sheet as extending infinitely in two dimensions, each tetrahedron resting on a triangular face in this infinitely extending plane and sharing the oxygens at all three corners with three other tetrahedra. This plane is referred to by some as the siloxane surface. The fourth, or apical, oxygen points upward in the direction normal to the base. Assembled without distortions, this sheet is hexagonal (Fig. 4.1A) with the Si–O bond distance about 1.62 Å and the O–O distance about 2.64 Å. Al may replace up to half of the Si, and when it does the dimensions of the sheet increase because Al^{3+} is larger than Si^{4+}. The Al–O distance is about 1.77 Å.

Octahedral Sheets

The octahedral sheet can be thought of as two planes of closest-packed oxygen ions with cations occupying the resulting octahedral sites between the two planes. When we connect the centers of the six oxygen ions packed around an octahedral cation site, we have an octahedron. Sharing of neighboring oxygen ions forms a sheet of edge-linked octahedra, again extending infinitely in two dimensions (Fig. 4.1B). The cations are usually Al^{3+}, Mg^{2+}, Fe^{2+}, or Fe^{3+}, but all the other transition elements (except Sc, which hasn't been reported as far as we know) and Li have been identified in cation sites of the octahedral sheet. The structure of the macroscopic minerals gibbsite [$Al(OH)_3$] and brucite [$Mg(OH)_2$] would fit the description for an octahedral sheet except that they are composed of two planes of closest-

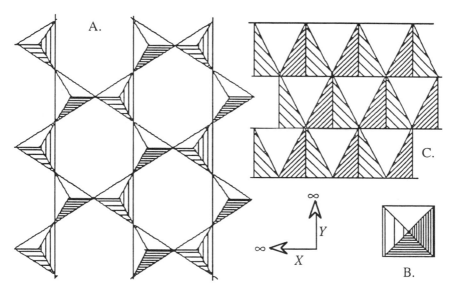

Fig. 4.1. A. Part of a sheet of corner linked tetrahedra. B. An isolated octahedron viewed down the 4-fold axis perpendicular to the page. C. Part of a sheet of edge-linked octahedra. These octahedra are sitting on a triangular face with a 3-fold axis perpendicular to the page.

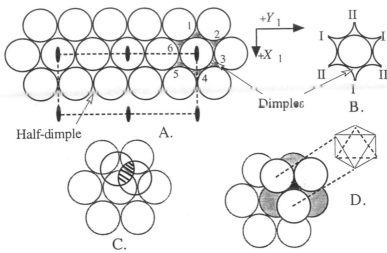

Fig. 4.2. Coordination in the octahedral sheet. A. Looking down on a plane of oxygen or hydroxyl ions (a plane in the *X-Y* dimension) with a cell imposed on it. Note that each anion has six nearest neighbors in this closest-packing arrangement and each is surrounded by six dimples (shaded) numbered 1 to 6. B. The six dimples make two sets, I and II, one set of which will be octahedral sites, the other tetrahedral (the latter are almost never occupied). C. The lighter circles represent oxygen or hydroxyl ions in the lower plane of anions. We assume the upper anions come to rest in the dimples of the lower sheet, consistent with closest packing. See that only three of the six dimples can be occupied. If adjacent dimples were occupied, both anions would have to occupy the volume represented by the shading in C. D. Note that a cation in dimple 2 would be surrounded by six oxygens or hydroxyls, three above it (lightly shaded) and three below (more heavily shaded). This site is, therefore, an octahedral site because if the centers of the six anions are joined by straight lines, an octahedron resting on a triangular base (the dashed lines) is created around the cation in the octahedral site.

packed hydroxyls, instead of oxygen ions. But the diameter of the hydroxyl ion is so similar to that of oxygen that there is essentially no difference. Therefore the structures of these two minerals serve as analogs in two ways: (1) For the octahedral sheets in clay minerals, the cation-to-anion ratio determines whether the mineral is referred to as dioctahedral or trioctahedral (see the next section); and (2) for the dimensions of the hydroxyl sheet within the 2:1 or 1:1 layer (see the section *Joining the Sheets*). The terms *gibbsite-like* and *brucite-like* should be used when discussing these analogs in the structure of clay minerals.

Dioctahedral and Trioctahedral

In layer silicates, octahedral sheets are either gibbsite-like $Al(OH)_3$ or brucite-like $Mg(OH)_2$ without too much overlap; there is limited solid solution between the two end members. In the brucite-like sheet the cation-to-anion ratio is 1:2. If you picture a plane of closest-packed hydroxyls and remember

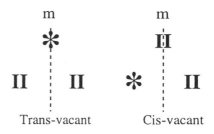

Fig. 4.3. Trans- and cis-vacant configurations for dioctahedral layer silicates. ✳ = vacant site, m = pseudomirror.

that they are essentially the same size as oxygen ions, you will recall that each hydroxyl ion has six nearest neighbors making six "dimples" (Fig. 4.2A) around it. When the next plane of closest-packed hydroxyls is fitted into the dimples of this first plane of hydroxyls, the six dimples will become three octahedrally coordinated and three tetrahedrally coordinated cation sites. For an octahedral sheet, we will use only the octahedral sites. By counting the octahedral sites in Fig. 4.2, you see that there is the same number of them as there are hydroxyl ions in a single plane or half as many sites as there are in two planes of hydroxyl ions. So, for the brucite-like sheet with divalent cations, all three octahedral sites around each hydroxyl must be filled to have electrical neutrality. These three sites are equivalent, in the crystallographic sense. Layer silicates with this arrangement are called *trioctahedral*. (You should find some sort of balls, Styrofoam, wooden, Ping Pong, or whatever, and try packing them so you can see the tetrahedral and octahedral sites.)

In layer silicates with gibbsite-like octahedral sheets, the cation-to-anion ratio in the sheet is 1:3. Here, to have electrical neutrality, only two Al^{3+} cations are needed instead of three Mg^{2+} ions. Therefore only two of every three octahedral sites around each hydroxyl need to be filled. Layer silicates with this arrangement are called *dioctahedral*. Trioctahedral and dioctahedral 2:1 clay minerals can usually be referred to the space groups C2/m or C2. The mirror, or pseudomirror in the case of C2, bisects one of the octahedral sites. This site is called M(1). The two sites on either side of the mirror are the M(2) sites. The M(2) sites are equivalent to one another, but not to the M(1) site. Until recently, it was assumed that the vacant site in the dioctahedral sheet of mica-like minerals was always the M(1) site. However, Tsipursky and Drits (1984) concluded that many smectites have an occupied M(1) site, and one occupied and one vacant M(2) site. Dioctahedral sheets with vacant sites located on a mirror are called *trans-vacant*, i.e., the mirror *"transverses"* the vacant site (how's that for a mnemonic device?); those with an occupied site on the mirror are called *cis-vacant* (Fig. 4.3). We will come back to this point in Chapter 10 when we discuss *trans-vacant* and *cis-vacant* dioctahedral sheets in illite polytypes.

Joining the Sheets

Now we must imagine these modular units, the tetrahedral and octahedral sheets, linked if we are to visualize the structure of clay minerals. (At the end of the chapter there is an exercise that gives some directions for making models of the layer silicates.) The oxygen-to-oxygen or hydroxyl-to-hydroxyl

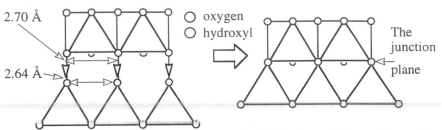

Fig. 4.4. Joining an octahedral sheet and a tetrahedral sheet into a 1:1 layer silicate at the junction plane. Dimensions from different sources differ, but these illustrate the general case. This makes a tetrahedral-octahedral-tetrahedral sandwich.

ionic dimensions of the octahedral the tetrahedral sheets are approximately the same (Fig. 4.4), so the apical oxygens of the tetrahedra can be seen as replacing two out of three of the hydroxyl ions in the lower plane of the octahedral sheet. This assemblage of one tetrahedral sheet and one octahedral sheet is called a *1:1 layer silicate structure.* Notice that one-third of the hydroxyl ions were not replaced but fit into, in the ideal hexagonal pattern of the tetrahedral sheets, the hole in the hexagonal ring made by the apical oxygens of the tetrahedral sheet (combined view of Figs. 4.1 and 4.4). A *2:1 layer silicate,* two tetrahedral sheets to one octahedral sheet, is formed by inverting a tetrahedral sheet, bringing it down on top of the 1:1 layer in Fig. 4.4, and again replacing two-thirds of the hydroxyls with apical oxygen ions. This assemblage makes a tetrahedral-octahedral-tetrahedral sandwich.

Not all, or even most, fits are as neat as that portrayed in Fig. 4.4. The lateral dimensions of the tetrahedral sheet are usually larger than those of the octahedral sheet because the larger Al^{3+}, or sometimes Fe^{3+}, has been substituted for the smaller Si^{4+}. Therefore, distortions or adjustments in one or both sheets are necessary if they are to combine at a common plane, called the *junction plane* by Bailey (1984). By referring the three sheets to X and Y coordinates in a plane parallel to their sheet dimensions, we can assign approximate values to unit cell dimensions in this plane (Fig. 4.2A, in which a is the unit distance along X and b along Y). The size of the b dimension discriminates trioctahedral from dioctahedral layers, the latter being the smaller. The way in which we approximate the b dimensions for such sheets is to use the values of those we find in the free state, and they are, for gibbsite, 8.64 Å (8.67 Å for bayerite, a polymorph of gibbsite), and for brucite, 9.43 Å. For an ideal tetrahedral sheet with no Al^{3+} substituted for Si^{4+}, $b = 9.15$ Å. We can see that such a tetrahedral sheet has dimensions intermediate to dioctahedral and trioctahedral sheets: the dioctahedral sheet is about 6% smaller and the trioctahedral sheet is about 3% larger. As Al^{3+} is substituted into the tetrahedral sheet, its b dimension increases in response to the larger Al^{3+} ion. With half of the Si^{4+} replaced by Al^{3+}, the b dimension becomes about 9.55 Å, or about 2% larger than that of the trioctahedral sheet.

Adjustments can be made in either the tetrahedral sheet or the octahedral sheet to accommodate a fit between them.

It might help to see this structure in three dimensions. Figure 4.5 shows the imaginary process of fitting two tetrahedral sheets to a dioctahedral sheet. We use the term *imaginary* so that you won't get the idea that this is how the crystals grow in nature! The triangular planes (shaded) on the top of the tetrahedral sheet are indicated as are the triangular planes (shaded) on the top of the gibbsite-like octahedral sheet. As this arrangement is put together, visualize that the apical oxygens indicated by arrows replace the hydroxyls of the octahedral sheet that are at the other ends of the arrows. But two hydroxyls stay put, so to speak, and they are indicated by the black balls at the center of the upper and lower oxygen-hydroxyl planes of the octahedral sheet. These occupy positions just below the centers of the ideally hexagonal holes in the outer oxygen planes of the tetrahedral sheets. You can see that these two black balls are displaced from each other with respect to an axis normal to the sheets. The direction of this displacement in the *X-Y* plane is along the -*X* crystallographic axis, and the two tetrahedral sheets are similarly displaced. This displacement converts what could have been trigonal symmetry to monoclinic symmetry. Ideally, the amount of this displacement for the 2:1 clay minerals (and micas) is 1/3 of the *a* dimension.

The lateral dimensions of the tetrahedral sheet can be reduced in three ways: (1) by rotating adjacent tetrahedra in opposite directions on axes through their apices and perpendicular to the (001) plane (Fig. 4.6), (2) by thickening the sheet, and (3) by tilting the tetrahedra. Tetrahedra are never, to our knowledge, rotated as far as they are shown in Fig. 4.6, but they do frequently rotate enough to form fat triangles. Changing the sheet thickness can be thought of as changing the ideal angle for a tetrahedron. (The ideal angle, measured from the center of the apical oxygen ion to the center of the tetrahedral cation to the center of one of the basal oxygen ions, equals 109° 28'.) The sheet thickens when the angle increases, thereby reducing the lateral, or *a-b*, dimensions of the sheet. Tilting of pairs of tetrahedra toward one another along the hinge line *r* in Fig. 4.6A and away from another along *s* results in corrugations of the basal plane of the tetrahedral sheet. Dioctahedral micas frequently show this corrugation (Bailey, 1984).

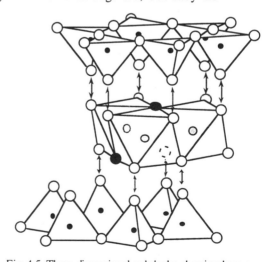

Fig. 4.5. Three-dimensional polyhedra showing how a 2:1 clay mineral can be imagined assembling. Large black balls are hydroxyls; small black balls are tetrahedrally coordinated cations; shaded balls are octahedrally coordinated cations; and unshaded balls are oxygens.

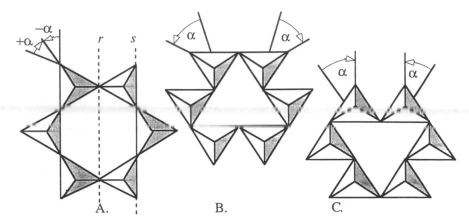

Fig. 4.6. All viewed in the *X-Y* plane. A. An ideal hexagonal ring. B. The result of rotating them away from one another. C. The result of rotating pairs of tetrahedra toward one another. (After Zoltai and Stout, 1984, p. 329.)

Tilting of successive rows of tetrahedra in the same direction results in the sheet forming a tube. We will see this type of distortion in the 1:1 layer silicates, which often have fiber-like or tube morphologies.

Adjustments in the octahedral sheets range from rather simple in the trioctahedral to more complex in the dioctahedral types. In trioctahedral sheets, distortion from the ideal is visualized by pulling each anion in three directions along three shared edges toward the cation plane with the resultant force inward (Fig. 4.7). The result is a thinner sheet with larger lateral dimensions.

In dioctahedral sheets, every third cation site is vacant so there are distortions in addition to those necessary for articulation with a tetrahedral sheet. Some of the distortions of dioctahedral sheets are shown in Fig. 4.8. Notice that the six anions around the vacant site all move away from one another and towards occupied cation sites. As a result of this movement, the edge length of the octahedra around the vacant sites increases from an ideal dimension of about 2.7 Å to about 3.2 to 3.4 Å. The anions in upper and lower planes moving toward one another shorten shared edges with the same effect as in the trioctahedral sheet: the sheet thickness decreases. Movement of anions around vacant sites has two results: reduction of sheet thickness, and combined but opposite movements for the upper and lower triangular faces of an octahedron. The upper face rotates counterclockwise and is pulled

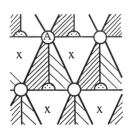

Fig. 4.7. Thinning of trioctahedral sheet. Open circles = upper plane of anions; X = central plane of cations; shaded half circles = lower plane of anions. Anion A is pulled along shared edges toward three closest anions in lower plane. This causes thinning of sheet.

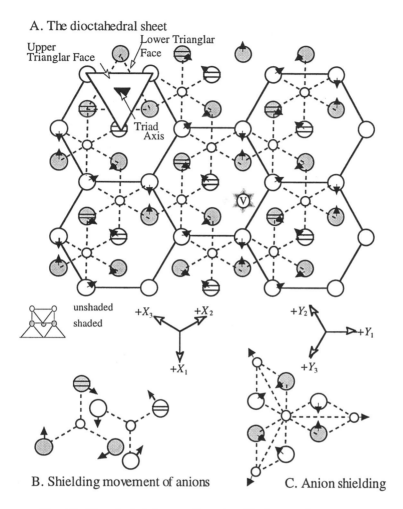

A. The dioctahedral sheet

Upper Trianglar Face

Lower Trianglar Face

Triad Axis

unshaded
shaded

$+X_3$ $+X_2$

$+X_1$

$+Y_2$ $+Y_1$

$+Y_3$

B. Shielding movement of anions   C. Anion shielding

Fig. 4.8. Dioctahedral sheet configuration. The larger spheres are oxygens or hydroxyls; the ones with the horizontal bars are the hydroxyls. The unshaded ones are in the upper plane and the shaded ones are in the lower plane. The smaller spheres are aluminums. How has the upper plane shifted with respect to the lower plane along $-X_3$.? The oxygens of the upper plane are arranged in a hexagonal net as shown by the solid lines drawn from oxygen to oxygen. There is a hydroxyl in the center of each hexagonal cell. This pattern is repeated in the lower plane except the oxygens are not joined by lines. The oxygens of both the upper and lower planes may also be thought of as apical oxygens attached to tetrahedral sheets. A. General view. Only one vacant octahedral site is marked with a v. How many others can you pick out? B. Shielding movement of anions results in rotation in opposite directions around the triad axis of upper and lower faces of the octahedra (upper and lower anions offset for clarity). C. Anions tend to shield and therefore shorten shared octahedral edges: O–O and O–OH ~ 2.3-2.5 Å, whereas unshared edges ~ 2.7-2.9 Å (modified from Bailey, 1967, with permission).

downward, while the lower face rotates clockwise and is pulled upward. Because a pair of upper and lower anions move closer together and are between two octahedrally coordinated cations, this movement is sometimes referred to as shielding one cation from another (see Fig. 4.8B and C). The combination of these distortions changes the sheet thickness from that for an ideal sheet, about 2.6 Å, to 2.04 to 2.14 Å.

We have emphasized that clay minerals are small, < 2 μm or < 4 μm. Perhaps the misfit of the tetrahedral and octahedral sheets at the junction plane (Fig. 4.4) contributes to the smallness of their grain size.

Stacking the Layers

Disorder is caused by unsystematic rotations and/or translations of one layer with respect to the one under it in a stacking sequence along Z, but polytypes are generated if these displacements are systematic, that is, if they are distributed according to some pattern, and if their magnitudes are *special* (Brindley, 1980) with respect to the symmetry of the silicate layers.

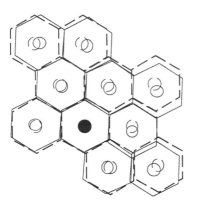

Fig. 4.9. Misalignment of 2:1 interlayer K coordination centers caused by a layer rotation of 3°.

Turbostratic disorder is the most extreme kind of stacking disorder. It is found in halloysite and in most smectites. For the smectites, the low layer charge permits only a low concentration of hydrated cations in the interlayer space, and the weak mutual attraction between them and the adjacent 2:1 layers leads to relatively large distances across the interlayer. Thus there is no "keying" effect between layers, and their relative positions and orientations are haphazard. A given layer can be displaced by any amount in the X-Y plane, and it can be rotated by any amount. These have been termed *arbitrary* displacements/rotations by Brindley. There is no layer regularity along Z, and the mineral is an assemblage of two-dimensional crystallites, although the basal diffraction pattern is sharp because the layers are perfectly parallel to one another. An example of such an arbitrary rotation is given in Fig. 4.9, which shows a lower hexagonal net (dashed lines) and an upper net that has been rotated 3° to the right. K atoms are drawn in the centers of the hexagonal holes. The center of rotation lies through the K atom shaded black, and you can see that the farther you go from that center, the poorer the alignment of the K sites will be with respect to the nets. Now a typical illite crystallite surface has dimensions of perhaps 40 by 40 hexagonal cells, so no matter where you pick the center of rotation, the distal hexagons are going to be poorly aligned indeed. The alignment required for three-dimensional diffraction is so demanding, that a 3°, or even 1°, rotation is enough for the X-ray machine to see the result as fully turbostratic. The nature and origin of the so-called two-dimensional diffraction patterns produced by such structures is discussed in Chapter 10.

We now turn our attention to the mica polytypes. The attractive force between 2:1 layers and the interlayer cation exceeds the dehydration energy of the interlayer cation because of the high layer charge. The layers collapse about the interlayer cations (usually K) that are located in the hexagonal or trigonal holes in the surface planes of the tetrahedral sheets. Because of the regular spatial distribution of these holes, the adjacent 2:1 layers are keyed into positions that make any layer displacements and arbitrary rotations impossible in the *X-Y* plane. The only stacking irregularity allowed is the special rotational kind, and if all the K sites are going to line up between adjacent layers, such interlayer rotations must be integral multiples of 60°. By far the most common special rotations are integral multiples of 120° because of the trigonal (3-fold) symmetry of the surface oxygen planes, though rarely, rotations of *n*60° (*n* is odd) do occur (see Chapter 10).

A layer rotation in the micas causes a shift in the *X-Y* plane of the unit cell center (Fig. 4.10A). Two 2:1 layers are shown in the *1M* configuration, i.e., they have identical orientations. For each of them, the upper and lower K sites are displaced in the -*X* direction by the amount *a*/3, caused by the slant in the octahedral sheet. Any rotation, which must be an integral multiple of 60°, takes place about an axis normal to (001) and passing through the interlayer K ion. Study Fig. 4.10A and satisfy yourself that such a rotation must result in a shift of the center of the unit cell and the upper K in the *X-Y* plane. Because rotations shift the position of the layers with respect to the idealized *1M* geometry, the diffraction pattern will not be identical to that of the *1M* structure. Table 7.7 lists the diffraction data for the polytypes generated by the various layer rotation schemes, and Chapter 10 discusses the diffraction theory and the results that apply to micas in which the sites of the special rotations are distributed randomly, leading to disordered structures. But here, we discuss the various regular patterns of rotations and the polytypes that are produced.

Figure 4.10B shows, schematically, the *1M* stacking. All the layers are identically oriented. In this example, and the ones that follow, an arrow depicts the orientation of the vector corresponding to the intralayer shifts of *a*/3 along -*X*. Figure 4.10C shows the ±120° rotations that define the *2M₁* polytype. Figure 4.10D depicts the spiral axis produced by successive 120° rotations that generate the *3T* polytype. The *a*/3 shifts cancel for a complete cycle, so the final structure is not monoclinic, but hexagonal, or more accurately, trigonal. Not shown on Fig. 4.10 are two other rare mica polytypes. One is the *2Or*, or 2-layer orthorhombic structure that is formed by alternate rotations of plus and minus 180°, and the last is the *2M₂*, which has alternate layers rotated by plus and minus 60° with respect to each other. The

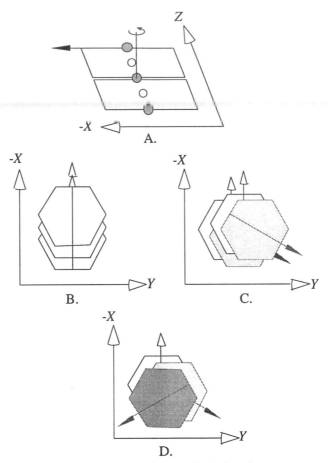

Fig. 4.10. A. Demonstration that a layer rotation in the micas causes a translation of the unit cell in the *X-Y* plane. Viewed normal to the (001) plane are B. the *1M* stacking, C. plus and minus 120° rotations forming the *2M₁* polytype and D. successive 120° rotations producing the *3T* polytype. Translation magnitudes are not to scale.

latter two polytypes have rotations that are odd-integral multiples of 60°, and as we discuss in Chapter 10, these are energetically unfavorable for the oxygen coordination about K and are rare probably for that reason (Bailey, 1975). There are many other possible polytypic stacking schemes, but they have, so far, not been found in nature.

Layer stacking constraints are different for kaolinite and chlorite because these minerals have no naked cation "keys" in the interlayer spaces, and bonding of the silicate layers is between a hexagonal oxygen surface on one side and a close-packed hexagonal hydroxyl surface on the other. The alignment of two adjacent silicate layers in the *X-Y* plane must be consistent with the spatial requirements of hydrogen bonding between these two surfaces (Bailey, 1980b).

Figure 4.11 is a cartoon of the stacking schemes for the kaolin minerals kaolinite, dickite, and nacrite viewed normal to (001). The shifts are not to scale, but have been exaggerated for clarity of presentation. The three octahedral sites are shown (A, B, and C) only two of which are occupied (circles). The unoccupied site (square) is almost always the B position in the kaolin minerals (Bailey, 1993).

The interlayer shift along -X is close to $a/3$ for the same reason that applies to the micas, namely, that the shift is due to the slant of the octahedral sheet, the upper atomic plane of which is displaced from the lower plane by $a/3$. For kaolinite this produces a simple *1M* stacking pattern. In dickite, alternating B and C site occupancy leads to a *2M* structure. The structure of nacrite is more complicated. Here, alternating 1:1 layers are rotated by 180° with respect to the zero degree datum, and the unit cells are displaced along b by the distance $b/3$. If you think about this arrangement, perhaps you can see that the structure will repeat on the sixth layer (three times $b/3$ times two because of the alternate 0-180 rotations). This produces a six-layer polytype,

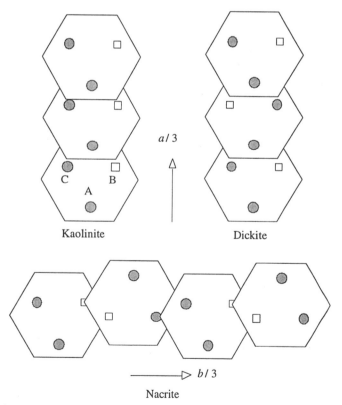

Fig. 4.11 Stacking sequences in kaolinite, dickite, and nacrite viewed normal to the (001) plane. Filled circles represent occupied octahedral sites. Squares depict the vacant site. Translation magnitudes have been exaggerated for clarity.

but as Bailey (1993) has demonstrated, the axes can be redefined so that X becomes Y, Y becomes X, and then Z lies in the plane that is normal to the page and includes the new X axis (the old Y). If this is done, then the pattern repeats every two layers. Nacrite is a 2-layer polytype, but one that is very different from the dickite $2M$ structure.

Stacking disorder in kaolinite is caused by an apparent random distribution of octahedral cation vacancies among A, B, and C positions (Fig. 4.11, but disorder not shown), although there is still discussion concerning the nature of the crystallographic operations necessary to produce this disorder. The work of Plançon and his co-workers on disordered kaolinite is summarized and discussed by Giese (1988), who concludes that such minerals are physical mixtures of moderate-defect and low-defect phases. Diffraction data for the kaolin polytypes are given in Table 7.6.

Different chlorite polytypes are produced by different translations of 2:1 layers with respect to the hydroxide sheets and by the direction of slant of the hydroxide octahedral sheet compared to the slant of the octahedral sheet in the silicate layers. If the octahedral sheets slant in the same direction, the polytype is called type I, and if they slant in opposite directions (caused by a 180° rotation of the hydroxide sheet), type II is the proper designation (see Fig. 4.12). Any translations of 2:1 layers in the X-Y plane must be consistent with the geometric constraints of hydrogen bonding between O and OH surfaces, and the two possibilities, a and b are shown by Fig. 4.13. Position a locates the hydroxide sheet cations directly over the tetrahedral Si positions in the underlying 2:1 layer (the *Iaa* polytype in Fig. 4.13). Position b superimposes the hydroxide cations over the octahedral 2:1 sheet cations (Fig. 4.13 *Ibb*).

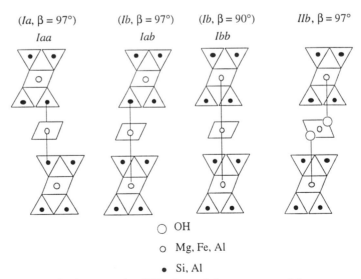

O   OH

o   Mg, Fe, Al

•   Si, Al

Fig. 4.12   Projection onto the (010) plane of the structures of four common chlorite polytypes. Modified from Shirozu and Bailey (1965).

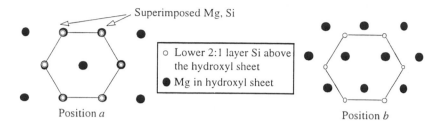

Fig. 4.13. Projection onto the (001) plane of the positions of the octahedral cations of the hydroxide sheet (Mg) and the Si positions in the tetrahedral sheet of the lower 2:1 layer.

Shirozo and Bailey (1965) concluded that the *a* position is energetically unfavorable because of cation-cation (Mg-Si) repulsion over the relatively short distance involved, and indeed, such polytypes are rare compared to the *b* types. Powder diffraction data for the chlorite polytypes are given in Table 7.5.

PROPERTIES

The properties of minerals are directly or indirectly related to their primary property, their structure. Clay minerals are no exception. In what follows, however, we will be concerned with properties that result from the interaction of clay minerals with other substances, mainly water. These properties depend on the nature of water, the ions that exist in the water, and the size, character, and distribution of charges on the surfaces of the minerals. Important aspects of the ions are their size, valence, electronegativity, and hydration energy. As Newman (1987b, p. 237) recalled, Cairns-Smith, in his discussion about the origin of life, described clay as "the story of sloppy, sticky, lumpy, and tough," indicating the stages clay goes through as it dries. This description indirectly acknowledges the overwhelming importance and presence of water. Many of the properties of clay minerals and other clay-sized materials are related to water attached to edges and to (001) surfaces. This includes such characteristic properties as plasticity, nutrient-holding and nutrient-exchange capacities, as well as bonding, compaction, and suspension capabilities. Understanding the different "states" suggested by Cairns-Smith, and reversing the process so that samples can be prepared for study, requires an understanding of the properties of clay mineral particles, or more generally, colloidal properties. The ones we will take up here are layer and surface charge, the electrical double layer, exchangeable ions, the nature of water when it is adsorbed on edges or in the interlayer space, and interaction with organic compounds. We will finish this chapter with a section on classification of the clay minerals. In the discussion of some of these properties, we include the group of clay-sized metal oxides and hydroxides called *soil oxides*.

Soil oxides are important because they offer insights into the pedogenic conditions prevailing during their formation and because of their contribution to the retention and release of nutrients. They frequently are intimately overgrown on the surfaces of other minerals. If we are to be able to read paleosols for the information they contain, we must understand both the soil oxides and the minerals with which they are associated. Taylor (1987) and Schwertmann and Taylor (1989), to whom we defer for a more extensive discussion, have offered thorough reviews of this important group. We offer only the briefest of introductions to these properties, and suggest references where you may pursue them. The classical works on the colloidal state of clay minerals are Marshall (1977) and van Olphen (1977).

Total Charge, Layer or Permanent Charge, and Variable Charge

Clay minerals and clay-sized minerals have charges on their surfaces. These determine ion-exchange capacities (see the section below on ion exchange); the dispersion/flocculation behaviors (next section); the transport and fate of solutes; and govern the rates of chemical weathering and the erodibility of the land surface. There are two kinds of charges at the surfaces of clay-sized minerals. One arises from substitution of a cation with one less valence charge than the one it replaces in the structure of the mineral. When the tetrahedral and octahedral sheets are assembled into layers, they may be electrically neutral or they may be negatively charged. Electrical neutrality exists if an octahedral sheet contains R^{3+} cations in two out of three octahedral sites, or R^{2+} cations in all octahedral sites, combined with the tetrahedral sheets containing Si^{4+} in all tetrahedral sites. Such minerals, other than kaolinite, are not particularly common, at least not in the clay-sized fractions of soils and sedimentary rocks. More often, there is some deviation from the ideal number of sites occupied exclusively by a single kind of cation resulting in a *layer charge*. Most commonly, Al^{3+} is substituted for Si^{4+} in the tetrahedral sheet, which, unless compensated for in the octahedral sheet, gives the layer a net negative charge. Two other ways of getting a negative layer charge are Mg^{2+} substitution for Al^{3+} in a dioctahedral sheet or substitution of vacancies in octahedral positions. There are many variations on this theme. The maximum layer charge for most 2:1 clay minerals is about -1.00 based on T_4O_{10}, or 4 tetrahedral cations and 10 oxygens. This is a formula unit or half of the unit cell of a layer silicate.

Neutrality is restored by having either single ions or ionic groups in the space between the layers, called, not too surprisingly, the *interlayer space*. Potassium (as in muscovite and biotite), sodium, and calcium are the most common interlayer cations. Complete and incomplete sheets of octahedrally coordinated cations occupy the interlayer space in chlorite and vermiculite. These sheets seem to be a consequence of hydrated cations, usually Fe^{2+}, Mg, Al, or Fe^{3+}, that organize OH ions or water molecules around them in the same configuration as in the brucite- or gibbsite-like sheets. Ammonium ions

NH_4^+ and organic molecules may enter this space also, and if the bonds are electrostatic, they neutralize the negative charge on the silicate part of the clay mineral.

An understanding of layer charge is important for at least two reasons: (1) it is used as one of the criteria for classification, as you will see in a section that follows shortly, and (2) it controls how the (001) surface of a 2:1 clay mineral interacts with the rest of the world. It seems possible that the tetrahedral sheet that faces the outside world may have a different charge than interior tetrahedral sheets. Some things for you to think about: (1) Do you suppose that the distribution of charges seen by a cation in the interlayer space is the same from a layer in which the charge arises from tetrahedral substitution as that from a layer with octahedral substitution? (2) What might be the consequences for the distribution of charges, and therefore the distribution of cations in the interlayer space, if the tetrahedra have had to rotate or tilt to fit with an octahedral sheet? Current thinking is undergoing some reconsideration. Güven (1992) has given a summary of the old and the new views, crediting Bleam (1990a and 1990b) for stimulating renewed discussion on the topic of layer charge. Complexes formed with different materials in the interlayer space are summarized and discussed by MacEwan and Wilson (1980).

The other charge is at edges of mineral particles, the boundaries where structural patterns end as broken bonds. Here, the chemical composition and structure cannot be maintained without additional ions, usually H^+ or OH^-, to satisfy the unsatisfied bonds. Some refer to the charge that results from the interaction of broken bonds and H^+ or OH^- as the *variable charge* and distinguish it from the layer charge that is referred to as the *permanent charge*. *Total charge* on a particle is then the sum of variable and permanent charges. For minerals with permanent charge, variable charge is a small fraction of the total, usually <1%, but becomes more important as layer charge gets smaller. Variable charge is important for the oxides, the hydroxides, and aluminosilicates imogolite, allophane, and perhaps kaolinite. It's called variable because it is a function of the pH of the medium in which the clay-sized particles are immersed. For minerals without permanent charge, ions that neutralize the variable charge are the potential-determining or net-charge-determining ions. Changes in pH, and therefore in net charge, control dispersion and flocculation behavior (see section on the double layer). These changes are measured and described in terms of *isoelectric points* or *points of zero charge* (McBride, 1989). Some authors consider these two slightly different properties. Some also use the term zero point of charge ZPC instead of point of zero charge (Taylor, 1987). When there is a surplus of H^+ in the surrounding environment, mineral particles have a positive net surface charge because H^+ ions have attached themselves to the surface, and, surprise, they have the opposite surface charge in the presence of excess OH^- ions. The pH at which negative variable charge equals positive variable charge ranges widely for different minerals. Manganese oxides reach this point, the isoelectric or point of zero charge, at a pH of about 1.5, goethite (α-FeOOH) and boehmite (γ-AlOOH) reach the point of neutrality in the vicinity of pH = 10, and kaolinite at pH = 4.7.

Electric Double Layer

The number of theories to explain the interaction between surfaces of colloidal particles and ions of the fluid in which they are immersed clearly points out that the finer points of this subject are not fully agreed upon or understood. The basic features of the double layer are, however, pretty straightforward, and are shown in Fig. 4.14. Given that the plane of atoms facing the outside world of a clay mineral is made up of anions (O^{2-} or OH^{-}),

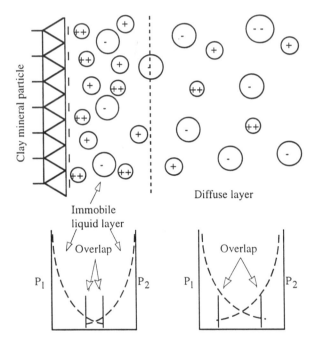

Fig. 4.14. The electrical double layer. The negative surface of the clay mineral attracts oppositely charged ions from the liquid in which the particle is immersed. This band of relatively tightly held positive ions attracts a less tightly distinguishable band of anions, which, in turn, attracts an even less distinguishable band of cations. These progessively less distinguishable bands around the particle form the diffuse layer part of the double layer. The effect of the charge or potential of the particle drops off along a line perpendicular to its surface. P_1 is the potential of one particle and P_2 that of a second particle. The two diagrams showing the dropping off of potential indicate that for thinner immobile layers, particles can approach one another more closely than when the immobile layer is thicker. The greater overlap of the P_1-P_2 pair on the right would result in greater repulsion than for the pair on the left.

it is easy to imagine that such surfaces would attract swarms of cations. First, however, such surfaces attract a layer of water that is relatively rigidly held (the water molecules still have thermal motions that can be detected by infrared spectroscopy). Within this layer, cations (in this context often called counterions) concentrate in a volume near the surface of the clay mineral. This, in turn, attracts anions, but less rigidly held. And these anions attract cations and so on until no segregation is apparent and electrical neutrality is achieved. Neutrality can be reached only in suspensions dilute enough in clay mineral particles. The importance of the concept of the double layer is to help explain dispersion and flocculation of clay minerals. When particles can approach one another closely enough, van der Waals forces come into play and bind the particles together into composites or clumps larger than colloid size, i.e., they *flocculate*. This causes them to settle because they become particles large enough that the random motions of Brownian movement can no longer keep them in suspension. When the cations in the surrounding fluid are divalent or trivalent, not as many are required to neutralize the clay mineral surface and the "immobile" layer is thinner, and therefore, particles can get closer to one another. In this case, they are more easily joined by van der Waals forces. On the other hand, univalent cations form a thicker immobile layer, keeping particles apart by a respectable distance, making it more difficult for the van der Waals forces to be effective (Fig. 4.14). In this case, clay mineral particles remain discrete entities, and their small size allows them to stay in suspension, i.e., to be *dispersed*. The thicknesses of the immobile layer and the diffuse layer also are affected by the concentration of ions in the fluid in which the clay mineral is immersed. As ions crowd together, the thicknesses of these layers decrease allowing closer approach of one particle to another, and therefore, more possibility for flocculation.

The double-layer interactions between particles can vary from particle pair to particle pair. Because the van der Waal's attraction can be different for attractions that are face to edge (FE), face to face (FF), or edge to edge (EE), and because pH can influence particularly edge charges, all three geometries can cause flocculation. The FE association is the arrangement that has led to the term *cardhouse structure*, and is the arrangement that has been the general perception of the way flocs form. However, scanning electron micrographs have recently called into question this arrangement. At least for smectites, clumps seem to form by faces attracted to faces (FF) (Güven, 1992). Newman (1987a) includes a full chapter on dispersion and flocculation (van Olphen, 1987). For additional information and a sense of different views on theoretical aspects of the double layer, see Greathouse et al. (1994), Low (1992), and Sposito (1992). In addition, J. W. Stucki, of the Department of Agronomy, University of Illinois, has written and made available a tutorial program entitled *An Introduction to Double Layer Theory*.

Exchangeable Ions or Cation-Exchange Capacity

Perhaps the earliest work on this reversible, low-energy transfer of ions between clay-sized material and surrounding liquid was by Way (1852) in his report on the power of soil to retain the nutrients of manure. Interest in exchange of ions is currently stimulated by four major needs: (1) to remediate radioactive wastes and other environmental contaminants either at disposal or accident sites; (2) to understand diagenetic processes; (3) to understand the properties of soils and the releasing and trapping of nutrients; and (4) to use the differences in cation-exchange capacity (CEC) qualities necessary for industrial techniques. In the everyday life of the clay mineralogist, exchangeable cations are important for at least three reasons, all of which are important in studying and preparing samples, they: (1) neutralize the layer charge; (2) have a physical influence (e.g., engineering properties and X-ray data for clay minerals vary with the nature and number of the exchangeable cations); and, (3) have a chemical influence. The distribution in interlayer space of exchangeable cations is probably controlled by the distribution of charges on the adjacent silicate layers that they neutralize. When there is Al^{3+} substituted for Si^{4+} in some of the tetrahedra, and the Al^{3+} may be regularly or irregularly distributed, those pseudohexagonal holes with more Al^{3+}-substituted tetrahedra around them are preferred spots for the location of cations (see Fig. 4.1). Do you see why? (See discussion on this topic in Güven, 1992, and his reference to Bleam, 1990a and b.) The mystery of why the ocean is Na^+-rich and not K^+-rich is probably best explained by an understanding of CEC and the average clay mineral's selectivity for these two ions.

The consequences of the colloidal size, and therefore the extremely large ratio of surface area to volume, cannot be overestimated. We, as large bulky objects, have surface charge, but it is quite inconsequential compared to our volume. It in no way influences our behavior. On the other hand, one important consequence of so much surface area on the clay minerals is the charge on their surfaces, the great majority of which is on the (001) surfaces, with the remaining charges on the much smaller surface of broken, unsatisfied bonds on edges (see sections on layer and surface charge). Ions and molecules, water most commonly, are attracted to and held by the charges on clay mineral particles. In most cases, with some important exceptions, cations are attracted to the (001) surfaces, anions to the edges. We are interested primarily in the cations. Especially in those clay minerals that expand, cations may be exchanged when the clay minerals holding them come in contact with a solution rich in other cations. Cation exchange needs to be viewed as a dynamic process governed by the law of mass action, at least in a first approximation, especially in the laboratory where the clay minerals can be exposed to strong concentrations of single-cation solutions. However, in the case of clay minerals exposed to solutions of low concentrations of a mix of cations, predictions are quite difficult to make. Cations in the fluid and those on the (001) surface interchange at some rate reflecting temperature, concentration, pH, the size and charge of the cation, the energy of hydration of the cation, and the amount and distribution of the layer charge of the clay mineral. The relative ease with which one cation will replace or exchange another is seldom predictable. For a series of montmorillonites, Eberl (1980) collected data from the literature that suggested for this mineral, under the same conditions, the following series:

$$Li^+ < Na^+ < K^+ < Cs^+ < Rb^+$$

i.e., generally K^+ is more stable, or more firmly fixed, in the interlayer space than is Na^+, for this mineral and for the conditions used in the experiments as reported. This series will vary with conditions or type of clay mineral, e.g., in acidic conditions H^+ is included in the series. The relatively common NH^{4+} wasn't among the ions tested but, for similar conditions, seems to usually be about in the middle of this series. Measuring the amount of cations freed and replaced by others, when carefully done, is a measure of both the negative charge on a layer and the cation-exchange capacity (CEC).

Beyond these simple generalizations, there is much to learn. CEC values are not only partly a consequence of the method of measurement, there are three or four actively used units for this value. Layer charge values, the charges these cations are supposed to be neutralizing, are most often determined from structural formulas (see end of chapter), whereas CEC values are based on the oven-dried weight of samples. Nevertheless, for pure, fully expandable samples, it is generally agreed that CEC values and total charge values match closely (Laird, 1994). Some of the points of contention about theoretical and empirical points of view can be read in McBride (1989) and Laudelout (1987), who observed, "Regularities in behavior [of exchangeable cations] seem fairly easy to observe and *ad hoc* theoretical interpretations not too difficult to find. At present, the predictive value of the latter seems limited."

Interaction of Water with Clay Mineral Surfaces

You might think that the subject of water interacting with clay mineral surfaces, water in the interlayer space, would be pretty straightforward. Ha! In spite of having thrown almost all the technology of modern instrumentation at the problem, there remains confusion and disagreement about a number of the details, e.g., the density of water in the interlayer space or the presence of absence of the hydronium ion H_3O^+. There are a few things that seem to be generally accepted, and these are: (1) some clay minerals swell with water, others do not; (2) the influence of the layer charge of a clay mineral extends only a few water layers away from the surface (and there is little agreement on what is meant by a few); (3) water forms a coordination "sphere" or shell around most cations and this sphere is more tightly held by some cations than others; (4) water in the interlayer space is dynamic, i.e., as measured by infrared and NMR spectroscopy, it is rotating, vibrating, and stretching; that water held at or very near the surface of a clay mineral doesn't freeze at temperatures as low as -30°C; and (5) it is in a state somewhere between that of bulk liquid water and that of ice.

Swelling clay minerals seem to take in water as single, double, or triple layers of water. Swelling or expansion is exclusively along the [001] or Z^* direction; there is no variation in the *a* and *b* dimensions of the crystals. Crystallographic integrity is maintained in the process of expansion. Compared to the rest of the mineral world, this is a remarkable and unique feat. Water in the interlayer space of expandable clay minerals is controlled by three factors: (1) the polar nature of the water molecule; (2) size and charge of cations in the interlayer space; and (3) the value and localization of the charge on the adjacent silicate layers, as discussed in the section on total charge. Expanded clay minerals carry water into their depositional sites. During the process of diagenesis, this water is released, carrying with it cations and anions in solution. This may be one of the primary driving factors in diagenetic changes such as the neoformation of clay minerals and the formation of other cementing minerals. Bruce (1984) suggested that it is a factor in the accumulation of petroleum. Bird (1984), commenting on engineering properties of shales, noted that adsorption of water in smectite can increase volume 80 percent and decrease friction by a factor of three. And Bethke et al. (1988) implicated it as a source of overpressuring in oil reservoirs. Hall (1993) offered a wider review of overpressuring that included the expulsion of interlayer water during diagenesis.

In the interlayer space of expandable clay minerals, water apparently takes the form of two-dimensional structures, or sheets. Extrapolating from work done on Na-smectite (Moore and Hower, 1986), these sheets are always present in integral numbers, 0, 1, 2, or 3, as a function of relative humidity, i.e., the activity of water. Each hydration state has a discrete $d(001)$: 9.6, ~12.4, ~15.2, and ~18.0 Å, respectively. In response to the amount of water available, $d(001)$ varies continuously but not linearly. It does not vary in a step function because intermediate values of apparent $d(001)$ result from an ordered interstratification of two appropriate hydrates. For example, at the low end of the relative humidity scale, the dehydrated clay mineral is interstratified with the form that contains one sheet of water so that the apparent $d(001)$ is between 9.6 and 12.4 Å.

As an example of an unsettled question about water, Loucks (1991) argued that the hydronium ion H_3O^+ is a far more common occupant of the interlayer space than generally recognized. Jiang et al. (1994) however, countered this argument by suggesting that the environments in which clay minerals were formed would have to have quite low pH's to form this ion, and that, perhaps, the analytical data used by Loucks were from mixed phases rather than single minerals.

As another example, agreement concerning the details of the structure of water in the interlayer space isn't too good (Table 4.1). Some argue for an increase in its density relative to free water; some argue for a density decrease relative to free water. Data of Carman (1974) and Colton (1986) show experimentally (approximating ~13 and ~1.2 km deep in the earth's crust, respectively) that $d(001)$ of expandable 2:1 layer silicates remains about the same with increasing pressure. Under burial pressures, Le Châtelier's Principle (or, quantitatively, the Clapeyron equation) indicates that water would come out of the interlayer space if it were less dense than pore water, or pore water would move into the interlayer space if it would be more dense in the interlayer space. Therefore, we think water must have about the same density in the interlayer space as outside in pore spaces. Based on Monte Carlo simulations to determine time-averaged thermodynamic quantities, Skipper et al. (1993) concluded that the density of water in the interlayer space of Mg- and Na-smectites ranged in density from 0.85 at a pressure of zero to 1.91 at approximately 300 km depth. Throughout the increase in pressure, with enough water for two layers of water, they pictured the Mg ions arranged close to the plane midway between the two silicate layer surfaces, whereas the Na ions had a tendency to bind to the layer surface.

Vermiculite exists in macroscopic crystals, and its interaction with water in the interlayer space has, therefore, received more attention with more precise results than that for smectites. Walker (1958) offered a detailed analysis for the hydration and dehydration of vermiculite that has been little modified by subsequent work. He showed dramatic photographs of hydration in progress, and identified five phases with $d(001)$ values of 9.02, 11.59, 13.82, 14.36, and 14.81 Å. (For more on hydrogen bonding and the structure of ice and water, start with Cotton and Wilkinson, 1972, pp. 157ff.)

Since most sedimentary rocks, buried sediments becoming rocks, and soils are really aqueous systems, one would think that the problems surrounding the interaction of water and minerals, in our case clay minerals, would have been solved. That they are not is a signal to upcoming generations that there is no shortage of problems for them.

Table 4.1. Density of water in the interlayer space and free water

P, bars	Skipper et al. (1993) With interlayer cation[b]		Hawkins&Egelstaff[a] Na	Kennedy et al. (1958) Free Water[c]
	Mg	Na		
0	1.38	1.15		
1			1.05	0.9957
10	1.38	1.14		0.9962
8500	1.38	1.08		1.2066[d]
100000	1.91	1.84		

[a](1980); [b]26.84°C [c]at 30°C [d]This value from Burham et al. (1969)

Interaction with Organic Compounds

A look at journals dedicated to clay mineralogy indicates that an understanding of organic compounds will be helpful. In this section, we will try to provide a toe-hold, a finger-grip, a starting place. We can, because of our own limitations and those of time and space, offer only an introduction to the simplest of compounds, a combined nomenclature-glossary, and their broad categories. We didn't find complete consistency in the vocabulary used by different disciplines, so you may expect some variation on the meanings of the terms we use here. We hope to offer enough to give you a starting point for such work as that of Mortland (1970), MacEwan and Wilson (1980), Rausell-Colom and Serratosa (1987), and Oades (1989).

In general, when geologists think of organics in nature, they think of fossil fuels. Waples (1985) has offered a clear, concise discussion of organic compounds associated with petroleum. Krauskopf (1979) has given a particularly clear introduction to the chemistry of organic compounds. From the standpoint of those interested in clay minerals, a much broader view of organic chemistry is necessary. Incentives for the study of interactions of organics with clay minerals stem from such applications as: (1) their importance to the formation of, exploration for, and recovery of petroleum; (2) the use and misuse of organic herbicides and pesticides; (3) the property of the clay minerals to act as oxidants (Surdam et al., 1989), catalysts, and templates for synthesizing by the polymerization of organic molecules; (4) clay minerals as barriers to and adsorbers of organics in landfills; (5) the possibility of using clay mineral-organic derivatives for improving the physical properties of plastics; and, as mentioned in the first chapter, (6) their role in prebiotic syntheses and storage of molecules of biological interest.

Organic chemistry is the chemistry of carbon in the same way that much of mineralogy is the mineralogy of silicon; organic chemistry is the chemistry of compounds associated with living organisms as opposed to inorganic chemistry associated with the nonliving world. The structural mineralogy or crystal chemistry of the geologists is essentially the same as the *stereochemistry* of the organic chemist; the polymorphs of mineralogy are the *isomers* of organic chemistry. Si and C are both group IV elements on the periodic table. Therefore, they have the same outer electron configuration. The most common bonds from C to other elements are the tetrahedrally distributed $2sp^3$ hybrid orbitals, just as are the $3sp^3$ orbitals for Si. And, again, just as for Si, these rigid directional bonds account for many of the important features of organic compounds. For Si, the element at the other end of these bonds is most often O, for C it is most often another C or H, thus the common name *hydrocarbons*. If all the atoms at the ends of the sp^3 orbitals were other carbons, the structure would be that of diamond.

Hydrocarbons, in the strictest sense, are compounds of C and H only. N, S, and O are the next three most common elements that occur in organic compounds. Therefore, an organic compound containing elements in addition to C and H is called a *heterocompound*, or an *NSO* compound. The hydrocarbons are classified in the same manner as the silicates, i.e., by structure. The broadest categories are as chains and rings. The chains are called *aliphatics* or *paraffins*. Rings come in two varieties: those with the prefix *cyclo-* and the *aromatics*. If all C-to-C bonds are single bonds in an aliphatic compound, it can then hold a maximum number of hydrogens. It is called, therefore, a *saturated hydrocarbon*. Chains can be straight, branched, or in rings. The prefix *n-* indicates that the chain is straight, *iso-* that it is branched, and *cyclo-* if it forms a ring. Aliphatic compounds can have C atoms linked by single, double, or triple bonds. When they do, they are no longer saturated, but are *unsaturated hydrocarbons*. The aromatic hydrocarbons have unique, flat, 6-membered rings of carbons, the bonds of which cannot be represented as either single or double but are best represented as a single bond between each pair of C atoms in the ring plus three nonlocalized bonds shared equally by all six carbons. The simplest of the aromatics is the benzene ring, with which you're probably familiar.

The structural details of organic compounds will be important when you think about organic compounds in the interlayer space of clay minerals. The structure of many organic compounds can be illustrated by a consideration of the simplest organic compound, methane, CH_4. It is represented by a tetrahedron as surely as is silica except that it doesn't polymerize in the same manner. In order for the C to polymerize, one of the four hydrogens must be dropped and the sp^3 orbital it shared is then shared by another C atom. In so-called "straight chains" of tetrahedrally coordinated carbons, the tetrahedral bonding angle tends to be preserved. A zigzag chain would be a more accurate name. This also holds for the cyclo-aliphatics, in which the ring is not flat but puckered. This differentiates it from the flat ring of benzene, which has bonding more like that of graphite than like that of diamond.

Aliphatic complexes introduced into the interlayer space of clay minerals have been divided into α- and β-complexes; the chains of α-complexes are parallel to (001), whereas those of β-complexes are perpendicular or at steep angles to (001). Furthermore, it is often possible to determine if the plane of the zigzag is parallel or perpendicular to the (001), e.g., ethylene glycol (which we routinely introduce into our samples to diagnose the degree to which expandable clay minerals are present) orients the plane of its zigzag chain perpendicular to the (001) surface (Reynolds, 1965).

The aliphatic compounds are classified based on the number of carbon atoms in the chain and on whether they have single, double, or triple bonds (*alkanes*, *alkenes*, or *alkynes*, respectively). The bonding character is designated by the suffix *-anes*, *-enes*, and *-ynes*; the number of carbon atoms by the Greek prefix for the number, except for the first four compounds (see Table 4.2). These first four in the table make up most natural gases; a mixture from hexane to decane is gasoline; and those from $C_{12}H_{36}$ to $C_{22}H_{46}$ are the constituents of lubricants. The roots of the names for these groups provide the

basis for much of the vocabulary of organic chemistry. We will not deal
further with the alkenes or alkynes.

With the simple vocabulary developed so far and the compounds listed in
Table 4.2, you can picture groups added to the chains, groups that replace the
hydrogens. We can change any of the compounds listed in Table 4.2 to groups
called *radicals* by taking away a H and changing the name by substituting *-yl*
for *-ane*. There are four rules for giving names: (1) select the longest possible
continuous chain of carbons and consider this as the base from which the
compound in question has been derived; (2) number in succession the carbons
in the chain so that the substitutions for hydrogens will be attached to the
lowest numbered C possible; (3) use the prefixes *di-, tri-, tetra-, penta-*, etc.
are used to indicate the presence of two, three, four, etc. groups on the parent
C chain; and (4) when several groups are present, list them in either increasing
complexity or alphabetically. For example, picture hexane, a straight aliphatic
alkane chain with six carbons. One of the simplest things we can substitute for
a H is the methyl group, CH_3^+. This would be methylhexane. To indicate
which C the methyl radical was attached to, the carbon atoms are numbered. If
the methyl radical were attached to the second C, we would have 2-
methylhexane; if there were 2 methyl radicals attached to this carbon, it would
be 2,2-dimethylhexane. You could probably diagram the structure of this
molecule, right?

In addition to radicals, other groups called *functional groups*, also can be
added to chains. A functional group is a specific group of elements that can be
attached to the molecules such as those in Table 4.2. The presence of one of
these groups confers on its host a chemical reactivity characteristic of that
functional group, regardless of the form of the structural framework of C
atoms of the host molecule. For example, for all alcohols the hydroxyl group,
-OH, is bonded to a hydrocarbon framework, and the names have the suffix
-ol, e.g., hexanol. All the alcohols have similar properties. There are only
about six or seven functional groups that occur frequently. An organic acid
has an -OH and an O bonded to a terminal C; the O is double bonded to the C;
an *amino acid* is an $-NH_2$ group attached to an organic acid. This latter is a
simple example of a *polyfunctional compound*, i.e., has more than one

Table 4.2. Naming the alkanes

Name	Formula	Name	Formula
Methane	CH_4	Heptane	C_7H_{16}
Ethane	C_2H_6	Octane	C_8H_{18}
Propane	C_3H_8	Nonane	C_9H_{20}
Butane	C_4H_{10}	Decane	$C_{10}H_{22}$
Pentane	C_5H_{12}	Pentadecane	$C_{15}H_{32}$
Hexane	C_6H_{14}	In general	C_nH_{2n+2}

functional group. The greatly simplifying feature of these groups of atoms is that the majority of chemical reactions involve changes of the functional group only. Any organic chemistry textbook will show you the common functional groups and how they are structurally attached to the carbon frameworks.

Some of the more complex straight-chain hydrocarbons that are important in sediments are *isoprenoids*, *steranes*, and *triterpanes*. Regular isoprenoids consist of a straight chain of carbons with a methyl branch on every fourth carbon; steranes contain three six-carbon rings and one five-carbon ring; and triterpanes contain five, or sometimes six rings, one of which is a five-carbon, the others six-carbon rings.

Turning to the aromatic compounds, all have at least one benzene ring (C_6H_6), some are a series of benzene rings linked together. There is a systematic nomenclature for this family of compounds, but it is difficult to find anyone who uses it. If one is to work with these, one just has to learn the formula and structure of benzene, toluene, anthracene, naphthalene, etc. Radical and functional groups are attached to aromatics just as they are to aliphatics. The points of attachment are numbered clockwise with carbon number 1 at 12 o'clock. For example, a benzene ring with two bromine atoms substituted for two hydrogens could be 1,2-dibromobenzene, 1,3-dibromobenzene, or 1,4-dibromobenzene. Can you picture these? The terms *ortho*, *meta*, and *para* are also used to describe points of attachment. Orthodibromobenzene would indicate that the bromines are attached to carbons next to one another; meta-, that they were separated by one carbon; and para-, separated by two carbons. Benzene, with an OH replacing one of the hydrogens, is the most common of the group called *phenols*. For aromatic compounds, if elements other than C are present in the rings, they are referred to as *heterocyclic* compounds. Multiple aromatic rings bound together, called *polycyclic aromatic hydrocarbons PAH*, are important geologically.

The study of the interactions of organic compounds with clay minerals began in the 1930s. Gieseking (1939) demonstrated that inorganic cations in the interlayer space could be replaced by organic cations; MacEwan (1944) and Bradley (1945) showed that uncharged polar molecules, glycerol and glycol, could enter the interlayer space without displacing cations. From an enormous number of studies made in the intervening time, a few generalizations can be made.

The geometric relations between organic compounds and clay minerals can help understand properties and behaviors of both. There are several ways to explain the bonding of organic compounds to clay minerals. As with inorganic, exchangeable cations, organic cations are attracted to the layer charge of the siloxane surfaces exposed in the interlayer space of 2:1 clay minerals. Neutral organic molecules can be adsorbed by forming coordination complexes with transition metal cations that have previously been introduced. Hydrogen bonding of organic molecules to clay minerals can be important, especially for molecules containing OH, NH_2, and NH_3 functional groups. Hydrogen bonding also apparently forms between the organic molecules, stabilizing whatever configuration they have assumed in the interlayer space.

The relatively few and relatively simple organic compounds initially introduced into geologic environments become many and generally more complex through diagenetic processes. The temperature and pressure increases that accompany burial of sediments transforms, through the loss of O, N, and S, the mixture of organic compounds into a complex substance called *kerogen*, the insoluble portion of organic matter in sedimentary rocks. The soluble part is called *bitumen*. Some geochemists recognize three types of kerogen and some four (see Waples, pp. 31ff). One relation not yet settled is the importance of clay minerals as catalysts in the formation of petroleum. Ungerer (1990) argued that clay minerals are relatively unimportant, whereas Johns and McKallip (1989) argued for the effectiveness of diagenetic illite in the natural pyrolysis of kerogen that yields petroleum.

CLASSIFICATION

Now that we have covered the properties and the rudiments of the structural features that help us identify clay minerals, we turn to their classification. We follow Bailey (1980b) and use layer type 1:1 or 2:1, as the main criterion for establishing divisions (Table 4.3). Within each division we use layer charge or charge per formula unit as the criterion for classification. Within these subdivisions, we make subgroups based on whether they are trioctahedral or dioctahedral. The other criteria seem real and distinct, whereas distinction on the basis of layer charge is approximate and somewhat arbitrary. It seems the least in accord with natural features of the minerals. This set of criteria will work moderately well except for mixed-layered clay minerals and for the transitional boundaries between micas and illites, illites and vermiculites, vermiculites and chlorites, and vermiculites and smectites. Table 4.3 shows this classification scheme. Mixed-layered clay minerals are not included in Table 4.3 because we assume that the components forming them are represented in this table.

Bowen (1928, Chapter XVIII, pp. 321-22) stated that a classification scheme will be of use and will endure to the extent that it coincides with real qualities of the nature of the things being classified. In addition, communication within a group of people with shared interests is effective in direct proportion to the extent that the group understands and agrees on the definitions covered by the classification scheme. This is somewhat abstract, so we'll give a concrete example or two. Perhaps the most useful classification scheme (and the oldest) is the binomial nomenclature for plants and animals. Still, when the boundaries between species are critically examined, some are blurred and are the basis for controversy. The problem is worse with minerals (and worse yet with rocks). The limitations of our understanding become apparent as we try to pin down definitions of minerals. It is certainly convenient to think of them as individual minerals. But in so many ways, most of them behave like members of a series, their properties grading from one to another. Some of them are more subtle in this regard than others. For example, consider illite and smectite. We are sure that muscovite is a discrete mineral. We are relatively sure illite-$2M_1$ is a distinct, discrete mineral. We are not so sure that illite-$1M$ ever occurs without some minimum amount of interlayered smectite, if there is such a mineral as smectite. Or perhaps what has been called smectite is really two minerals. We make do with perfect end-member species that probably don't exist but do seem to pin down both ends

Table 4.3. Classification of phyllosilicates, emphasis on clay minerals

Layer type	Group	Subgroup	Species
1:1	Serpentine-kaolin ($z{\sim}0$)	Serpentines (Tr)	Chrysotile, antigorite, lizardite, berthierine, odinite
		Kaolins (Di)	Kaolinite, dickite, nacrite, halloysite
2:1	Talc-pyrophyllite ($z{\sim}0$)	Talc (Tr)	
		Pyrophyllite (Di)	
	Smectite ($z{\sim}0.2\text{-}0.6$)	Tr smectites	Saponite, hectorite
		Di smectites	Montmorillonite, beidellite, nontronite
	Vermiculite ($z{\sim}0.6\text{-}0.9$)	Tr vermiculites	
		Di vermiculites	
	Illite ($0.6 > z < 0.9$)	Tr illite?	
		Di illite	Illite, glauconite
	Mica ($z{\sim}1.0$)	Tr micas	Biotite, phlogopite, lepidolite
		Di micas	Muscovite, paragonite
	Brittle mica ($z{\sim}2.0$)	Di brittle micas	Margarite
	Chlorite (z variable)	Tr,Tr chlorites[a]	Common name based on $Fe^{2+}, Mg^{2+}, Mn^{2+}, Ni^{2+}$
		Di,Di chlorites	Donbassite
		Di,Tr chlorites	Sudoite, cookeite (Li)
		Tr,Di chlorites	No known examples
2:1	Sepiolite-palygorskite	Inverted ribbons (with z variable)	

Based on Bailey (1980a, b), Brindley (1981), Hower and Mowatt (1966), and Środoń (1984) .

[a]2:1 layer first in name of chlorite; Tr = trioctahedral and Di = dioctahedral; z = charge per formula unit.

of this series. In general, most agree with the definitions of the two ends, but many disagree on the details.

In the end, we use a classification system that is based partly on logic and partly on intuition. The only "real" classification that comes to mind is the periodic law or periodic table. An element is either sodium or magnesium because a species between would require a fraction of a proton, and that apparently is impossible. In addition, there are no green or pink sodium atoms. All atoms of sodium in the universe are indistinguishable from each other except for isotopic variations, which, again, are distinct species. Do you understand that we are not satisfied that we understand everything about minerals? If that is clear, let us proceed to the next chapter in which individual clay minerals are considered. (If these last two paragraphs have left you cold, perhaps you should consider some more exact discipline.)

Box 4.1. Nomenclature

This is an aside to emphasize some recommended terms that apply to clay minerals. The first two paragraphs of what follows are from reports of the Nomenclature Committee of the Association Internationale pour l'Etude des Argiles (AIPEA) chaired by S. W. Bailey (Bailey, 1980a, 1982).

In the classification of clay minerals (Table 4.3), *smectite* is the accepted group name for clay minerals with layer charge between 0.2 and 0.6 per formula unit and that swell in the presence of water. Chlorite is a 2:1 layer type rather than a 2:2 or 2:1:1, and the interlayer hydroxyl sheet is to be treated like other interlayer material.

When discussing structures it is important to note that *lattice* and *structure* are not synonymous. Lattice is the abstract, invented pattern, whereas structure is the concrete, real mineral material. Keep this analogy in mind: Lattice is to the mind as crystal structure is to the brain. For using the terms *plane*, *sheet*, *layer*, and *unit structure*, a single *plane* of atoms, a tetrahedral or octahedral *sheet*, and 1:1 or 2:1 *layer* are the recommended designations. Plane, sheet, and layer refer to increasingly thicker arrangements. The assembly of one or more layers plus interlayer material is referred to as a *unit structure*. Brucite and gibbsite sheets are not acceptable terms. Either *hydroxide* or *interlayer sheet* for interlayer material or *brucite-like* or *gibbsite-like* should be used for indicating the trioctahedral or dioctahedral nature, respectively, of the interlayer material.

In referring to the composite peaks of mixed-layered clay minerals, we put the peak of the mineral with the smaller $d(001)$ first, e.g., the illite/smectite 002/003 peak. To designate specific proportions, we use the familiar scheme for other silicate minerals like Ab_{37} but use the format that follows: chlorite(0.85)/smectite. A distinction is also made between the terms *clay* and *clay mineral*. *Clay* is a general term for material < 2 μm (sometimes <4 μm) in size; it is used to indicate rock, whereas *clay mineral* refers to a specific mineral (most of which, conveniently, occur in the clay-sized fraction).

We also include here the minimum necessary terms for discussing *diagenesis*, which we take to be the sum of all chemical and physical changes in minerals during and after their initial accumulation, a process limited on the high-temperature, high-pressure side by the lowest grade of metamorphism. Diagenesis involves addition and removal of material, transformation by dissolution and recrystallization or replacement, or both, and by phase changes. We include *pedogenesis*, the changes and formation of minerals in the soil environment, within diagenesis. Retallack (1983, p. 829) discussed the problems soil scientists have with the term and argued for including soil-forming processes within the concept of diagenesis. *Authigenic* refers to minerals formed in place. It is also applied to minerals that are clearly the result of new crystal growth on older crystals of the same kind, e.g., K-feldspar overgrowths are referred to as authigenic overgrowths. We use authigenesis as a subprocess of diagenesis, and in most cases diagenesis is the better term. Of the two definitions offered in Bates and Jackson (1980), our use of the term diagenesis is closer to the second definition, diagenesis [sed], even though the first definition is framed in terms of clay minerals. We think that a convenient boundary between diagenesis and metamorphism is Weaver and Brockstra's (1984) boundary: that point at which all illite-IM has been converted to $2M_I$. Neither this definition nor the one for diagenesis adequately covers the changes in hydrothermal, geothermal, or simple thermal situations.

The terms *neoformation* and *transformation* are widely used. Neoformation is new formation from solution, and transformation is remodeling of an existing structure in which parts of the parent mineral are retained. Both are diagenetic changes. Examined on finer and finer scales, these distinctions begin to blur, as seen in an example from Smith et al. (1987). They showed that olivine may weather to a mixture of saponite (Mg-rich, trioctahedral smectite) and goethite, both showing an orientation related to the structure of the parent olivine as seen in transmission electron micrographs (Fig. 5.20, p. 188). Two terms we will find useful in discussing secondary and host minerals are *topotaxy* and *epitaxy*. They are defined in the report of the Joint Committee on Nomenclature of the International Mineralogical Association—International Union of Crystallography (Bailey *et al.*, 1977): *Topotaxy* is the phenomenon of mutual orientation of two crystals of different species resulting from a solid-state transformation. *Epitaxy* is the phenomenon of mutual orientation of two crystals of different species, with two-dimensional lattice coincident (net or mesh in common), usually, though not necessarily, resulting in an overgrowth. So topotaxy is a more general term. The Committee recommended that the adjectival forms use the endings *-taxic* or *-tactic*, but not *-axial*. *Ostwald ripening* is another term related to crystal growth. It's a recrystallization process in which, within a solution containing the material of the crystals with which it is in contact, the smallest crystals are dissolved and their material is added to the larger crystals of the same phase (Ostwald, 1900; Baronnet, 1982; Eberl et al., 1990; and especially see Fig. 5.11 from Jahren, 1991). The driving mechanism is a shift to lower surface free energy. And then there are terms that have come into use recently enough so that they are used differently by different groups of workers. Until the Nomenclature Committee can act on the following terms, you will have to take their meanings from the context in which they appear: *fundamental particle, interparticle diffraction*, and *MacEwan crystallite*.

A few crystallographic terms seem appropriate. In X-ray crystallography, it has become a convention to use the unbracketed symbol to denote a family of structural planes and to bracket the index to denote the actual crystal face. Thus, 001 "reflections" arise from structural planes parallel to the face (001). 001 also is the first basal plane. For a *1M* polytype, the structural plane or actual face symbol is (001), and for a *2M* polytype, it is (002). d(001) is the repeat distance along a line perpendicular to the 001 face, or along Z^*. (We have chosen to continue to use Ångstroms rather than nanometers, although we see the change coming.) Square brackets, [321], enclose the coordinates of a zone axis, i.e., denote a line perpendicular to family planes 321 (after Phillips, 1971, p. 42, and Juster et al., 1987). Bailey's Committee (1977) recommended that X, Y, Z or [100], [010], [001] be used for directions of crystallographic axes, u, v, w or x, y, z are used for fractional coordinates of atom positions within the unit cell, and a, b, c for the repeat distances along these axes. Comparable directions and dimensions in reciprocal space are indicated with an asterisk * following any of these same symbols.

Mackenzie (1963) reviewed nomenclature from the depths of history if you are interested in the twists and turns terms associated with our discipline have taken.

REFERENCES

Bailey, S. W. (1975) Cation ordering and pseudosymmetry in layer silicates: *Amer. Minerl.* **60**, 175-87.

Bailey, S. W. (1980a) Summary and recommendations of AIPEA nomenclature committee: *Clays and Clay Minerals* **28**, 73-78.

Bailey, S. W. (1980b) Structures of layer silicates: in Brindley, G. W., and Brown, G., editors, *Crystal Structures of Clay Minerals and Their X-Ray Identification*, Monograph No. **5**, Mineralogical Society, London, 1-123.

Bailey, S. W. (1982) Nomenclature for regular interstratifications: *Amer. Minerl.* **67**, 394-98.

Bailey, S. W. (1984) Classification and structures of the micas: in Bailey, S. W., editor, *Micas*, Vol. **13** in Reviews in Mineralogy, Mineralogical Society of America, Washington, D. C., 1-12.

Bailey, S. W. (1993) Review of the structural relationships of the kaolin minerals: in Murray, H. H., Bundy, W. M., and Harvey, C. C., editors, *Kaolin Genesis and Utilization:* Special Publication No. 1, The Clay Minerals Society, P.O. Box 4416, Boulder Colorado, 25-42.

Bailey, S. W., et al. (1977) Report of the International Mineralogical Association (IMA)– International Union of Crystallography (IUCr) Joint Committee on Nomenclature: *Acta Cryst.* A33, 681-684.

Baronnet, A. (1982) Ostwald ripening in solution. The case of calcite and mica: *Estudios geol.* **38**, 185-98.

Bates, R. L., and Jackson, J. A. (1980) *Glossary of Geology*: 2nd ed., American Geological Institute, Falls Church, Va., 751 pp.

Bethke, C. M., Harrison, W. J., Upson, C., and Altaner, S. P. (1988) Supercomputer analysis of sedimentary basins: *Science* **239**, 261-67.

Bird, P. (1984) Hydration-phase diagrams and friction of montmorillonite under laboratory and geologic conditions, with implications for shale compaction, slope stability, and strength of fault gouge: *Tectonophysics* **107**, 235-60.

Bleam, W. F. (1990a) Electrostatic potential at the basal (001) surface of talc and pyrophyllite as related to tetrahedral sheet distortions: *Clays and Clay Minerals* **38**, 522-26.

Bleam, W. F. (1990b) The nature of cation-substitution sites in phyllosilicates: *Clays and Clay Minerals* **38**, 527-36.

Bowen, N. L. (1928) *The Evolution of Igneous Rocks*: Princeton Univ. Press, Princeton, N.J., 334 pp.

Bradley, W. F. (1945) Molecular associations between montmorillonite and organic liquids: *Jour. Amer. Chem. Soc.* **67**, 975-81.

Brindley, G. W. (1980, 1984) Order-disorder in clay mineral structures: in Brindley, G. W., and Brown, G., editors, *Crystal Structures of Clay Minerals and Their X-Ray Identification*, Monograph No. 5 (1st and 2nd printing), Mineralogical Society, London, 125-95.

Brindley, G. W. (1981) X-ray identification (with ancillary techniques) of clay minerals: *in* Longstaffe, F. J., editor, *Short Course in Clays for the Resource Geologist*: Mineralogical Association of Canada, Toronto, 22-38.

Bruce, C. H. (1984) Smectite dehydration—its relation to structural development and hydrocarbon accumulation in northern Gulf of Mexico Basin: *AAPG Bull.* **68**, 673-83.

Burnham, C. W., Holloway, J. R., and Davis, N. F. (1969) The specific volume of water in the range 1000 to 8900 bars, 20° to 900°C: *Amer. Jour. Science* **267-A**, 70-95.

Carman, J. H. (1974) Synthetic sodium phlogopite and its two hydrates: stabilities, properties, and mineralogic implications: *Amer. Minerl.* **59**, 261-73.

Colton, V. A. (1986) Hydration states of smectite in NaCl brines at elevated pressures and temperatures: *Clays and Clay Minerals* **34**, 385-89.

Cotton, F. A., and Wilkinson, G. (1972) *Advanced Inorganic Chemistry*: Wiley, New York, 1145 pp.

Eberl, D. D. (1980) Alkali cation selectivity and fixation by clay minerals: *Clays and Clay Minerals* **28**, 161-72.

Giese, R. F., Jr. (1988) Kaolin minerals: Structures and stabilities: in Bailey, S. W., editor, *Hydrous Phyllosilicates (exclusive of micas)*, Vol. **19** in Reviews in Mineralogy, Mineralogical Society of America, Washington, D.C., 29-62

Gieseking, J. E. (1939) Cation exchange in smectites: *Soil Science* **47**, 1-14.

Greathouse, J. A., Feller, S. E., and McQuarrie, D. A., (1994) The modified Gouy-Chapman theory: Comparisons between electrical double layer models of clay swelling: *Langmuir* **10**, 2125-30.

Güven, N. (1992) Molecular aspects of clay-water interactions: in Güven, N., and Pollastro, R. M., editors, *Clay-Water Interface and its Rheological Implications*: Workshop Lectures, Vol. **4**, Clay Minerals Society, Boulder, CO, 1-79.

Hall, P. L. (1993) Mechanisms of overpressuring: an overview: in Manning, D. A. C., Hall, P. L., and Hughes, C. R., editors, *Geochemistry of Clay-Pore Fluid Interactions*: Chapman & Hall, London, 265-315.

Hawkens, R. K., and Egelstaff, P. A. (1980) Interfacial water structure in montmorillonite from neutron diffraction experiments: *Clays and Clay Minerals* **28**, 19-28.

Hower, John, and Mowatt, T. C. (1966) The mineralogy of illite and mixed-layer illite/montmorillonite: *Amer. Minerl.* **51**, 825-54.

Jiang, W-T., Peacor, D. R., and Buseck, P. R. (1994) Chlorite geothermometry? –Contamination and apparent octahedral vacancies: *Clays and Clay Minerals* **42**, 593-605.

Johns, W. D., and McKallip, T. E. (1989) Burial diagenesis and specific catalytic activity of illite-smectite clays from Vienna Basin, Austria: *AAPG Bull.* **73**, 472-82.

Juster, T. C., Brown, P. E., and Bailey, S. W. (1987) NH_4-bearing illite in very low grade metamorphic rocks associated with coal, northeastern Pennsylvania: *Amer. Minerl.* **72**, 555-65.

Kennedy, G. C., Knight, W. L., and Holser, W. T. (1958) Properties of water. Part III. Specific volume of liquid water to 100 C and 1400 bars: *Amer. Jour. Science* **256**, 590-95.

Krauskopf, K. B. (1979) *Introduction to Geochemistry*, Second Edition: McGraw-Hill, New York, 617 pp.

Laird, D. A. (1994) Evaluation of the structural formula and alkylammonium methods of determining layer charge: in Mermud, A. R., editor, *Layer Charge Characteristics of 2:1 Silicate Clay Minerals*: CMS Workshop Lectures, Vol. **6**, Clay Minerals Society, Boulder, Colorado, 79-104.

Laudelout, H. (1987) Cation exchange equilibria in clays: in Newman, A. C. D., editor, *Chemistry of Clays and Clay Minerals*: Monograph No. **6**, Mineralogical Society, London, 225-36.

Loucks, R. R. (1991) The abound interlayer H_2O content of potassic white micas: Muscovite-hydromuscovite-hydropyrophyllite solutions: *Amer. Minerl.* **76**, 1563-79.

Low, P. F. (1992) Interparticle forces in clay suspensions: Flocculation, viscous flow and swelling: in Güven, N., and Pollastro, R. M., editors, *Clay-Water Interface and its Rheological Implications*: Workshop Lectures, Vol. **4**, Clay Minerals Society, Boulder, CO, 157-90.

MacEwan, D. M. C. (1944) Identification of the montmorillonite group of minerals by X-rays: *Nature* **154**, 577-78.

MacEwan, D. M. C., and Wilson, M. J. (1980) Interlayer and intercalation complexes of clay minerals: in Brindley, G. W., and Brown, G., editors, *Crystal Structures of Clay Minerals and Their X-Ray Identification*: Monograph No. **5**, Mineralogical Society, London, 197-248.

Mackenzie, R. C. (1963) De natura lutorum: *Clays and Clay Minerals*, Vol. **13**, The Macmillan Co., New York, 11-28.

Marshall, C. E. (1977) *The Physical Chemistry and Mineralogy of Soils*, Vol. II: *Soils in Place*: Wiley, New York, 313 pp.

McBride, M. B. (1989) Surface chemistry of soil minerals: in Dixon, J. B., and Weed, S. B., editors, *Minerals in the Soil Environment*, 2nd Edition, Soil Science Society of America, 35-88.

Moore, D. M., and Hower, John (1986) Ordered interstratification of dehydrated and hydrated Na-smectite: *Clays and Clay Minerals* **34**, 379-84.

Mortland, M. M. (1970) Clay-organic complexes and interactions: in Brady, N. C., editor, *Advances in Agronomy*, Academic Press, New York, 75-117.

Newman, A. C. D., editor, (1987a) *Chemistry of Clay and Clay Minerals*: Monograph No. **6**, Mineralogical Society, London, 480 pp.

Newman, A. C. D. (1987b) The interaction of water with clay mineral surfaces: in Newman, A. C. D., editor, *Chemistry of Clays and Clay Minerals*: Monograph No. **6**, Mineralogical Society, London, 237-74.

Oades, J. M. (1989) An introduction to organic matter in mineral soils: in Dixon, J. B., and Weed, S. B., editors, *Minerals in the Soil Environment*, 2nd Edition, Soil Science Society of America, 89-159.

Ostwald, W. (1900) Über die Vermeintlich Isomerie des roten und gelben Quecksilberoxyds und die Oberflächenspannung fester Körper: *Z. Phys. Chem. Stoechiom. Verwandsch.* **34**, 495-503.

Phillips, F. C. (1971) *An Introduction to Crystallography*, Fourth Edition: John Wiley & Sons, New York, 351 pp.

Rausell-Colom, J. A., and Serratosa, J. M. (1987) Reactions of clays with organic substances: in Newman, A. C. D., editor, *Chemistry of Clay and Clay Minerals*: Monograph No. **6**, Mineralogical Society, London, 371-422.

Retallack, G. J. (1983) A paleopedological approach to the interpretation of terrestrial sedimentary rocks: The mid-Tertiary fossil soils of Badlands National Park, South Dakota: *Geol. Soc. Amer. Bull.* **94**, 823-40.

Reynolds, R. C., Jr. (1965) X-ray study of an ethylene glycol-montmorillonite complex: *Amer. Mineral.* **50**, 990-1001.

Schwertmann, U., and Taylor, R. M. (1989) Iron oxides: in Dixon, J. B., and Weed, S. B., editors, *Minerals in the Soil Environment*, 2nd Edition, Soil Science Society of America, 379-438.

Shirozu, H., and Bailey, S. W. (1965) Chlorite polytypism: III. Crystal structure of an orthohexagonal iron chlorite: *Amer. Minerl.* **50**, 868-85.

Skipper, N. T., Refson, K., and McConnell, J. D. C. (1993) Monte Carlo simulations of Mg- and Na-smectites: in Manning, D. A. C., Hall, P. L., and Hughes, C. R., editors, *Geochemistry of Clay-Pore Fluid Interactions*: Chapman & Hall, London, 40-61.

Smith, K. L., Milnes, A. R., and Eggleton, R. A. (1987) Weathering of basalt: formation of iddingsite: *Clays and Clay Minerals* **35**, 418-28.

Sposito, G. (1992) The diffuse-ion swarm near smectite particles: Modified Gouy-Chapman theory and quasicrystal formation: in Güven, N., and Pollastro, R. M., editors, *Clay-Water Interface and its Rheological Implications*: Workshop Lectures, Vol. **4**, Clay Minerals Society, Boulder, CO, 127-55.

Šrodoń , J. (1984) X-ray identification of illitic materials: *Clays and Clay Minerals* **32**, 337-49.

Surdam, R. C., Crossey, L. J., Hagen, E. S., and Heasler, H. P. (1989) Organic-inorganic interactions and sandstone diagenesis: *AAPG Bull.* **73**, 1-23.

Taylor, R. M. (1987) Non-silicate oxides and hydroxides: in Newman, A. C. D., editor, *Chemistry of Clays and Clay Minerals*: Monograph No. **6**, Mineralogical Society, London, 129-202.

Tsipursky, S. I., and Drits, V. A. (1984) The distribution of octahedral cations in the 2:1 layers of dioctahedral smectites studied by oblique-texture electron diffraction: *Clay Miner.* **19**, 177-93.

Ungerer, P. (1990) State of the art of research in kinetic modelling of oil formation and expulsion: *Org. Geochem.* **16**, 1-25.

van Olphen, H. (1977) *An Introduction to Clay Colloid Chemistry* 2nd ed.: Wiley, New York, 301 pp.

van Olphen, H. (1987) Dispersion and flocculation: in Newman, A. C. D., editor, *Chemistry of Clay and Clay Minerals*: Monograph No. **6**, Mineralogical Society, London, 203-224.

Walker, G. F. (1958) Reactions of expanding lattice minerals with glycerol and ethylene glycol: *Clay Miner. Bull.* **3**, 302-13.

Waples, D. W. (1985) Geochemistry in Petroleum Exploration: International Human Resources Development Corp., Boston, Massachusetts, 232 pp.

Way, J. T. (1852) On the power of soils to absorb manure: *Jour. Royal Agric. Soc. England* **13**, 123-43.

Weaver, C. E., and Brockstra, B. R. (1984) Illite-mica: in Weaver, C. E., and associates, *Shale Slate Metamorphism in the Southern Appalachians*, Elsevier, Amsterdam, 67-199.

Zoltai, T., and Stout, J. H. (1984) *Mineralogy: Concepts and Principles*: Burgess, Minneapolis, Minn., 505 pp.

*We dedicate this chapter to S.W. (Bull)
Bailey (1919-1994). A brief glance at the
Reference section will indicate why.*

Chapter 5
Structure, Nomenclature, and Occurrences of Individual Clay Minerals

Finally! We can turn our attention to actual clay minerals. You will find that they will be discussed in terms of what has been presented in the preceding chapters, especially Chapter 4. Before you can begin to use clay minerals as puzzle solvers, you will need to understand them as individuals, and as members of mineral assemblages. One of the most curious aspects of this pursuit is that we will describe individual clay minerals, and then you will spend the rest of your professional life trying to explain why the mineral at hand doesn't quite match the neat, individual descriptions we've offered. You will recognize that we are faced with a dilemma: when should we be discussing individual clay minerals and when mixed-layered clay minerals. From the other side, looking back, you will say, "It seems as if every clay mineral can transform, one way or another, to any other clay mineral." In any number of ways, it is artificial to single them out and treat them one by one. This will become more clear when we discuss mixed-layered clay minerals. The degree to which we characterize a clay mineral and differentiate it from others, you will find to be closely related to how it has been analyzed, or, to say it another way, it will depend on the resolution limits of the analytical instrument used. Warren and Ransom (1992) and Jiang et al. (1994) offered helpful discussions about the limits of analysis related to the kind of instrumentation used. But we delay. Let's turn to the individual clay minerals.

THE INDIVIDUAL CLAY MINERALS

The 1:1 Layer Type
The 1:1 layer type usually has no layer charge or a very small layer charge because the tetrahedral cation sites (see Fig. 5.1) usually are all occupied by Si^{4+} and the octahedral sites by all Al^{3+} or all Mg^{2+}. If there is substitution in one sheet of a 1:1 layer silicate, there almost always is a compensating substitution in the other sheet so that neutrality is maintained.

There are dioctahedral and trioctahedral varieties of 1:1 layer silicates, and within these subgroups there are individual species. Most species differ from one another in the manner in which the layers are stacked, i.e., they are different polytypes. See Bailey (1988a) for comprehensive information on polytypes of 1:1 layer silicates.

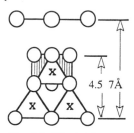

Fig. 5.1. The 1:1 layer.

Serpentine minerals, $Mg_6Si_4O_{10}(OH)_8$ The serpentine minerals are usually larger in particle size than clay minerals, but they do cross the boundary into the clay-size range. They are found as silky asbestos fibers and as smaller fibers in more massive or splintery material (chrysotile), as platy material (lizardite and amesite), or as the iron-rich variety berthierine. Some material is only apparently platy. Upon investigation it proves to be a series of alternating waves with a wavelength of 30-100 Å (antigorite). These structures are called *modulated*.

Differences in morphology are attributed to adjustments of the smaller tetrahedral sheet of chrysotile, and that of antigorite to the larger octahedral sheet. In these two varieties, the tetrahedra are tilted slightly on an axis parallel to the (001) plane, spreading the apical oxygens enough to match the anion sites of the octahedral sheet. This results in the layers forming tubes in one case and alternating waves in the other. There is enough Al^{3+} substitution in the tetrahedral sheet in lizardite and amesite to increase its dimensions so that it can join with the octahedral sheet without bending. See Wicks and O'Hanley (1988) for additional information on serpentine minerals.

Berthierine $(Fe^{2+}_3Mg_{0.75}Al_2)(Si_{2.5}Al_{1.5})O_{10}(OH)_8$ is a representative formula) is a mineral that has been recognized for a long time (a thin section made by Sorby in 1856 shows oolites of berthierine), but has often been overlooked. Its identity also was confused by the misunderstanding of the nature of chamosite, a name, at times, used for a 7 Å, Fe-bearing mineral. Note that it contains fewer than the ideal 6 octahedral cations. Berthierine also is easily mistaken for kaolinite in XRD patterns from oriented aggregates. Its occurrence was sufficiently uncommon that Brindley (1982) published chemical data on essentially all 14 of the known specimens of relatively pure berthierine. It apparently can form in shallow marine environments rich in iron, is commonly present in unmetamorphosed sedimentary iron formations, and can form in pedogenic environments (Young and Taylor, 1989). It has been found as a flint clay, and in this form, has been used by Native Americans for utilitarian and ritual artifacts (Hughes et al., 1990). The pace at which berthierine is being recognized has increased remarkably in the past decade.

Odinite $(Fe^{3+}_{1.6}Mg_{1.6}Al_{1.1}Fe^{2+}_{0.6})(Si_{3.6}Al_{0.4})O_{10}(OH)_8$ (a representative formula)   Odinite is a second 1:1 mineral similar in composition to berthierine but with more ferric than ferrous iron and currently forming on continental shelves and in reef lagoons. Bailey (1988d) described it as an Fe^{3+}-rich green clay that occurs widely in shallow marine tropical seas, especially at points of entry of rivers carrying Fe in solution. Odinite is part of the verdine facies, a group of green marine clay minerals extensively studied and reported on by Odin (1988). Either berthierine or odinite apparently may be a precursor to diagenetic chlorite (Bailey, 1988b, d), however, berthierine is considered the leading candidate for the 7 Å layer interstratified with chlorite in a 7 Å/chlorite mixed-layered clay mineral (Reynolds et al., 1992).

Kaolin minerals, $Al_4Si_4O_{10}(OH)_8$ In general, there is no recognized substitution in any of the species of kaolin minerals: kaolinite (in various degrees of disorder), dickite, nacrite, 7 Å halloysite, and 10 Å halloysite. However, if analyzed in detail, there may be from 0.1 to 10% vermiculitic, micaceous, or smectitic layers, or some combination, present in kaolinite (Talibudeen and Goulding, 1983a). This is probably the explanation for the negative charge found on some kaolinites. The dominant structural feature of these minerals appears to be from the regularly distributed cation site vacancies in the octahedral sheet and the distortions these vacancies create with respect to joining the octahedral and the tetrahedral sheets (Fig. 5.1). Kaolinite and the two halloysites are single-layer structures, whereas dickite and nacrite are double-layer polytypes, i.e., the repeat distance along the direction perpendicular to (001) for dickite and nacrite is two 7 Å layers, or 14 Å.

There is confusion over whether there are two halloysites or only one, and what the relation of halloysite is to kaolinite. There is a 10 Å variety that

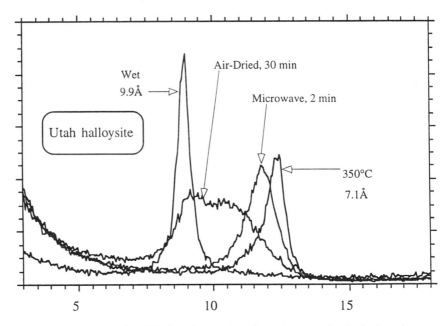

Fig. 5.2. An example of halloysite moving from the completely hydrated to dehydrated. Sample courtesy of R. E. Hughes, Illinois State Geological Survey.

clearly has a layer of water about 2.9 Å thick between the 7 Å layers, giving a 10 Å spacing. The water may leave spontaneously or can be driven out by very gentle heating or by putting the sample in a vacuum (Fig. 5.2). It commonly occurs as a tube formed by rolling of the layers, although it has been observed in other morphologies. Brindley (1980, p. 154) suggested that halloysite is a highly disordered form of kaolinite, disordered enough to take water into the interlayer space. Minerals clearly identified as kaolinites show varying degrees of disorder. Geise (1988) argued that though the 1:1 layer of halloysite is essentially the same as the other minerals of the kaolinite group, its structure is different in detail from the others. Bailey (1989) offered the hypothesis that halloysite is different in structure and composition from kaolinite; that halloysite has small layer charges that are compensated by interlayer cations; and that these cations attract water. These various hypotheses need testing. Can you suggest tests? We do not see a need at this time for recognizing a separate 7 Å variety of halloysite.

The nature of halloysite and its relation to kaolinite is a particularly intriguing problem. Kaolinite, in contrast to halloysite, has not been observed to swell in water, but treatment with some organic chemicals, e.g., formamide or dimethyl-sulfoxide, will open up kaolinite, or cause it to swell. Several workers studying the transition steps from the fresh silicate minerals of igneous rocks to their weathering products have identified the tube morphology and the subdued X-ray diffraction pattern of halloysite. Robertson and Eggleton (1991) presented compelling TEM images of muscovite altering to kaolinite, and, in turn, kaolinite to halloysite, in an Australian kaolinitic granite (Fig. 5.3).

As for the petrology of kaolinite, it is probably the most ubiquitous aluminosilicate mineral in soils and permeable bedrock in warm, moist regions, forming as a residual weathering product, or sometimes by hydrothermal alteration, of other aluminosilicates, especially of feldspars. It is part of several diagenetic sequences, and is found fairly commonly as a diagenetic mineral filling pore spaces; e.g., kaolinite with excellent structural order is found in geodes and vugs in Mississippian carbonate outcrops on the Illinois-Iowa-Missouri border. Dickite and nacrite are much less common in nature, normally restricted to hydrothermal occurrences. However, it is impossible, we think, to distinguish kaolinite from dickite and nacrite on the basis of basal XRD reflections of an oriented aggregate. For this reason, it has been suggested that we should use kaolin or kaolin-group mineral unless we can identify the polytype (Ehrenberg et al., 1993).

Large, sedimentary, or secondary, lacustrine, deltaic, or lagoonal deposits of kaolinite of great commercial value occur, e.g., in the southeastern United States (See **Box 5.1** on the history of the use of kaolinite.) They are of great value because they contain only coarse-grained quartz, which is easily removed. Quartz is intolerable if the kaolinite is to be used to coat paper. Neither the Brazilians nor the Japanese have been able to find a way to separate fine-grained quartz from the kaolins occurring in their countries (Murray, 1988).

Kaolinite and dickite have been used in working out the history of petroleum basins. Kaolinite transforms to dickite at about three kilometers below the seafloor, at about 120°C, in sandstone reservoirs on the continental shelf of Norway. This origin is in agreement with calculations of the electrostatic bonding of kaolinite, nacrite, and dickite made by Giese (1973) that indicate that dickite is more stable than kaolinite at standard temperature and pressure; it just takes the 120°C to push the metastable kaolinite into transforming. Ehrenberg et al. (1993) suggested that the evidence was good enough to allow its use as a paleogeothermometer. There also are several instances cited of kaolinite being the precursor to illite in sandstone reservoirs.

Four names for clay materials are associated with kaolinite: ball clay, fireclay, flint clay, and underclay. These names come from the people that mine them and use them commercially. There is considerable overlap and flexibility in the properties and uses associated with these names. *Ball clays* yield a white product when fired. They have the plasticity and the green and dry strength for making sanitary ware, ceramic tiles, etc. They frequently contain small amounts of other minerals and organic matter. *Fireclays* are similar to ballclays but are usually found immediately beneath coals. When underclays have been referred to as fireclays, they are kaolinite-rich underclays. They are used to make high strength and refractory bricks and cements, and are used to bond molding sands in foundries. *Flint clay* also can

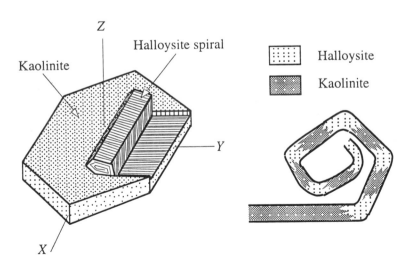

Fig. 5.3. Halloysite and kaolinite. Sketched from TEM photos, it seems difficult to argue for anything other than the halloysite forming from the kaolinite and inheriting a major part of the structure of the kaolinite. Robertson and Eggleton (1991) interpret the kaolinite as forming from muscovite by the loss of one tetrahedral sheet. Modified from Robertson and Eggleton (1991) with permission.

be used to make refractory items. This term is more specific and applies to a clay that is hard, smooth, breaks with conchoidal fracture, and develops no plasticity when ground up and wetted as other clays do, i.e., it does not slake in water. The kaolinite crystals making up this rock are tightly interlocked, reminiscent of the texture of an igneous rock. The interlocking crystals are the most likely explanation for its physical properties. One type of occurrence is in dish-shaped bodies that suggest the shape of the bottom of a pond. One origin suggested for it is crystallization from a gel of kaolinite composition formed in the bottom of perhaps a karst lake. Another theory is that flintclay has recrystallized from an intrusive body and has retained both the original texture and structure of its igneous precursor. Older tonsteins that, according to Bohor and Triplehorn (1993), have gone through a solution-precipitation reaction, are flint clays. (Tonsteins are altered ash layers that have fallen into coal swamps. We'll say more about them when we discuss bentonites.)

Kaolinite is prominent in various *underclays*, but underclays do not necessarily have kaolinite in them. An underclay is the unit immediately

---

Box 5.1. Uses of Kaolinite

The term *kaolinite*, its use in porcelain, and the making of paper all originated in China centuries ago. Papermaking and ceramics continue to be the two chief uses of kaolinite.

The word *kaolin* is from the Chinese Kau-ling or high ridge, specifically a hill east of Jingdezhen City, Jiang Xi Province, Central China, where the earliest of clay sent to Europe was obtained by François Xavier d'Entrecolles (1662-1741), a French Jesuit missionary. This rock, which is composed essentially of the clay mineral kaolinite, had been used for centuries by the Chinese for ceramics in general and, for the nine or ten centuries before d'Entrecolles, for porcelain in particular. Europeans were taken with the translucence and the esthetic quality of Chinese porcelain, which became one of the main items of trade in the first commercial interchanges between China and Europe. In an effort to duplicate the Chinese porcelain, there was much unsuccessful experimentation in Europe beginning in the late 17th century. The first kaolin deposits in Europe used in attempts to duplicate the porcelain imported from China, or china clay as it was called in reference to its source, were probably those near Meissen, Germany initially exploited about 1707. Almost as soon as attempts to make porcelain were getting underway in Europe, it also was being attempted in America. "1741—Porcelain clay was discovered in or near Savannah (Georgia) by Mr. Duchet, and china cups made. The trustees gave him fifty pounds sterling to 'encourage him in his enterprise'" (Sholes, 1900, p. 1, Smith, 1929). Legend also has it that kaolins from Georgia were being shipped to the famous Wedgewood Pottery works in England before the discovery of the deposits in Cornwall in about 1750.

Although pottery is almost as old as sites of human habitation, porcelain, made from kaolin and ground feldspar, was probably first made in China in a primitive form during the 7th century. Porcelain is a particularly hard kind of pottery, usually white and translucent with a pleasant and characteristic ring when tapped. The development of this form of ceramics depended on designing a kiln that could produce the 1450°C temperature necessary for the vitrification of kaolinite and feldspar.

Papermaking began about 105 A.D., when Ts'ai Lun, a Chinese court official, developed the idea of forming a sheet of paper from the macerated bark of trees, hemp waste, old rags, and fish nets. In the 8th century, knowledge of papermaking spread to Japan and to Arabs occupying Samarkand (check your atlas), who, four centuries later, brought papermaking to the European continent when they invaded Spain and Sicily. In Europe, the first "loading" or filling of paper with clay is recorded as the English patent granted to Samuel Pope in 1731. It is unclear how long before 1731 the Chinese had been using clay to fill paper. Again in imitation of the Chinese, the first patent in Europe for coating of paper was in England and granted to George Cummings in 1764. Today, approximately one-half of the kaolin produced in the world is used to coat and fill paper. As filler mixed with the cellulose fiber, it forms an integral part of the paper sheet giving it body, color, opacity, and the ability to hold a sharp edge for printing with colored inks. As a coating held with starch, latex, or casein, it is plated on the surface from a slurry that gives paper gloss and adds to the properties gained by using it as a filler. Kaolinite is also used as an ingredient in fiberglass; as a filler in plastics, adhesives, rubber, and pharmaceuticals; as an extender in paints; for making catalysts and synthetic zeolites; and many other minor uses.

Before kaolinite can be used in the making of paper, it must be processed, or beneficiated, as they say in the mining business. It cannot contain any grit, usually quartz and mica, because that would ruin the paper-coating machines. It must meet standards for viscosity in a slurry, whiteness, and brightness. As mined, kaolinite may contain some Fe-bearing and Ti-bearing minerals and have a grain size that is too large. The amount of impurities and the particle size usually need to be reduced. Meeting such requirements, and competition from other materials that can be used in a similar way, drives the need for continuing research on the properties of kaolinite, its geologic setting, and its origin.

The United States produced, from open-pit mines, more than seven million tons of kaolin in 1991, valued at $1.1 billion, 90 percent of which came from deposits along the Fall Line in Georgia and South Carolina. Western Europe is the second largest producer, with other producers scattered around the globe (Prasad et al., 1991).

below a coal bed that was the soil upon which the vegetation grew that formed the coal. The British refer to these beds as *seat earths*. The mineralogy of the underclays must represent substantial alteration of detrital material by growing vegetation (Hughes et al., 1987; Wnuk and Pfefferkorn, 1987). In the

Illinois Basin, the underclay beneath the No. 2 coal is kaolinite-rich in a relatively broad band around the margin of the Basin. Such kaolinite-rich bands are progressively narrower in underclays of the successively younger coals above the No. 2 coal. The origins of this pattern are ascribed to two mechanisms. One is based on grains of kaolinite being larger than that of most other clay minerals; i.e., it is part of the detrital pattern in which the coarser-grained minerals are deposited closer to shore and the finer-grained ones farther out (Parham, 1964). The other idea is that depositional patterns near shore are due to differential flocculation, differential settling, and differential transport of mixed and individual clay minerals (Whitehouse et al., 1960). A definitive explanation is not yet available.

Allophane and imogolite, r.Al_2O_3/*s*.SiO_2/*t*.H_2O (*r*, *s*, and *t* are rational numbers)   We put allophane and imogolite tentatively in this section because they seem to be incipient 1:1 minerals both structurally and chemically. Wada (1977) has thoroughly reviewed what is known about these semiordered materials, which are sometimes described as amorphous or noncrystalline. They seem to be two distinct materials that give indications of having characteristic properties, but are too variable to qualify as full-fledged minerals with three-dimensional order. They seem chemically distinguishable from one another, but there is not general agreement on this point, partly because there are materials that are intermediate between allophane and imogolite, on the one hand, and hydrous oxides of Al, Fe, and Si, on the other. The ratio of SiO_2 to Al_2O_3 is usually 1.3 to 2.0 for allophane, but has been reported as low as 0.83. Imogolite does not seem to vary as much. It gives a ratio of SiO_2 to Al_2O_3 of 1.05 to 1.15. Part of the problem has to be that as imperfect, tentatively ordered materials they respond to the treatments used in analytical procedures, so that part of the data gathered represents response to these treatments. However, there does seem to be more Al in imogolite than in allophane. Furthermore, it appears that up to half of the Al in allophane is in fourfold coordination, whereas all Al in imogolite is in sixfold coordination.

The morphology of imogolite seems to be cylindrical, whereas allophane forms hollow spherules. The structural model of imogolite that best explains the available data has been presented by Cradwick et al. (1972). They began their model from the structure of gibbsite and substituted silicate ions for hydrogens around the vacant cation sites on one side (top or bottom) of the gibbsite sheet. The fit of the silica tetrahedra on one side causes a contraction because the O–O distances are smaller for the silica tetrahedra than for those around the vacant octahedral sites of the gibbsite. This contraction would explain the tubular shape of imogolite as seen in electron micrographs. These tubes have an apparent diameter of 18.3 to 20.2 Å and a distance of 21 to 23 Å between the axes of tubes.

Allophane seems to be best described as hollow spherules 35 to 50 Å in diameter. The walls of the spherules may be constructed of a modified imogolite or from an incomplete kaolinite-like or halloysite-like material.

The most common occurrence of allophane and imogolite is in soils formed on volcanic ash, although under tropical, humid conditions they have been found in soils derived from basalts. In ash or pumice beds, imogolite typically is found as gel films on individual particles, whereas allophane is found within the grains. This common association makes the se materials of particular importance to an understanding of the formation of the minerals composing bentonites. Disordered material on the surface of any primary aluminosilicate mineral that has been subjected to weathering may well be included in the allophane-imogolite category, further adding to the importance of these materials.

The 2:1 Layer Type, z = 0

The minerals of these 2:1 layer types without layer charge are important because they serve as prototypes for the discussion of the structures of other clay minerals that do have a layer charge. The z in the title of this section refers to the amount of layer charge per formula unit: $T_4O_{10}(OH)_8$ for 1:1 layer silicates and $T_4O_{10}(OH)_2$ for 2:1 layer silicates (Figs. 5.1 and 5.4) [z the sum of layer charge from the tetrahedral sheet x and that from the octahedral sheet y ($z = x + y$)]. The latter is the talc-pyrophyllite group, which contains trioctahedral and dioctahedral members respectively [talc = $Mg_3Si_4O_{10}(OH)_2$ and pyrophyllite = $Al_2Si_4O_{10}(OH)_2$]. Ideally, these minerals have no tetrahedral or octahedral substitution, no layer charge, and no interlayer material. Natural materials often have small amounts of substitution, however, which gives a small amount of ionic attraction between layers that supplements van der Waals bonding, the main force that holds these layer silicates together. This weak force is probably the reason these minerals are soft, have excellent cleavage, a quality called slipperiness, and exhibit varying degrees of stacking disorder. See Evans and Guggenheim (1988) for additional information.

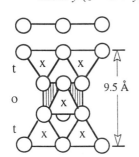

Fig. 5.4. The 2:1 layer with 0 layer charge. The x is a cation site.

The 2:1 Layer Type, z ~ 1

We will have a layer charge of -1 if we imagine the substitution of one Al^{3+} for one out of every four Si^{4+} ions in both the talc and the pyrophyllite prototype structures, if there are no other compensating substitutions. This charge is neutralized by a univalent cation in the interlayer space, called an *interlayer cation*. These layer silicates are referred to as the true micas, and they include the familiar species muscovite, which is dioctahedral, and biotite, which is trioctahedral (Fig. 5.5). There is little intergrading between dioctahedral and trioctahedral micas; they tend to be near one end or the other. Members of the mica group, especially muscovite and biotite, do sometimes occur in the clay-size fraction of sedimentary rocks and soils. However, we need to consider them here primarily because, in their macroforms, they serve as structural and chemical models for the 2:1 clay minerals. Most clay minerals cannot be studied macroscopically because they do not occur in grains large enough for single-crystal X-ray diffraction studies.

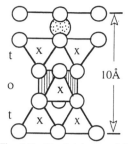

Fig. 5.5. The 2:1 layer with interlayer cation. The x is a cation site.

When workers studying soils or sedimentary rocks say that fine-grained micas frequently are the most abundant component of the clay-size fraction, they are blurring the distinction between the 2:1 layer silicate with $z \sim 1$ and those with $z < 1$. Almost certainly they are blurring it because it is blurred in nature. Micas lose some of their K due to weathering processes. If they lose enough, they are partly transformed to expandable 2:1 minerals so that the mica is interstratified, or occasionally they are found as a core surrounded by zones of expandable minerals. According to Fanning et al. (1989), transition to expandable 2:1 minerals takes place by release of K in layer weathering and edge weathering. The latter gives frayed edges or wedges around a mica core. Micas thus serve as precursors for other 2:1 layer silicates, especially illite and vermiculite. These layer silicates are discussed later.

The trioctahedral subgroup, $(K,Na)_{x+y}(Mg,Fe^{2+},R^{3+})_{3-y}R^+_y(Si_{4-x}Al_x)O_{10}(OH)_2$ Phlogopite, biotite, and annite are the three main species in this subgroup. Mg dominates the octahedral sites in phlogopite, Fe dominates them in annite, and Fe and Mg occur subequally in biotite. There is commonly substitution of R^{3+} ions in the tetrahedral position beyond the 1:3 ratio. This extra substitution is compensated for by substitution of R^{3+} ions in the octahedral sites. Only rarely is Fe^{3+} found substituting for Si^{4+} in the tetrahedral sites. In all other cases it is Al^{3+}. In the octahedral sites, Fe^{2+} and Mg^{2+} are the common cations, with Mn^{2+}, Li^+, Al^{3+}, and Fe^{3+} found occasionally. The interlayer spaces in pseudo-12-fold coordination are sites for K^+, Na^+, NH_4^+, and Ca^{2+} (rarely Rb^+, Cs^+, or Ba^{2+}).

The dioctahedral subgroup, $(K,Na)_{x+y}[(Al,Fe^{3+})_{2-y}(Mg,Fe^{2+})_y](Si_{4-x}Al_x)O_{10}(OH)_2$ Muscovite and paragonite are the most familiar species of this subgroup. They apparently differ only in the interlayer cation; K in muscovite and Na in paragonite. Si-rich varieties are called phengite if they have a Si:Al ratio in tetrahedral sites of about 3.5:0.5, and celadonite if they have almost all Si. These latter two maintain a layer charge of about 1 by substituting R^{2+} cations in the octahedral sites. We will need these as models when we consider illite and smectite.

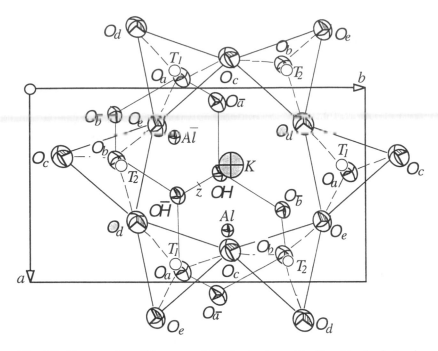

Fig. 5.6. Upper portion of the muscovite-$2M_1$ structure with ellipsoids of anisotropic thermal vibration for each atom. The octahedral vacancy is in the upper right and lower left of the a-b rectangle (shaded O_d's). You can see the other two Al. The tetrahedra have been rotated 11.4° about an axis perpendicular to this drawing. Letters with a bar, e.g., $O_{\bar{a}}$, indicate that this atom is related by the inversion center Z to the atom Oa (modified from Güven, 1971).

As with the other subgroups, there are interesting structural complications because there is an octahedral sheet distorted by vacancies. The trioctahedral micas show little stacking variation or polytype variation. The dioctahedral micas, on the other hand, show several polytypes and serve as the model for determining polytypes of illite. The most common are the $1M$, $2M_1$, and $3T$ polytypes. This terminology for polytypes may be applied to illites. Techniques for identifying the polytypes are considered in Chapters 7 and 10.

Figure 5.6, although it represents the structure of muscovite, is as close as we will get to a detailed look at the illite structure. Notice the interlayer K^+ within the ditrigonal cavity.

The 2:1 Layer Types with z< 1

Collected into this group are illite, glauconite, smectite, vermiculite, and chlorite. Although we use separate names for these minerals, implying that they are distinctly different from one another, they are in many ways transitional one to another. Furthermore, they are so often interstratified, with one another and with 1:1 clay minerals, that it is difficult to consider them one at a time. Examples of illite with no detectable expandable layers are rare.

And then there is the question of whether the expandable layers in mixed-layered clay minerals are the same as the smectite in bentonite, or the smectite from soil. If anything is understood about this group of minerals, it is that smectite is transformed to illite under several different sets of circumstances. The conversation about the mechanism(s) for this change is vigorous and current.

Illite and glauconite are, almost always, dioctahedral. Smectite and vermiculite are found in dioctahedral and trioctahedral forms, and, though macroscopic vermiculite is apparently only trioctahedral, clay-size vermiculite of soils and sediments may be either dioctahedral or trioctahedral. Chlorites are usually trioctahedral, but can be mixed dioctahedral/trioctahedral with the interlayer hydroxyl sheet and the interior hydroxyl sheet with different octahedral site occupancy patterns. Let's consider these minerals one at a time keeping in mind that we will return to them when we discuss mixed-layered clay minerals.

Illite   Compositionally, muscovite appears to be on one side of illite, and smectite on the other side (or perhaps pyrophyllite should be the other side). Illite is generally agreed to have more Si, Mg, and H_2O but less tetrahedral Al and less interlayer K than muscovite. Hower and Mowatt (1966) suggested that discrete, diagenetically formed illite has 0.75 K ions fixed in its interlayer space per $T_4O_{10}(OH)_2$ formula unit. The same polytypes identified for muscovite are found in illite. Some workers would allow small amounts of expandable layers (up to 10%) in illite; others would allow none. There is general agreement that, operationally, it is characterized by a rational series of basal reflections with $d(001) = 10$ Å (Fig. 5.5). Środoń and Eberl (1984) would have us add, based on an air-dried XRD tracing compared to that of a glycolated tracing of an oriented aggregate, that it has a peak-height-intensity ratio (Ir) of 1, for

$$Ir = \frac{I(001)/I(003)_{\text{Air dried}}}{I(001)/I(003)_{\text{Glycolated}}} = 1$$

A ratio of one indicates the complete absence of expandable layers [$I(001)$ = intensity of the 001 peak], while an Ir > 1 indicates the presence of expandable layers. Unfortunately, his relation is unusable in the presence of quartz. Do you see why?

More recently, Środoń et al. (1992) determined the following average structural formula for illite based on extrapolation from 163 analyses of I/S samples:

$$VI = -0.10 \qquad IV = -0.8$$
$$\text{Fixed Cations}_{0.89}Al_{1.85}Fe^{3+}_{0.05}Mg_{0.10}Si_{3.20}Al_{0.80}O_{10}(OH)_2$$

They concluded that illite has unique chemical characteristics different from those of muscovite and phengite, and that 0.89 is the layer charge for all illite layers. This formula gives us a reference point. You can see from examining this formula that the tetrahedral sheet has a charge of -0.8 [(4 x 3.2) + (3 x 0.8) = 15.2 and (15.2 - 16) = -0.8]. The 16 comes from the eight oxygens assigned to the tetrahedral sheets, with the remainder of the negative charge assigned to the octahedral sheet. Compare this with the zero charge on the tetrahedral sheets in pyrophyllite. The octahedral sheet of the illite of Środoń et al. (1992) has a charge of -0.10 [(3 x 1.85) + (3 x 0.05) + (2 x 0.10) = 5.90, and 5.90 - 6.0 = -0.10]. The 6 comes from 2 oxygens and 2 hydroxyls. Again, compare this charge to the zero charge on the octahedral sheet in pyrophyllite. The sum of the charges on the tetrahedral and octahedral sheets, (-0.8) + (-0.10) = -0.90, is almost exactly balanced by the +0.89 charge of the interlayer cations. When you fully understand the source of the layer charge, it is easy to see what fraction of the four tetrahedral sites are occupied by Al and realize that this fraction is the same as the amount of layer charge originating in the tetrahedral sheet. In similar fashion, for a dioctahedral mineral, the fraction of the octahedral positions that are occupied by divalent cations is the same as the contribution of the octahedral sheet to the layer charge. (At the end of this chapter, we describe a method for devising structural formulas and their layer charges from chemical analyses, and note a few caveats.)

Not all illites have K as the primary fixed cation in the interlayer space. NH_4^+-illite may be commonly overlooked because the $d(001)$ value is just a bit larger (~10.3 Å) than regular illite (~10.0 Å), and it is difficult to analyze for N. An illite 001 peak with $d(001)$ slightly larger than normal is routinely identified as illite-rich illite/smectite. Daniels and Altaner (1990) showed that NH_4^+-illite could be clearly distinguished from K-illites when they occur together by using the separation of their fifth order basal reflections. Sterne et al. (1982) reported NH_4^+-illite from black shales associated with an ore deposit in Alaska. Juster et al. (1987) and Daniels and Altaner (1990) both studied the minerals in the coal and shales of the anthracite region in northeastern Pennsylvania. Each recognized important amounts of NH_4^+-illite. However, they have some interesting contradictory conclusions about the polytypic form of NH_4^+-illite, the reactions by which it formed, and the relation of the amount of it to increasing grade of metamorphism or coal rank. The only NH_4^+-illite found to date is associated with rocks with high organic content. Therefore, it is assumed the organic material is the source of the ammonium.

Grim et al. (1937) introduced the name *illite* (for Illinois) as a term for the clay-size mica-like minerals commonly found in argillaceous rocks. Bravaisite, degraded mica, hydromica, hydromuscovite, hydrous illite, hydrous mica, K-mica, micaceous clay, illinois-geological-surveyite, and sericite are some of the terms that have been used approximately synonymously, reflecting the variable and heterogeneous nature of this material. Much of the variability reflected in this list of names originates in views from different disciplines, people with different purposes, as well as different types of alteration processes. (Discussions about the naming of minerals, this one in particular, can be vigorous and even acrimonious.) Bailey (1966), in summarizing a consensus of work to that time, said, "sedimentary illite is . . . a heterogeneous mixture of detrital $2M_1$ muscovite, detrital mixed layer micaceous weathering products, detrital weathering products partly reconstituted by K-adsorption or by diagenetic growth of chloritic interlayers, plus true authigenic $1Md$ and $1M$ micas—some having mixed layering also."

When used in the original sense of Grim et al. (1937), the term *illite* is ambiguous. On the one hand, it is used in the general sense for the micaceous material in the clay-size fraction. On the other hand, it is used for a specific mineral. The problem with using it in the latter sense is that there is little general agreement on the nature of the specific mineral. In the treatment here, we will use the term *illitic material* to cover the original, general intention of Grim et al. (1937), and the term *illite* when referring to a specific mineral.

Our definition of illite is as follows. As in the conventional treatment of a mineral compositional series, it is an end-member of a series. For example, albite is the end-member of the plagioclase series, and, although its formula is written $NaAlSi_3O_8$, it is understood that up to 10% of the mineral designated albite may be anorthite ($CaAl_2Si_2O_8$). Because less than 5% of interstratified material is difficult to detect by conventional X-ray methods, illite, in the sense we use it as a specific mineral, may contain up to 5% of an interstratified component. This component will be the other end of a compositional series. It is most commonly smectite but can be vermiculite, or perhaps chlorite. Although the conclusion of Środoń et al. (1992) seems sound, we are not quite ready to commit to -0.89 as the layer charge for all illite layers.

On a more speculative side, the two tetrahedral sheets of each illite layer may have the same layer charge except for the one that is exposed to the outside world. This tetrahedral sheet has a reduced layer charge, a smectite-like layer charge. This has yet to be proven. However, we think the evidence points to this as a distinct possibility. If this is so, it leads to some interesting implications about illite and smectite. Perhaps there is only one mineral, not two, and it behaves more like illite in thicker stacks, and more like smectite in thinner stacks. Perhaps you will be able to think of a way to test this idea. (You won't be too surprised to learn that shortly after writing this paragraph, the current issue of the *American Mineralogist* arrived carrying a study by Jakobsen et al. (1995) in which they reported on two rectorites that have scattering domains that average from eight to ten 2:1 layers. They described one rectorite as having an illite-like high-charge on its outer tetrahedral sheets, and one as having smectite-like low-charge on the outer tetrahedral sheets.) We'll return to this intriguing problem in the section on illite/smectite.

As for origin, illite the mineral is elusive. Therefore, we will consider illitic material in the sense described by Grim et al. (1937). Illitic material, especially when the illite component of interstratified illite/smectite (I/S) is included, is the most important, or at least the most abundant, clay mineral in sedimentary rocks. It is formed at, and stable or metastable at, the earth's surface. Some of the illitic material in sedimentary rocks is recycled, some is formed in pedogenic and weathering processes, and some is formed diagenetically. This is indicated by K-Ar ages from successive size fractions of shales; coarser-grained fractions usually yield dates that are older than the stratigraphic age of the unit, whereas finer-grained fractions yield dates that are younger. Illitic material can also form in weathering, in hydrothermal, and in metamorphic environments. Its formation from smectite in deep burial diagenetic conditions is discussed in the section on mixed-layered clay minerals.

Polytypes also reflect environmental conditions. Several workers have suggested that the sequence *1Md-1M-2M₁* correlates with conditions of increasing temperature and pressure, an extension of the classical Barrovian-type isograds into temperatures and pressures below metamorphic thresholds. In a study of a metamorphic sequence in the southern Appalachian Mountains, Weaver and Brockstra (1984) characterized metamorphic zones in terms of the illite polytypes *1M* and *2M₁*; they recognized no *3T*. They characterized the diagenetic zone by the absence of the *2M₁* polytype (for diagenetic illite, not detrital). Appearance of the *2M₁* polytype coincides with the beginning of the lowest-temperature (280 to 360°C) metamorphic zone, the *anchizone*. It increases in abundance across this zone. Where the *2M₁* polytype constitutes 100% of all polytypes present, this defines the boundary of the next zone, called the *epizone*. This zone could be viewed as the high-temperature part of the smectite-to-illite transition. It represents a change from illite into true mica with a layer charge of one. Grathoff et al. (1994) used polytypes to distinguish diagenetic from detrital illite, and illite crystallinity, polytypes, and K/Ar dating to distinguish the Ordovician Maquoketa Group illites from illites in the underlying units. In our opinion, there may be no full *1Md-1M-2M₁* sequence, but rather separate *1Md-2M₁* and *1M-2M₁* sequences. This is another puzzle for future clay mineralogists.

Illite crystallinity rightfully belongs in the section on illite/smectite (I/S), but we will consider it briefly here. Various forms of a sharpness ratio or crystallinity index for the illite 001 peak are well established (e.g., Kübler, 1964; Eberl and Velde, 1989). Most commonly, the width of the illite 001 peak, measured at half of the peak height above the background, is measured on tracings of an oriented, < 2μm fraction of a sample. Środoń and Eberl (1984) and Eberl and Velde (1989) discussed the theoretical basis for the relation between the width of the 001 peak and the crystallinity index. They concluded that peak width is controlled by at least two crystal-chemical factors, X-ray-scattering-domain size and percent expandable layers, i.e., the smaller the scattering domain, the wider the peak, and the more expandable layers present, the wider the peak. Based on these two assumptions, the width of the 001 peak can be used to track depth of burial or degree of tectonic stress, e.g., Duba and Williams-Jones (1983). (For illites from sedimentary rocks, apparently strain as a cause of peak broadening can be ignored.)

Illites can express the environment in which they formed in other ways, some of which we think we are beginning to understand. Illites forming in soils seem different from those forming in sandstones and shales. Changes due to weathering have been well documented: from muscovite flake to flakes with edges frayed like an old paint brush, with bristles of illite (or perhaps vermiculite), to completely replaced muscovite flakes (Fanning et al., 1989). Further weathering often results in the illite changing from partly interlayered with expandables to being fully expandable (i.e., smectitic)[1], a reversal of the smectite to illite transition due to burial (Willman et al., 1989). Perhaps the differences between the illites of sandstones and shales are as simple as differences in permeability and space to grow. The hairy illite that has grown in the pores of sandstones is immediately recognizable. In shales, pore space is apparently too small, or fluid flow too low, for hairy illite to form. Additionally, K-Ar ages of illites in sandstones develop differently than those in shales. Frequently, the K-Ar ages of illites in sandstones increase with increasing depth, whereas those of illites from shales are often older for stratigraphically equivalent units. Lee et al. (1989) concluded that for illitic clay minerals in the Rotliegende Sandstone in the North Sea: (1) K/Ar dates give the time of diagenesis, and this timing is related to tectonic events; and (2) by combining the K/Ar age with burial-history curves, the burial depth at the time of diagenesis can be discerned. Illites from shales may get younger, older, or remain constant with increasing depth (Matthews et al., 1994). Pevear (1992) and Grathoff and Moore (1996) have suggested methods for extrapolating to the ages of diagenetic and detrital illites in shales. By plotting the K/Ar ages of at least three size fractions of illite against the percent of the illite that is detrital, extrapolation to zero detrital illite gives the age of diagenesis and extrapolation to zero diagenetic illite gives the age of the detrital illite.

You will have noticed that we couldn't keep smectite, AKA, expandables, out of the discussion. We will discuss their relation again in the section on mixed-layered clay minerals. You may expect further developments in the general understanding of the nature of illite, e.g., the substantiation or rejection of some of the ideas offered here.

Glauconite   Glauconite is another clay mineral name that has been used without any precise, generally agreed-upon meaning. As was suggested for illite, the term glauconite is ambiguous. The term for a specific mineral,

---

[1]We just cannot break loose from our ambivalence about expandable and smectitic. We are not sure whether they are the same or different.

glauconite, and the term for materials similar to or rich in glauconite, *glauconitic*, are useful. This is the way we will use them: Glauconite, the mineral, has been used for greenish grains or pellets associated with microenvironments of reduction in an otherwise oxygenated marine environment of sedimentation on shallow shelves. Glauconitic should be used until it has been established that the sample is monomineralic. Glauconite the mineral is the primary component of Odin's (1988) glaucony facies. As instrumentation has improved, and the mineralogical nature of glauconite has come to be better understood, the nomenclature has sharpened. The AIPEA Nomenclature Committee (Bailey et al., 1979; Bailey, 1980a; and 1982) recommends the term be restricted to: ". . . an Fe-rich dioctahedral mica with tetrahedral Al (or Fe^{3+}) usually greater than 0.2 atoms per formula unit and octahedral R^{+3} correspondingly greater than 1.2 atoms . . . and . . . $d(060) >$ 1.510 Å . . ." and "Mode of origin is not a criterion, ..." a criterion that doesn't seem appropriate based on the material presented by Odin and co-workers (1988). It is implied that octahedral $Fe^{3+} >>$ Al. However, Ireland et al. (1983) presented data showing a continuous variation from an end-member that would normally be called glauconite, with Fe^{3+} in the octahedral sheet, to one with Al in the same site so that it would normally be called illite. They do recognize two distinct differences between illite and glauconite: (1) Illite has a better inverse relation between Mg and Al in the octahedral sheet than glauconite; and (2) glauconites have relatively higher interlayer cation occupancies. Not everyone agrees with this second point (e.g., see Thompson and Hower, 1975). Glauconite has been found in only the *1M* and *1Md* polytypes.

Now for the interpretive part. Glauconite seems to be forming today in marine sediments between about 50°S and 65°N with maximum accumulation at about the divide between the outer shelves and continental slopes, areas of very slow accumulation of detritus. Consequently, a mineral forming here spends a relatively long time at the sediment-water interface, a region transitional from oxidizing above to reducing below. It is frequently associated with fecal pellets, which can be microenvironments of reduction. The presence of organic material seems to be an essential feature in the diagenesis of glauconite. Ireland et al. (1983) suggested that, given the stability relations among common iron-bearing minerals, this situation is optimal for the formation of glauconite by precipitation or by transformation from other minerals. They also concluded that glauconite requires a long time to form, whereas nontronite (Fe^{3+} dioctahedral smectite that also forms on the ocean floor, especially near smokers on spreading ridges) can form rapidly. Odom (1984) and Odin (1988) both concluded that glauconite forms initially only by direct precipitation and not by transformation. The initial mineral to form is apparently smectite-like. From this stage, still at the sediment-water interface, it gains K and becomes more ordered as it transforms toward an illite-like mineral. All of Odin's observations are on modern and Late

Cenozoic glauconites. He and his co-workers always find their glauconites to be Fe^{3+}-rich. Ireland et al. studied Cretaceous glauconites. They found some quite rich in octahedral Al and speculated that once buried in a reducing environment, because the Al-rich glauconites were found in association with pyrite, Fe^{3+} was replaced by Al. Odin concluded that the environment in which modern glauconite is forming is at temperatures $<15°C$, pH about 8, and an Eh right at the oxidation-reduction boundary. This is in water depths slightly cooler and deeper than those in which berthierine and odinite make up the verdine facies.

Smectite   As illite is the yeoman, smectite is the princess of the clay minerals. Long considered the most interesting of the clay minerals and a curiosity, it now shares center stage with illite as research attention focuses on the smectite-to-illite transition. Smectite is really the name for a group of minerals, both dioctahedral and trioctahedral, all of which display the startling property of being able to expand and contract their structures while maintaining two-dimensional crystallographic integrity. When smectite expands, the interlayer cation may be replaced by some other cation (i.e., an exchange takes place). Accordingly, the cation-exchange capacities (CEC) of smectite (and of vermiculite, discussed in the next section) are high compared with clay minerals that do not expand. Expansion takes place as water or some polar organic compound, such as ethylene glycol, enters the interlayer space. Although smectite is more commonly found as a discrete mineral than is illite, it too should be viewed as the end-member of a compositional series. For example, smectite can be found without illite interlayers in many Tertiary and Mesozoic bentonites. Again, we defer discussion of the nature of this series to the section on mixed-layered clay minerals when we can discuss illite and smectite together.

To consider the compositions of individual species of this group, refer to the talc and pyrophyllite formulas, $Mg_3Si_4O_{10}(OH)_2$ and $Al_2Si_4O_{10}(OH)_2$. Within the dioctahedral smectite group, montmorillonite has the origin of the layer charge primarily in the octahedral sheet. An ideal formula is

$$VI = -0.33   IV = 0$$
$$R_{0.33}{}^{+}(Al_{1.67}Mg_{0.33})Si_4O_{10}(OH)_2 \text{ (montmorillonite)}$$

in which VI is used to indicate the octahedral sheet and its charge (VI because of the 6-fold coordination), IV the tetrahedral sheet and its charge, and R is the exchangeable cation in the interlayer space. The smectite varieties in which the layer charge is the result of substitution primarily in the tetrahedral sheet are beidellite and nontronite. They have the following ideal formulas

$$VI=0   IV = -.33$$
$$R_{0.33}{}^{+}Al_2(Si_{3.67}Al_{0.33})O_{10}(OH)_2 \text{ (beidellite)}$$
$$R_{0.33}{}^{+}Fe^{3+}{}_2(Si_{3.67}Al_{0.33})O_{10}(OH)_2 \text{ (nontronite)}$$

The layer charge arises primarily in the octahedral sheet for montmorillonites and primarily in the tetrahedral sheet for beidellite and nontronite. Compositions intermediate between montmorillonite on one end of the series and beidellite and nontronite on the other are common. Most compositions are on the montmorillonite side, whereas those near the beidellite end-member are rare (Brindley, 1980, p.170), except in soils.

In trioctahedral smectites, hectorite resembles montmorillonite with dominantly octahedral substitution and has an ideal formula

$$VI = -0.33 \quad IV = 0$$
$$R_{0.33}{}^{+}(Mg_{2.67}Li_{0.33})Si_4O_{10}(OH)_2 \quad \text{(hectorite)}$$

Saponite is a bit different from the other smectites in that it has a positive charge on the octahedral sheet that partially compensates for a large negative charge on the tetrahedral sheet, so the ideal formula would be

$$VI = +0.33 \quad IV = -0.66$$
$$R_{0.33}{}^{+}(Mg_{2.67}R_{0.33}{}^{3+})(Si_{3.34}Al_{0.66})O_{10}(OH)_2 \quad \text{(saponite)}$$

(Is it dioctahedral or trioctahedral?)

Though these ideal formulas are convenient, they bypass the question of just what the layer charge of smectite must be to be called smectite. Some claim that the layer charge ranges from approximately 0.2 to 0.6 and cation proportions vary accordingly. Środoń et al. (1992), based on a study of 163 superior analyses of illite/smectite, concluded that smectite, at least in illite/smectite, has a fixed and constant layer charge of 0.40.

Members of the smectite group differ due to the source of the layer charge, and within particular species of smectites there can be significant variation of properties. For example, workers have detailed variations in the properties of Na- and Ca-saturated montmorillonites. Talibudeen and Goulding (1983b) measured slight differences in weight loss, surface area, CEC, and surface charge density for three samples of the same size fraction of a Wyoming smectite from the same location but probably from different pits. Based on the amount of energy released as Ca replaces K, they concluded that their samples had six different exchange sites. They generalized and suggested that all smectites have exchange sites with a mixture of energies and that these sites are probably distributed unevenly on the surfaces of the adjacent silicate layers.

The unique property of swelling or expansion is almost certainly due to the relatively small layer charge of smectites. Perhaps the primary cause is that there is not a large enough attraction from the interlayer cations to keep the layers together compared with the larger layer charge of illite. In contradiction to this suggestion is the observation that pyrophyllite, with no layer charge at all, doesn't expand a bit. Why would that be? Perhaps the interlayer space of smectites expands because the interlayer cations are attracted more to water than to the relatively small layer charge. In the presence of water, the interlayer cations pull water into the interlayer space. An observation that supports this is that the swelling behavior is a function of both the size and the charge of the interlayer cation present.

Early studies (Nagelschmidt, 1936; Bradley et al., 1937) indicated a continuous variation of $d(001)$ with available water or relative humidity. Other workers (Mooney et al., 1952) asserted that $d(001)$ changed discontinuously. Moore and Hower (1986) showed that for Na-smectite the variation is continuous but not linear. The change in apparent $d(001)$ results from an ordered interstratification of continuously changing proportions of successive hydrates; i.e., as relative humidity is increased from 0 to 100%, Na-smectite has discrete thicknesses of 9.6, ~12.4, ~15.2, and ~18 Å. The value 2.8 Å is taken to correspond to the thickness of one layer of water. Any two successive hydrates of this series can be mixed in an ordered interstratification to reach equilibrium with the available water and to provide an intermediate $d(001)$. In apparent disagreement with Moore and Hower, Ransom and Helgeson (1994) concluded that aluminous, dioctahedral smectite, in its various states of hydration, should be treated as a regular solid solution of hydrous and anhydrous thermodynamic components that are randomly interstratified. They stated that smectites in partial states of dehydration should yield XRD tracings indistinguishable from those of randomly interstratified I/S. The higher K content of I/S is unexplained.

The origin of smectite is a puzzle that is not fully solved to the satisfaction

---

Box 5.2. Alteration of Ash-Fall Layers

Volcanic ash falls in a blanket over landscape and seascape. Such blankets are important geologically because they are distinct time boundaries that often can be dated. This can be invaluable for stratigraphic correlations and for establishing rates of organic and inorganic change (see e.g., Huff et al., 1992).

Depending on the environment into which the ash falls, it can change into different minerals, or be preserved. For alteration, water is needed for leaching away some of the ions that are freed in the alteration reactions. If the environment is acid, as it would be in a coal swamp in which organic acids have been generated by the breakdown of plant material, the ash usually alters to kaolinite. This forms a rock called a tonstein (Bohor and Triplehorn, 1993). If the environment is mildly alkaline, as it would be in a marine environment, the ash usually alters to smectite. This forms a rock called a bentonite. If the environment is more highly alkaline, as it would be in a playa lake or hypersaline inland sea, the ash usually alters to a zeolite or a combination of zeolite and K-feldspar.

Bohor and Triplehorn (1993) offered a conclusive argument for the volcanic origin of tonsteins and bentonites, along with a thorough discussion of nomenclature and rare circumstances in which bentonites are kaolinitic and tonsteins are smectitic.

of most workers in the field of clay mineralogy. An important source of smectite is the alteration of volcanic glass with relatively high silica content (more about this in Box 5.2 on ash-fall layers). Nadeau and Reynolds (1981) argued that much of the smectite in the Cretaceous marine shales of the western interior of North America can be accounted for by alteration from volcanic ash.

In addition to forming from volcanic glass, smectite seems to precipitate directly in pore spaces of sandstone and apparently forms in weathering environments characterized by very slow movement of water whether in swampy lowlands or in arid to semiarid regions (Berner, 1971). Smectites carry indications of the chemical composition of the parent from which they have formed by weathering (Fig. 5.7). Adding to the diversity of situations in which smectite is found is the startling report of Yamada et al. (1995) that they formed large (>10 μm), well-crystallized smectite at pressures of 2 to 5.5 GPa and temperatures of 700 to 1000°C (equivalent to mantle conditions 65 to 175 km deep!). These crystals formed in an assemblage of coesite, kyanite, and jadeite, an assemblage clearly recognized as a high-pressure, moderate-temperature assemblage. Scattered

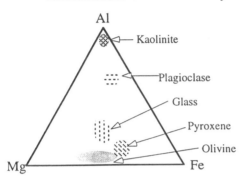

Fig. 5.7. Triangular Al:Fe:Mg diagram showing the compositions of smectites produced by weathering of minerals and glass in basaltic andesite from the Abert Lake, Oregon, drainage basin. Modified from Banfield et al. (1991) with permission.

through the literature on soils and weathering are interpretations of smectite forming from illite, from kaolinite, and from chlorite. Some workers in soils make no distinction between vermiculite and smectite. They use the terms *high-charge* and *low-charge* expandables for clay minerals that behave like vermiculite or smectite or as an intermediate. Here is fertile ground for some careful work. Are there two kinds of smectite: one produced by the kind of weathering found in soils and one by neoformation such as in the alteration of volcanic glass? Based on differential thermal analysis (DTA), Glass et al. (1992) noted that smectites from the Great Plains of North America lose their hydroxyls near 550°C, whereas those from bentonites lose them at 650 to 700°C.

We still have more to learn about smectite, and, as with illite and chlorite, the nature of smectite is closely related to other specific minerals with which it is interstratified and to primary minerals from which it formed (Fig. 5.20).

Vermiculite   We can represent the composition of vermiculite by thinking of it as a biotite, $K(\underline{Mg},Fe^{2+})_3(Si_3Al)O_{10}(OH)_2$, in which some Fe^{2+} has been oxidized to Fe^{3+}, thus reducing the layer charge. (Underlining of an ion indicates it is the dominant ion when more than one occupant in a structural site is possible.)

$$[\sim(\underline{Mg},Fe^{3+})_3(Si_{4-x}Al_x)O_{10}(OH)_2]^-$$

The remaining layer charge is often balanced with hydrated ions, commonly Mg, in the interlayer space

$$(4H_2O(Mg)_{0.5})^+$$

Such a simple representation is, however, deceptive. If smectite is the princess and illite the yeoman of the clay minerals, then vermiculite is the bad boy. It is difficult to identify positively because of its variable characteristics, which are difficult to explain, compared with similar minerals. It can be seen as a member of three different compositional series: (1) from biotite to trioctahedral vermiculite to trioctahedral smectite; (2) from muscovite to dioctahedral vermiculite to dioctahedral smectite; and (3) from chlorite to vermiculite, most likely both trioctahedral. Due to the nature of their origin, they commonly contain some interlayers of their precursor as interstratified layers, i.e., they may be mixed-layered muscovite/vermiculite, biotite/vermiculite, illite/vermiculite, or vermiculite/chlorite. These changes, assumed to have occurred in nature, have been duplicated in the lab. In clay-size material, vermiculite may be trioctahedral or dioctahedral, whereas in macroscopic varieties, it is trioctahedral with platy morphology like that of the trioctahedral micas. A puzzle yet to be resolved is how the change is made from trioctahedral to dioctahedral, as seems to be the case in some situations (Weaver, 1989, p. 167). Vermiculite apparently has the same pattern of layer charge distribution as saponite, i.e., a positive octahedral charge reducing some of the negative tetrahedral charge. A likely origin for vermiculites involves partial oxidation of ferrous iron in the octahedral sheet, leading to a decrease in layer charge. The decreased charge allows the stripping of K ions and their replacement by (usually) hydrated Mg ions. One of the largest vermiculite mines in the United States, near Libby, Montana, is an augite pyroxenite body that has been altered to biotite, hydrobiotite, and vermiculite, a sequence that must represent the path of alteration.

The chlorite to vermiculite alteration can proceed in the other direction. The interlayer hydroxyl sheet in chlorites can be replaced by hydrated cations. The NEWMOD© program can represent this nicely by using the option of indicating that the hydroxyl layer is complete (= 1) or incomplete to some degree (< 1). As this value decreases, the pattern more and more resembles that of vermiculite. Most commonly, hydroxy-Al complexes form in the interlayer space of vermiculite replacing the cations that are coordinated to water molecules, but hydroxy-Fe and hydroxy-Mg complexes have been observed as well. The Mg complexes are the dominant ones in evaporitic environments. If the hydrated cations are completely replaced, the mineral is essentially a chlorite.

The AIPEA nomenclature committee (Bailey, 1980a) set the range of layer charge for vermiculite from 0.6 to 0.9, so it would vary from one that would be high for smectite to one that would be high for illite. (How we will perceive layer charges for expandable minerals in light of the idea of a fixed charge of 0.89 for illite is yet to be seen.) Conventionally, the identification of vermiculite is made first by Mg-saturation and then by solvation with glycerol after which it should expand to about 14.5 Å compared to about 18 Å if it were smectite (Walker, 1958). These expansion dimensions indicate that vermiculite has accepted one layer of glycerol and smectite two layers. Upon gentle heating, ideally the Mg variety should collapse in stages. Minerals identified as vermiculites show expandabilities to all of the values between 14.5 and 18 Å; they also collapse to varying degrees of "completeness," which seems to indicate that, rather than hydrated cations in the interlayer space, there are some cation-hydroxyl complexes that are not developed into complete sheets such as those found in chlorite. Douglas (1989, p. 653), among others, suggested that these sheets are hydroxy-Al interlayers. The variation in the extent to which these sheets are complete or incomplete seems to be the chief reason for the inconsistent collapse behavior of vermiculite.

Malla and Douglas (1987), working with both reference clay minerals and soil clay minerals, did not find the AIPEA definition of vermiculite adequate. They suggested glycerol solvation was not an adequate basis for distinguishing vermiculite from smectite. Further, Malla and Douglas suggested that vermiculite could be better defined and more clearly distinguished from smectite by first estimating the layer charge by the n-alkylammonium ion exchange technique (Lagaly, 1992) and then glycerol solvation of K-saturated, heat-treated samples. (The n-alkylammonium ion technique takes much practice and cannot be considered a routine procedure.) Their tests showed that K-saturation causes collapse to 11.2 to 12.8 Å if the layer charge is less than 0.57 and to 10 to 10.6 Å if the layer charge is at least 0.63. Glycerol solvation caused expansion to 18 Å if the layer charge was not greater than 0.36, to 14 Å if the layer charge was 0.39 to 0.43, and showed no expansion if the layer charge was at least 0.63. They concluded that some vermiculites have two different layer charges based on the appearance of peaks representing spacings of both 14 and 18 Å from the same mineral. They also achieved expansion after Mg^{2+}-saturation and ethylene glycol solvation for minerals with layer charge as high as 0.72. Based on a review of the features of macroscopic vermiculites, de la Calle and Suquet (1988) suggested that low-charge vermiculites and saponite belong in the same family. These (Malla and Douglas, 1987; de la Calle and Suquet, 1988) complications should indicate the problems involved in distinguishing vermiculite from other minerals. They also clearly indicate another area inviting further work.

Soil scientists have to deal with this ambiguous creature more than other clay scientists. Vermiculites occur in variable amounts in all major soil groups but are most common in soils of temperate and subtropical climates; they are not a major component, except locally. Because of their importance in soils and their relative unimportance in sedimentary rocks, we will digress here for a moment to delve into a region with which we are not as familiar as we should be, i.e., soils. Because clay minerals in soils tend to be found mixed with more components than their counterparts in sedimentary rocks, tend to be more disorganized, more defective, and therefore have smaller diffracting domains, they are generally more difficult to characterize (e.g., Hughes et al., 1994).

Because expandables and illite are the most abundant clay minerals in the parent materials for soils in many areas, the weathering products of these 2:1 clay minerals dominate pedogenic profiles. When dealing with clay minerals in soils, in place of vermiculite or smectite, we use the terms expandables, collapsibles, low-, intermediate-, and high-charge. For example, for the mixed-layered mineral conventionally called I/S, in soils the expandable component may not be quite smectite nor quite vermiculite (Malla and Douglas, 1987; Suquet and Pézerat, 1988). In a given soil profile developed on glacial till, the nature of the first order peak of this expandable clay mineral will continuously change from the unaltered till to the A horizon. It will shift to lower and lower angles and become wider and wider, or "smear," as Frye et al. (1968) have described it. The soil profile on a Late Wisconsinan Richmond Loess in Illinois, illustrated by the XRD patterns in Fig. 5.8, shows this smearing of the expandables peak that is the product of weathering. (This use

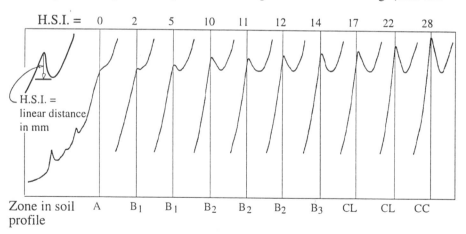

Fig. 5.8. Sequence of XRD curves of glycol-solvated samples showing progressive alteration of the 001 peak of expandables from the surface (on the left) through a soil profile developed on Richland Loess (from Frye et al., 1968). Vertical lines 0 to 28 are at the 17Å position. CL = C horizon, leached; CC = C horizon, calcareous; HSI = heterogeneous swelling index.

of smearing of XRD peaks should be distinguished from preparation of smear slides.) The heterogeneous swelling index (HSI, invented by H.D. Glass, Illinois State Geological Survey) measures the peak broadening or smearing of the 001 expandables peak, which, in turn, indicates the degree of weathering. Because we cannot tell when to stop calling this expandable component vermiculite and when to start calling it smectite, we refer to it as simply expandable. In other cases, it may be useful to describe low-, intermediate-, and high-charge vermiculite corresponding to clay mineral peaks at about 17, 16.5, and 14 Å after ethylene glycol solvation (Newman et al., 1990). The degree to which chlorite had altered to high-charge and low-charge vermiculites in weathering profiles was used by Newman et al. (1990) to identify and distinguish two superposed tills in the area of Boston, Massachusetts. [In the section on mixed-layered clay minerals, we will consider another mixed-layered clay mineral, the origin of which also seems to be restricted to soils and paleosols. We referred to it as kaolinite/smectite (K/S) or kaolinite/expandables (K/E).]

Another mixed-layered clay mineral, the origin of which also is restricted to soils and paleosols, so far as we know, referred to as kaolinite/smectite (K/S), should be described as kaolinite/expandables (K/E) because the 2:1 layer in this mineral can vary from a low-charge expandable through high-charge vermiculite (Hughes et al., 1993). Perhaps vermiculite would better be called the transitional expandable component TEC of soil mineralogy, and give it up as a discrete mineral, at least in the <2 μm size range.

In the study of sedimentary rocks, vermiculite is seen less often than the other three clay minerals of this group (illite, smectite, and glauconite). It is, however, an extremely common mineral in the well-drained soils of humid regions if the parent material contains micas such as the mica schists of the Appalachian Region. Therefore, the discontinuity between content of vermiculite in soils and the lack of it in shales is striking. Throughout the Appalachian Region from Canada to Alabama, the suspended load of rivers contains vermiculite. However, none is found in the coastal or estuarian sediments (Weaver, 1989, p. 168). This is one more enigma to cloud further the issue of whether or not vermiculite is really a clay mineral, or just a transitional expandable component in several weathering sequences; maybe it is only a mineral on the macroscopic scale.

Chlorite   Chlorite consists, ideally, of a negatively charged 2:1 layer,

$$[(R^{2+},R^{3+})_3(Si_{4-x}R^{3+}_x)O_{10}OH_2]^-$$

and a positively charged interlayer octahedral sheet (Fig. 5.9),

$$[(R^{2+},R^{3+})_3(OH)_6]^+$$

Chlorite was sometimes referred to as a 2:1:1 or a 2:2 layer silicate in the older literature, but the 1978 AIPEA Nomenclature Committee chaired by Bull Bailey (Bailey et al., 1980a) clearly stated that it should be seen as a 2:1 mineral with an interlayer sheet of cations octahedrally coordinated by hydroxyls. There is a considerable range of cation substitutions in chlorites, although the most common octahedral cations are Mg^{2+}, Fe^{2+}, Al, and Fe^{3+}. In addition, the great preponderance of chlorite specimens are trioctahedral in both octahedral sheets. One example is known that has both octahedral sheets dioctahedral, and it is called

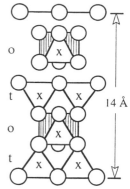

Fig. 5.9. The structure of chlorite. x = cation site.

dioctahedral chlorite. There are two, perhaps three, examples with a dioctahedral 2:1 layer and trioctahedral interlayer material. These di,trioctahedral chlorites are called sudoites. Recently it was reported in an Archean shale in the Witwatersrand Basin (Zhou and Phillips, 1994). Tri,dioctahedral chlorites are not yet known, but more examples of these exotic varieties should be expected (Table 4.3, p.131).

There have been many classification schemes, some of them quite complex, for the overwhelmingly common trioctahedral chlorites. So many schemes suggest the dilemma their variation presents to the student. The Nomenclature Committee (Bailey, 1980a) suggested the adoption of a simplified nomenclature for chlorites based on the dominant divalent octahedral cation. They recommend only the four species names clinochlore, chamosite, nimite, and pennantite for varieties dominated by the divalent ions Mg, Fe, Ni, and Mn, respectively. They state clearly that all other species and varietal names should be discarded for the clay-size chlorites. Perhaps there are other factors that need to be considered when classifying macroscopic chlorites. We will pass quietly by the fact that some workers still speak of 7 Å chlorite. In the older literature, the term chamosite often refers to a 7 Å mineral, probably berthierine. See Bailey (1988c) for a comprehensive review.

Almost all of the work on the properties of chlorites up to the decade of the 1980s was done on chlorites from igneous and metamorphic rocks or macrocrystalline chlorites. Those working with chlorite grains large enough for optical microscope study have the option of using optical properties for determining Fe/(Fe+Mg) (Albee, 1962). Albee generalized on the basis of over 200 samples that the composition of chlorites from igneous and metamorphic rocks generally matches the chemistry of the rock in which it is found. Characterization of diagenetic chlorites in sedimentary rocks has had quite a bit of attention in recent years (Curtis et al., 1984, 1985; Jahren and Aagaard, 1989; Humphreys et al., 1989; Jahren, 1991; Hillier and Velde, 1991, 1992). There clearly has been a distinction made between diagenetic chlorites and those from igneous and metamorphic rocks, a distinction we address below.

Chlorite is from the Greek *chloros*, green, and gives the color to the green schist facies. We will address the nature of chlorite in circumstances less intense than the greenschist facies and refer you to Laird (1988) for a review of chlorites in metamorphic rocks. The chlorites we deal with are found (1) in shales at the highest grades of diagenesis, (2) in soils, (3) as apparently *de novo* crystallization (i.e., neoformation) on the surfaces of sand grains in porous sandstones where they may be the alteration product of either odinite or berthierine (Bailey, 1988d; Ryan and Reynolds, 1996), (4) as a replacement of carbonate grains and matrix in carbonate rocks, and (5) in geothermal and low-temperature hydrothermal systems. We offer a suspicion for which we have little evidence other than the unusual circumstance that some chlorites form in near-surface conditions while others, in soils, weather more quickly than calcite (Droste, 1956). That suspicion is that chlorites, apparently not distinguishable on the basis of chemical composition or XRD data, from metamorphic rocks are different from those formed in sedimentary and soil environments. Can you think of a way to test this?

As is the case for illite, the chlorites in sandstones can usually be distinguished as diagenetic or detrital based on morphology, but this is a more difficult task in shales. Unlike illite, chlorite has no readily datable isotopes in the usual cations found in its structure to help discriminate between diagenetic and detrital origins. Walker (1987, 1993) concluded that most chlorites in shales and slates are the IIb polytype, and can form directly at temperatures well below 200°C without passing through any intermediate polytypes regardless of temperature. This is disappointing, because the suggestion by Hayes (1970) of a type I to type II conversion at some fixed temperature within the diagenetic-metamorphic sequence, if demonstrable, would have given shale petrologists a useful temperature datum for regional studies. Chlorite coatings on sand grains, and crystallizations in vugs in carbonate rocks, constitute an apparently anomalously low-temperature formation of chlorite. Type I chlorite is uncommon, and most of the recorded instances of its occurrence are these coatings and crystallizations (Walker, 1987).

In sandstones, chlorite often forms a microcrystalline coating of rosettes that produce spectacular SEM pictures (Fig. 5.10). These grain-coating chlorites are quite often a "funny" type of chlorite interlayered with a 7 Å 1:1 phase. We'll put off additional comments on this mineral to the section on mixed-layered clay minerals. The chlorite coatings seem to inhibit later quartz cementation, thus preserving primary porosity. Preservation of porosity makes an interesting story and one that is important to the oil industry. Chlorite also is found in high concentrations in certain, Mg-rich evaporite sequences, such as the Salina Group of New York State (e.g., Bodine and Standaert, 1977; Hluchy, 1988). It may be a replacement or it may be primary. The relative importance of diagenesis and fluid chemistry is not yet understood for this

Fig 5.10. The typical morphology of diagenetic, grain-coating chlorite. This sample is from the Cypress Sandstone in the Illinois Basin. The rosette on the right is about 10 μm in diameter. This SEM photo is courtesy of Beverly Seyler, Illinois State Geological Survey.

occurrence or for the grain-coating varieties.

As with problems determining diagenetic from detrital in sediments, formation of chlorite in soils is difficult to document because convincing demonstrations of valid genetic pathways are compromised by the obstacles to separating detrital and pedogenic forms. But one pathway that is well known and well understood consists of vermiculite that has developed Al-hydroxy interlayers. Operationally, this phase is indeed chlorite because it is a 14-Å phase that is unaffected by polar solvents and mild heat treatments. It differs from "normal" chlorite, however, in that the Al-hydroxy interlayer can be removed by sodium citrate extraction (Tamura, 1958; Sawhney, 1960). After such treatments, the mineral behaves like vermiculite with respect to heat and solvation. This mineral is important because it is thought to represent Al mobilized at the low soil pH values that are caused by acid rain (April et al., 1986). It is a common constituent of the spodosols of the eastern and northeastern United States.

To make interpretive use of what we have learned about chlorite has been frustrating so far. de Caritat et al. (1993) and Jiang et al. (1994) reviewed the use of chlorite as a geothermometer. de Caritat et al. evaluated four approaches and concluded that chlorite geothermometry should be used cautiously and only with other corroborating methods of estimating paleotemperatures. Jiang et al. focused on the suggestion of Cathelineau and Nieva (1985) (and subsequent additional work) that the amount of tetrahedral

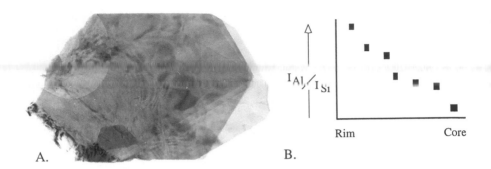

A.   B.

Fig. 5.11. A zoned crystal of chlorite and graph of microprobe analyses from the center of the crystal to the edge. Jahren offered this as an expression of Ostwald ripening (from Jahren, 1991, with permission).

Al should increase as temperature increases, and concluded that this was flawed because analyses had been made on volumes that were too large, a volume that included minerals other than chlorite. However, Jahren (1991) presented solid evidence that the Al/Si ratio increases from core to rim in a single diagenetic, zoned crystal of chlorite from a sandstone reservoir in the North Sea, which had grown by Ostwald ripening (Fig. 5.11). This trend suggests that the crystal grew as it was buried more deeply, and therefore subject to increasing temperatures. Attempts to use chlorite crystallinity, analogous to illite crystallinity, also have been made. Again, there are hints that it should work, but has not yet proved consistent. So, there are apparently systematic changes with increasing temperature and pressure, but we haven't figured them out yet. We will come back to the topics of chlorite geothermometry and crystallinity when we discuss chlorite as a member of mixed-layered minerals.

Hower et al. (1976) noted that chlorite appeared in Gulf Coast shales at the greatest depths studied (ambient temperatures near 150°C). They proposed that chlorite forms as part of the overall diagenetic reaction that converts smectite to illite (discussed later). Many other investigators have noted the presence or abundance of chlorite in old, deeply buried shales, so this mode of formation can be considered well established. It makes sense, too. Lynch and Reynolds (1985) noted that contact metamorphism of a Cretaceous shale by a Pliocene dike produced a smectite-to-illite reaction sequence much like that observed in the Gulf Coast burial diagenetic sequences. The highest temperature portion of the contact zone contained a chloritic mineral. They concluded that at least some of the K needed for illitization was derived by the reaction of K-feldspar with pore-fluid Na to produce albite. It is likely, then, that a final, ideal mineral assemblage present at the end of diagenesis and at the beginning of classical metamorphism consists of illite (muscovite)-chlorite-albite, which is precisely the mineral assemblage defined as the greenschist facies of regional metamorphism for pelitic rocks.

We believe that chlorite is a mineral worthy of serious and extensive future study by clay mineralogists and shale petrologists. There are the possible variables of tetrahedral Al content, Fe content, and the distribution of the Fe between the octahedral sites of the silicate and hydroxyl layers, the numerous polytype possibilities, and the chlorite mixed-layered minerals. In addition, based on the work of Odin and his colleagues as reported by Bailey (1988d) and recent studies by Ryan and Reynolds (1996), there is the intriguing possibility that chlorite is part of a diagenetic sequence that parallels the smectite-to-illite sequence. There must be much geologic information that can be coaxed out of a complete description of a given chlorite or, better yet, a suite of chlorite samples from well-documented sedimentary and diagenetic environments. But systematic studies are lacking, and we don't even know the questions yet, much less the answers. For details on structure and crystal chemistry, see Bailey (1988c); for a general review, see Weaver (1989).

Sepiolite and Palygorskite

The morphology of these fibrous clay minerals, generally found in association with carbonates in arid to semiarid soils, paleosols, and alkaline lake sediments, is a reflection of their ribbon-like structure. They are also found in marine sediments, but there is disagreement on whether they are detrital or diagenetic in this environment. They are 2:1 layer silicates because their tetrahedral sheets are linked essentially infinitely in two dimensions. But they are structurally different from the other clay minerals in two ways: (1) the octahedral sheets are continuous in only one dimension, in ribbons, and (2) the tetrahedral sheets also are divided into ribbons by inversion of every two (palygorskite) or three (sepiolite) rows of tetrahedra (Fig. 5.12). The channels between ribbon strips are ~4 Å by 6 Å and ~4 Å by 9.5 Å, the larger channel in sepiolite. They contain two kinds of water. One type of water is more tightly bonded because it is coordinated to the octahedral cations; the other, less tightly bonded water is referred to as zeolitic water. Again similar to zeolites, the channels may contain exchangeable cations. The nature of these channels makes these minerals important industrial raw materials for catalysts and molecular sieves. Crystallographers find the structural modifications of these two species interesting. They represent an example of a more general structural style called *modulation*, which is periodic inversion of tetrahedral sheets in layer silicates, or phyllosilicates (also a feature of some serpentines, p. 139).

Compositionally, they are both Mg silicates, but palygorskite is more aluminous and has more structural diversity than sepiolite. An ideal approximate formula for sepiolite is

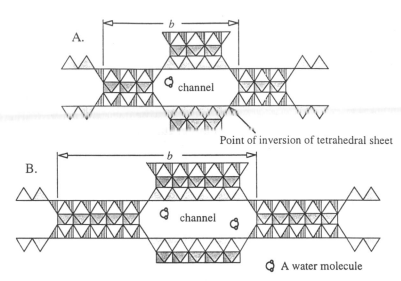

Fig. 5.12. Schematic structure diagrams: A. Palygorskite with b approximately 18 Å;
B. Sepiolite with b approximately 27 Å; Note the tetrahedral sheets maintain continuity
through the inversion point, whereas the octahedral sheets do not.

$$\sim Mg_8Si_{12}O_{30}(OH)_4(OH_2)_4 \cdot n(R^{2+}(H_2O)_8)$$

and for palygorskite,

$$\sim MgAl_3Si_8O_{20}(OH)_3(OH_2)_4 \cdot n(R^{2+}(H_2O)_4)$$

Sepiolite may also have Fe, Mn, Al, and Ni in octahedral positions.
Palygorskite, which is closer to a true dioctahedral layered silicate, may have
Na, Fe, and Mn in octahedral sites.

Although sepiolite and palygorskite are considered relatively rare, they are
quite abundant in the Mediterranean and Middle East Region. They appear to
be products of weathering reactions, but may form from smectite or convert to
smectite in marine environments that are either sub- or hypersaline. As
mentioned above, the genesis of palygorskite and sepiolite in marine
sediments is in question. They are unevenly distributed in time and in space;
occurrences seem restricted to latitudes ~ 40° N to 35° S; and the rocks of
some epochs contain notably more than others. The distribution offshore in
the direction of prevailing winds from arid regions suggests that careful
consideration be given to eolian transport as the source of these two minerals.
Velde (1987) offered a petrologic analysis of the distribution of palygorskite
and sepiolite in closed basins. Singer (1989) reviews these two minerals and
their importance in soils.

MIXED-LAYERED CLAY MINERALS

No aspect treated in this book has changed more since the first edition than that of the mixed-layered clay minerals. The transition of smectite to illite was an established axiom within the discipline of clay mineralogy. Now, we not only have a vigorous debate about the mechanism of this transition, we see many other transitional series: We see smectite going to kaolinite with an intermediate kaolinite/expandables (expandables because we're not sure the expandable layers are exactly smectite, Hughes et al., 1993); we see 7 Å/chlorite (7 Å because we're not sure whether the 7 Å phase is berthierine, kaolinite, odinite, or something else, Moore and Hughes, 1991; Reynolds et al., 1992); and we see mixes of chlorite, smectite (or at least an expandable layer), and something much like corrensite, or perhaps it is just chlorite and corrensite intergrown (Moore et al., 1989; Jiang and Peacor, 1994) that seem to be moving toward chlorite. As we look at the layering in detail through the marvelously revealing eyes of lattice fringe imaging (e.g., Banfield and co-workers, Eggeleton and co-workers, Peacor and co-workers, and Veblen and co-workers), the mixing and transition of one mineral to any other mineral seems the only limit on available possibilities. As an indication that the phenomenon of mixed layering is not restricted to clay minerals, we suggest that you see Veblen's (1991) review of polysomatism and polysomatic series. (Polysomes are structures created by combining two or more slabs or layer modules that are distinct structurally and stoichiometrically.) There may or may not be a difference between polysomatism and mixed layering.

Inevitably, as we begin to understand these transitions, we try to apply them to deciphering parts of the history of the earth. Several of the transitions seem to measure the depth of burial of, or expenditure of tectonic energy on, the sediments in which they are contained; the proportion of expandable layers in kaolinite/expandable seems to be a measure of the intensity or the duration of weathering in soils, or both. The empirical crystallinity indices are apparently based, at least in part, on mixed layering. Once we thoroughly understand these indices, they may be even more useful than we have so far found them to be. The *general thesis* that emerges from studies of mixed-layered clay minerals is that the collective changes form a trend from the metastable, highly imperfect, and disordered arrangement of all varieties of layered components of clay minerals to well-ordered, homogeneous, defect-free phases. These changes are driven by time, increased temperature and pressure levels, fluid movements, and shearing associated with tectonism. They are most clearly expressed in the transition from argillaceous sediments to metamorphosed pelitic rocks (e.g., Weaver and Brockstra, 1984; Lee et al., 1985; Lynch and Reynolds, 1985). Out of the effort to understand the transitions of one mineral to another through an intermediate of mixed layering is emerging a number of sequences that extend the classical Barrovian sequence of metamorphic-grade indicators into the low-temperature and low-pressure conditions of diagenesis.

Mixed-layering, interlayering, and interstratification

These terms are synonymous for clay minerals formed of two or more kinds of intergrown layers, not physical mixtures. Interstratifications of more than two components seem quite rare, or perhaps we have not learned to identify the third components in mixtures. The layers are stacked along a line perpendicular to (001) or Z^*. Component layers are found stacked in random, partially regular, or regular sequences. A large variety of mixed-layered clay minerals seem more susceptible to ordering than to random stacking, especially when one component is expandable and makes up half or less of the mixture.

Two or more clay minerals stacked together along the direction Z^* should not be too surprising. Recall that they are all constructed from the same two modules: an octahedral sheet and a tetrahedral sheet. One mineral growing on another with structures that are coincident in two dimensions is called an *epitaxic overgrowth*, i.e., you may think of their 2D nets as fitting together. This is common in clay minerals. We will not go over any other structural details of individual minerals here because we think we can safely assume that we can treat the components in mixed-layered clay minerals as if they were the same components when they occur as discrete or end-member minerals (a justifiable assumption?). Illite/smectite mixed-layered clay minerals are the most common, more common than either discrete illite or discrete smectite. The abundance of chlorite/smectite is just being recognized. The list of other mixed-layered clay minerals that would not ordinarily be anticipated continues to grow: e.g., celadonite/nontronite in modern sediments from the Red Sea rift zone (Butuzova et al., in Drits, 1987), or chlorite/saponite from marine basins off the coast of Brazil (Chang et al., 1986), or kerolite/stevensite (Eberl et al., 1982; Hay et al., 1995).

Table 5.2. Regularly stacked, 50/50 mixed-layered clay minerals

Aliettite	1:1	Talc/trioctahedral smectite
Corrensite	1:1	Two varieties, a trioctahedral chlorite/low-charge trioctahedral smectite and a trioctahedral chlorite/high-charge trioctahedral vermiculite
Dozyite	1:1	Serpentine/chlorite
Hydrobiotite	1:1	Biotite/vermiculite
Kulkeite	1:1	Talc/chlorite
Rectorite	1:1	Dioctahedral mica[a]/dioctahedral smectite
Tosudite	1:1	Dioctahedral chlorite/smectite

[a]The Nomenclature Committee recognized K-, Na-, and Ca-rectorites (S. W. Bailey, University of Wisconsin, written communication 5/14/1990).

Special names are assigned to regularly alternating sequences of components present in a fixed ratio. Examples of regular sequences that have been accepted by the AIPEA Nomenclature Committee, and a more recently described one, dozyite (Bailey et al., 1995), are given in Table 5.2. The Committee's criterion for a regular sequence is that it have at least ten 00*l* peaks that yield a mean value for the 001 spacing with a rioctahedral smectite and a trian 0.75% and whose peaks are all the same breadth (Bailey, 1982).

Turning to mixed-layered clay minerals that are not perfectly ordered, we noted at the beginning of the section on 2:1 layers with $z < 1$ the many ways these minerals are transitional one to another. Figure 5.13 is an attempt to show these transitions in compositional space. The clay minerals in these diagrams would be more accurately represented if we could make a three-dimensional diagram of a three-sided prism with the triangle for the dioctahedral minerals on one end and that for the trioctahedral on the other. Then the clay minerals would occupy the space within the prism rather than on the triangular surfaces. The contours indicating percent expandable material or smectite are adapted from Środoń and Eberl (1984, and Lippmann, 1982). There does not seem to be a similar relation between vermiculite and the trioctahedral smectites (or, maybe no one has looked for it or figured out how to look for it).

Of the many not perfectly ordered mixed-layered clay minerals, illite/smectite (I/S) and chlorite/smectite (C/S) are the most important, not only from a quantitative standpoint, but because we are learning to use them

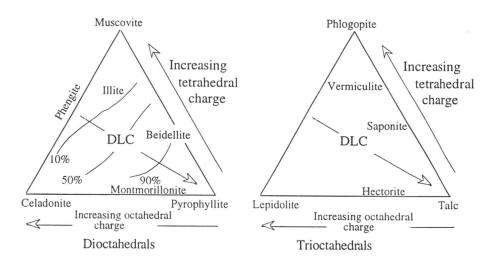

Fig. 5.13. Generalized chemical relations for the dioctahedral and trioctahedral clay minerals in relation to their macroscopic counterparts. 10, 50, and 90% label lines of equal expandable layer content in illite/smectite. DLC = decreasing layer charge.

for interpreting geologic history. In what follows, we will restrict ourselves to discussion of these two plus chlorite interstratified with a 1:1 layer (7 Å/C) and kaolinite/expandables (K/E). The I/S series has received a great deal of attention because it serves as an index of thermal maturity for organic material in the sediments of basins. Water from the interlayer space in smectite is released as the change takes place to illite. This water seems to at least partially explain how petroleum moves out of source rocks and towards reservoir rocks. It is also implicated in higher than hydrostatic pressures (overpressuring) of some oil fields. These relations are covered by Bruce (1984). Chlorite forms two-component interstratified minerals with all the other common phyllosilicate layer types, e.g., biotite/chlorite, illite/chlorite, talc/chlorite, wonesite/chlorite (wonesite is a Na-biotite), as well as with smectite and vermiculite. We will restrict ourselves to discussion of two examples of chlorite minerals, those interstratified with smectite (C/S), and those interstratified with a 1:1 layer (7 Å/C).

Illite/smectite (I/S)
Let us start with what seems most clear, the mixed-layered clay mineral that we think we understand best. Illite is generally understood to be more stable than smectite as conditions change to higher temperatures and pressures and to different chemical environments than those at the surface of the earth. We stated in the first edition that we would start with illite/smectite (I/S), and the transition from smectite to illite, because our understanding of it is broader than that of other mixed-layered clay minerals, although our understanding is not yet complete. Like many subjects one approaches, we have learned much about I/S in the intervening 7 years but full understanding now seems as least as far away as when we first wrote. People in our discipline have not yet agreed on whether illite and smectite are members in a solid solution series, separate phases, or even a whole series of phases, or whether this is one mineral behaving differently when in different particle sizes. (Recall our suggested definition from the section above on illite, p. 151.)

Early workers recognized that there is a change in mineralogy with increasing depth in Gulf Coast sediments (e.g., Burst, 1959; Powers, 1967) and these changes have since been recognized in basins throughout the world (Fig. 5.14). Just how the transition from smectite to illite/smectite to illite takes place is another problem on which there is yet no general agreement. The classic work on this problem is that of Hower and co-workers on Gulf Coast sediments (Perry and Hower, 1970; Reynolds and Hower, 1970; Hower et al., 1976). Hower et al. argued that illite forms from smectite under diagenetic conditions according to the reaction:

$$\text{K-feldspar} + \text{smectite} \rightarrow \text{illite} + \text{chlorite} + \text{quartz}$$

Keeping in mind that K is needed to change smectite into illite, they noted

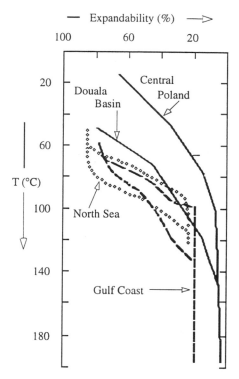

Fig. 5.14. The relation between depth and percent expandability in illite/smectite in shales from different sedimentary basins. The Douala Basin is in North Africa. (Modified from Środoń and Eberl, 1984.)

that while the amount of K_2O in the whole rock chemistry of the Gulf Coast sediments does not change with depth, the amount of K_2O in the >2 μm fraction decreased with depth as it increased in the <2 μm fraction. They concluded that the K was coming from dissolution of K-feldspar or other K-bearing minerals (Fig. 5.15). They also observed that for less than 50% illite layers, the two layers were apparently always stacked in random order. At some value, usually 55 to 65% illite, and usually at a depth of about 10,000 ft, stacking changed to nearest-neighbor ordering, or $R1$ (Box 5.3). At some value >75% illite layers, $R3$ ordering was observed, i.e., there were at least three illite layers between any two smectite layers and no smectite layers directly succeeded one another in the stacking sequence. Illite/smectite in the Gulf Coast sediments seldom exceeded 80% illite layers. Środoń and Eberl (1984) suggested that the Gulf Coast sediments stopped changing because detrital K-bearing minerals that had been the source of K had been consumed. They based their suggestion on some K-rich sediments from Poland that continue to change to about 10% smectite (Fig. 5.14). (As an aside, the depth

Box 5.3.   Reichweite or Ordering

We use a terminology for the stacking order based on the term *Reichweite*, *R* ("the reach back"), i.e., terminology for expressing the probability, given layer A, of finding the next layer to be B. This scheme, originally suggested by Jadgozinski (1949), is used in this book. Another way to think of the Reichweite is to ask how much influence does A have on what the next layer will be or on the next layer after that, i.e., what is the reach back of A? Flipping a coin is $R = 0$, i.e., there is no influence at all of one flip on another. The probability of getting a head depends only on the proportion of heads and tails, in this case 0.5. For perfectly ordered 50/50 illite/smectite (I/S), given an S or an I, the other must be next in line. Smectite, then, has a reach of one unit along the line of stacking or $R = 1$. (You could state this in terms of illite, also: Illite has a reach of one unit.) In a two-component system, containing layers of types A and B, the respective fractions of A and B are P_A and P_B so that

$$P_A + P_B = 1.$$

The probabilities of one type of layer occurring next to another are given by junction probabilities (Reynolds, 1980), e.g.,

$$P_{A.B} = 1 \text{ and } P_{A.A} = 0$$

for a 50/50 proportion of the two layer types for $R1$ ordering. If P_A and one of the junction probabilities ($P_{A.B}$ or $P_{A.A}$) are fixed, then the others can be calculated. For any two-component mixture, the relation $P_A + P_B = 1$ is read as: The sum of the probability of finding an A layer plus the probability of finding a B layer is 1. The relation $P_{A.B} = 1$ and $P_{A.A} = 0$ indicates perfect ordering or that the probability of finding an A layer followed by a B layer is 1 and that of an A layer following an A layer is zero. A perfectly ordered material with layers ABABABAB . . . is not a mixed-layered mineral, but a new mineral with a cell that can be defined as AB and whose *c*-axis dimension is the sum of the *c*-axis dimensions of A and B. These are the minerals in Table 5.2.

It is important to see that disorder occurs (some degree of randomness) even if R = 1 for compositions that are not 50/50 proportions of A and B. Suppose that $P_B = 0.7$ and R = 1, i.e., $P_{A.B} = 1$. This tells you that an A can never follow an A. Why? Consider the equation

$$P_{A.B} + P_{A.A} = 1$$

which is a mathematical way of saying that something must follow an A and the only possibilities are A and B. If you set $P_{A.B} = 1$, then $P_{A.A}$ must be zero. But, and this is the important part, R1 ordering does not mean that a B can never follow a B. This must be so because 70% of the layers are B so B layers must follow B layers—there are simply not enough A layers to sandwich between all B layers. R1 ordering for this case means that all A layers are separated by *at least one* B layer. Many A layers are separated by more than 1 B layer. In this 0.3/0.7 case, 40% of the B layers are still available to be randomly distributed after each A has been matched with a B layer. So you see that this condition is ordered with respect to one component but random with respect to the other. Order for both components occurs at exactly the 50/50 composition for R = 1.

Some minerals have patterns of ordering represented by the sequence ISII. This ordering is $R = 3$, meaning that, in this case, the first I has influence that extends or reaches three positions to the last I. Another way to think of this pattern is that each smectite layer is surrounded by at least three illite layers on each side. Current speculation differs on whether $R = 2$ ordering has been identified and is possible. Some workers believe $R = 2$ ordering is possible and that they have seen it in experimental diffraction tracings (Bethke et al., 1986; Bethke and Altaner, 1986); others suggest that $R = 2$ is not a probable ordering arrangement and that the experimental patterns are better explained as mixtures of $R = 1$ and 3 ordering (Šrodoń and Eberl, 1984). See Reynolds (1980, pp. 252ff) for detailed development of the probabilities of ordering. Bethke et al. (1986) and Bethke and Altaner (1986) extend the use of these probabilities to an analysis of smectite illitization. To describe structures by means of the Reichweite nomenclature, we use $R1$ to mean $R = 1$, $R3$ for $R = 3$, etc.

beyond which the percentage of illite layers showed no further increase coincided with the point at which no further increases of chlorite were noted. This is consistent with the conclusion of Hower et al. (1976) that Fe and Mg released by smectite were critical to chlorite formation.)

Not everyone has agreed that the reaction suggested by Hower et al. (1976) is the correct one, or at least, the only one. For example, Boles and Franks (1979) and Pollastro (1985) suggested that Al necessary in the illitization of smectite is derived from the dissolution or "cannibalization" of smectite rather than dissolution of K-feldspar. They cited as evidence the decreasing amount of total illite/smectite with depth for their sets of samples.

There are examples from all over the world of the smectite-to-illite transition corresponding to burial depth. The individual transitions vary both in descriptive details and in their interpretations, but in general, they seem to fit the scheme described by Hower et al. (1976). In the Niger Delta, the transition seems to have gone closer to completion at somewhat lower temperatures (Velde et al., 1986). The transition in Illinois Basin sediments has gone to about 90% illite, even though inferred temperatures were no greater than 80°C (Moore, 1982). There is an interesting exception off the east

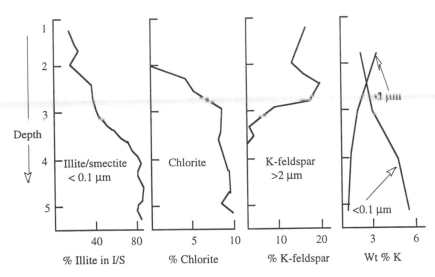

Fig. 5.15. Depth-dependent changes (depth given in thousands of meters) in the composition of illite/smectite, percent chlorite, percent K-feldspar in the >2μm fraction, and weight percent K in the greater than 2 μm and the less than 0.1 μm fractions from a Gulf Coast well (from Hower, 1981).

coast of Brazil. The smectite in the sediments is nontronite and does not change with depth (Anjos, 1986); why it does not is unclear.

The smectite-to-illite transition has been identified in three other geological situations and has been at least partly reproduced in the laboratory as well. Smectite-bearing shales intruded by dikes yield a smectite-to-illite transition series along a line perpendicular to the tabular surface of the dike. In a dike-intruded shale system investigated by Lynch and Reynolds (1985), the total amount of illite/smectite increased as the dike was approached, whereas in deep burial the total amount of illite/smectite seems to decrease with depth. Geothermal areas exhibit the transition series, but they are developed farther than can be found in burial diagenesis. And some hydrothermally altered zones around ore bodies yield a more complex but similar series (Horton, 1985; Vergo, 1984).

Three of the most apparent variables driving the smectite-to-illite transition, called by some the illitization of smectite, are the time involved, the temperature, and the availability of K. As with all chemical reactions, time and temperature are inversely related. Temperature, or perhaps better, heat energy, to stimulate this transition, can come from three sources: (1) burial and the geothermal gradient; (2) hydrothermal fluids; or (3) intrusion of an igneous body. Permeability also seems to make a difference in the sense that the smectite to illite transition appears to have gone farther in more permeable rocks in situations where there is the opportunity to compare rocks that are more and less permeable. Early studies like those of the Gulf Coast sediments

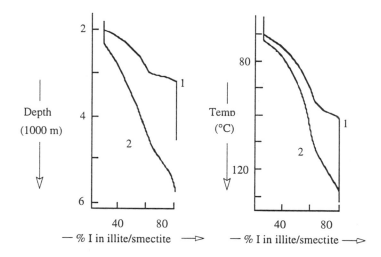

Fig. 5.16. Comparison of the depth and temperature influence on illite/smectite composition for two shale diagenetic sequences. Sequence 1 is largely from the Eocene, sequence 2 from Plio-Pleistocene (Hower, 1981)

cited above, attributed to the geothermal gradient the energy driving the transition.

Time is also an important factor in the diagenetic change from smectite to illite. Samples of sediments of Miocene age containing smectite but buried less than 200 m contain no illite (Środoń and Eberl, 1984). Figure 5.16 shows change in percent expandability (or smectite) as a function of depth and as a function of temperature for two sequences. In each case it is obvious that the younger rocks have not changed as much as the older ones, which suggests that the smectite-to-illite transition is kinetically controlled by some function of time and temperature rather than by temperature alone. Pytte and Reynolds (1989) and Altaner (1986) offered evidence that a sixth-order rate equation best explains the transition. Huang et al. (1993) experimentally quantified these kinetics and concluded that a simple rate law that is second order with respect to the fraction of smectite in I/S and first order with respect to K^+ concentration, best explains the smectite-to-illite conversion.

However, there are those who argue that at least some transitions happen abruptly, or, as the proponents have termed it, punctuated diagenesis (Morton, 1985; Ohr et al., 1991; Freed and Peacor, 1992). In punctuated or instantaneous diagenesis, all smectite is illitized at the same time, the degree to which the sediment is illitized being a function of depth, and therefore, temperature. Awwiller (1994), working with samples from the same age sediments from the same area in the Gulf Coast of Texas as those of Ohr et al. and Freed and Peacor, argued that his data do not show that diagenesis was instantaneous. In addition, although Morton's data from the <0.06 µm fraction apparently show no variation in time of diagenesis, those for Morton's <0.1 μm data do. With careful workers on both sides of a question such as this, it is obvious that additional work and insight are needed. Awwiller suggested that perhaps the only difference is semantics, that perhaps instantaneous transition and one that takes place in rapidly buried sediments may be indistinguishable on a geologic time scale.

An example that clearly shows the importance of the availability of K was shown by Altaner (1989) from a study of a zoned bentonite bed 2.5 m thick. The amount of K in the adjacent shale beds was depleted relative to background levels, and the percentage of illitic layers in the I/S of the bentonite bed decreased from both top and bottom towards more smectite-rich I/S in the center of the bed. Altaner concluded that the transition to illite was controlled by diffusion of K into the bentonite bed. There also has been untested speculation that the smectite-to-illite transition is triggered by the release of K as K-feldspar transforms into albite, a transition clearly shown for the Frio Formation in the Texas Gulf Coast (Land and Milliken, 1981).

Weathering of illite in an acid environment produced by the oxidation of pyrite seems to drive the sequence in reverse, i.e., illite-to-smectite transition (Rimmer and Eberl, 1982; Willman et al., 1989).

Models for smectite-to-illite transition

As we said in the first edition, attempts to explain the mechanisms of this transition are inextricably bound up with notions of the nature of illite and smectite. The four models we offered no longer seem suitable. The fundamental particle model and the MacEwan crystallite model now seem reconcilable. The question, "Do you see that if we begin with the notion that these are two discrete minerals, we have already limited our conclusions?" is still valid; and the two-solid-solution model of Inoue et al. (1987), because it is primarily descriptive, still seems valid. Clay mineralogists have reduced the kinds of mechanisms they envision for this transition to three, a solid-state transformation, neoformation, and an in-between one proposed by Altaner and Ylagan (1993). For this latter mechanism, the region surrounding a smectitic interlayer dissolves and reforms as an illitic interlayer without disturbing the remaining part of the structure. There may also be a fourth mechanism in which I/S simply precipitates directly with an illite percentage that represents the conditions at the time of formation, an I/S that has not gone through any changes. Some of the old questions have been answered, some remain, and new ones have been asked. Perhaps it would be worth your while to find a copy of the first edition and read about the four models for perspective.

MacEwan crystallite model   The MacEwan crystallite model is the one assumed for Reynolds's (1985) modeling scheme (see the Appendix and the exercise at the end of Chapter 3). In this model, layers of illite and smectite are pictured as intimately interlayered, stacked either randomly or regularly, into a fixed sequence. This sequence, acting as a coherent unit, scatters X-rays. This model imagines the repeat distance along Z^*, the same as other models, but the division between layers is through the center of the octahedral sheet, whereas the fundamental particle model divides them at the interlayer space (Fig. 5.17). In this model, the transition from a smectite to an illite layer is made by remodeling the chemistry and structure in place without disrupting the stack of layers, a solid-state transformation. Indications of solid-state transformation are a maintenance of morphology and polytype as the proportion of illite increases. It isn't clear whether particles undergoing solid-state transformation can increase the degree to which they are ordered, e.g., whether they can go from *1Md* to *1M* within the mechanism we describe as solid-state transformation.

Fundamental particle model   Nadeau et al. (1984) showed convincing evidence that they can measure individual sheet-shaped particles with the thickness of the sheets approximately integral multiples of 10 Å. This is pure smectite, and they call each 10 Å particle a fundamental particle. The fundamental particle shows the behavior we call expandable because it can adsorb water or organic molecules to both surfaces. Illite particles must be at least 20 Å thick and are composed of two 10 Å units held together by interlayer K. If two particles are joined by a (Mg, Fe) hydroxide sheet in the interlayer space, they become a chlorite particle with a thickness of about 24 Å. The surface of this particle will also adsorb water or organic molecules. An I/S mineral in this model that has illite particles that have grown to 30, 40, 50 Å, or more, would, in the MacEwan crystallite model, have 33%, 25%, and 20% smectite

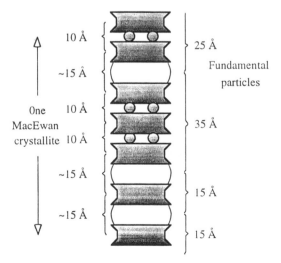

Fig. 5.17. Mixed-layered illite/smectite according to the MacEwan crystallite model (left) and the fundamental particle model (right). Expandable surfaces have two waterlayers. (Used with permission and modified from Altaner and Bethke, 1988.)

layers, respectively. Only the (001) surfaces of these packets of K-connected layers can adsorb water or organic molecules. This model suggests that the particles are not grown together epitaxially. Rather, they are simply stacked together into a group of particles that diffract X-rays. You might picture them as if they were a pile of wet playing cards, i.e., in close enough physical contact to act as a diffracting unit when interacting with X-rays.

Nadeau et al. have made physical mixtures of 10 Å particles that have yielded X-ray diffraction tracings exactly like those of materials that the MacEwan model would have taken to be fixed sequences of large (perhaps $N > 20$) numbers of layers. They referred to diffraction from their small ($N = 2$, 3, or 4), independent but stacked particles as *interparticle diffraction*. Formation of fundamental particles by direct precipitation rather than a rearrangement of the structure of precursor minerals is a tenet of this model, a dissolution-crystallization model. Indications of the dissolution-crystallization model are changes in morphology and polytype as the proportion of illite increases. Šrodoń et al. (1992) offered a more recent definition of a fundamental particle: "there are no smectite interlayers in a fundamental particle, but . . . its planar surfaces are smectitic, i.e., there is one smectite interlayer per fundamental particle (each surface counts as 1/2 of the smectite interlayer) and (N - 1) illite interlayers, where N is the total number of silicate layers in the particle."

Eberl et al. (1987, 1990) have argued for the validity of the fundamental particle model on the basis of the distribution of particle sizes within single samples. They suggested that fundamental particles grow by Ostwald ripening and that the shape of the distribution curves is another tool for deciphering crystallization history.

Figure 5.18 indicates that MacEwan and fundamental particle models can be reconciled. In Fig. 5.18A, there are fundamental particles with different numbers of layers, $N = 3, 6, 4, 7$, etc. If the tetrahedral sheets of these packets that face the outside world have layer charges that are smectite-like, they will be able to adsorb water or polar organic molecules such as ethylene glycol. A packet with $N = 12$ would have 11 illite layers and 1 smectite layer, i.e., it would appear from an XRD tracing to be I/S with $(1/12)X100 = \sim 8\%$ smectite; the packet with $N = 6$ would be $\sim 17\%$ smectite. These packets are quite MacEwan-like. The XRD beam would, of course, average the N of 10^6 to 10^9 packets so the XRD tracing would represent an averaging of the packet sizes.

Two-solid-solution model   We use this model to show that both the MacEwan model with solid-state transformation and the fundamental particle model with dissolution and crystallization are needed to explain a carefully recorded transition. In hydrothermally altered silicic volcanic glass, Inoue et al. (1987, 1988) identified three phases, each with a distinct morphology: two are solid solutions, and the third shows little variation in chemical composition. Of the two solid solutions present, one varies from 100% to 50%

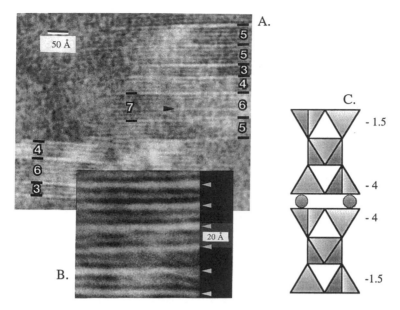

Fig. 5.18. Two TEM images and a 2:1 structure. A. Image showing variable spacings between smectitic surfaces. The digits, i.e., 5, 4, 7, etc. each represent the number of layers in a fundamental particle. Note that one changes from 7 layers to 6 layers. B. This shows *R*1 ordering of 20 Å fundamental particles. The dark parts or fringes are the expandable interlayer spaces. C. Our cartoon of one 20 Å fundamental particle with the layer charge values per tetrahedral sheet shown on the right. Photos used with permission of David R. Veblen.

expandable layers with random stacking (*R*0). The other varies from about 50% expandable layers, regularly to partially ordered, to pure illite. They characterized the first solid solution as smectite undergoing K-fixation, the second, as maturing illite. The third phase present has less than 5% expandable layers.

The first phase has a corn flake-like morphology and is found in samples that yield X-ray tracings showing the turbostratically stacked clay minerals to be between pure smectite and 50/50 I/S. The second has a lath-like morphology and is a *1M* polytype. The lath phase has from nearly 50% to as little as 5% expandable layers. The third phase takes the form of hexagonal plates of a *2M₁* polytype for samples that consist of almost pure illite. As soon as illite can be clearly detected by X-ray diffraction, a few tiny laths, perhaps 20 Å thick and 300-400 Å wide, may be seen among the smectite flakes. Moving toward the source of heat, the laths increase in thickness, length, and number as smectite flakes decrease in number. X-ray diffraction confirms that the percent of expandable layers decreases as the laths increase in size and the number of flakes decreases (Fig. 5.19). It seems easy to infer, as they have

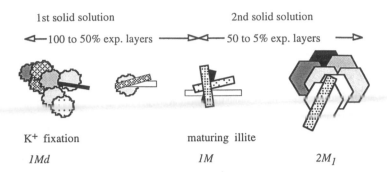

Fig. 5.19. Morphologies in the Inoue sequence from the cornflake shaped *1Md* smectite-rich I/S, to the *1M* laths of *R*1 I/S, to the hexagonal plates of I/S with <10% smectite.

done, that the illite laths are increasing in size by the addition of material from solution, and that the material is coming from the smectite flakes that are decreasing in number because they are dissolving. Supporting this inference are chemical analyses that show coexisting flakes and laths to have the same interlayer K content. Flakes are seldom visible for samples showing more than 50% illite. Inoue and coworkers proposed two different reactions for these two series. K-fixation takes place as a remodeling or solid-state transformation in the flakes as solids, whereas the laths grow because the flakes are dissolving and being reprecipitated as laths or as additions to laths, a dissolution-crystallization mechanism. In contrast to studies of the transitions in the chlorite/smectite system, Inoue et al. (1987, 1988) encountered no rectorite (Table 5.2). This is an important and illuminating example. The problem is that it is not so readily transferable to sandstones, shales, or soils. Perhaps to do so, we would be comparing apples and oranges.

The MacEwan model does not lend itself well to thermodynamic modeling because, although it sees two units with distinctly different compositions, it is not absolutely clear that they are separate phases because of the location of the boundaries within the crystallites (Fig. 5.17). The fundamental particle model sees the disordered part of the illite/smectite series as a mixture of two phases, but sees the ordered part of the series as a single phase. The two-solid-solution model compromises the MacEwan and fundamental particle models by having two physically separate phases, each of which is a solid solution. ^{29}Si-NMR (nuclear magnetic resonance) data convincingly show that different NMR signals are received from Si atoms bordering the interlayer space in smectite compared to those from Si atoms next to the interlayer space in illite because tetrahedral sites in illite have fewer Si and more Al atoms in them (Altaner et al., 1988).

It seems clear that, however many phases there are and by whatever mechanism they form, they are metastable in most of the circumstances we are likely to consider in soils and sedimentary rocks. It seems only somewhat less clear that there are different mechanisms. The differences, were we to have to put our money on the source of differences, seem most closely related to the permeability of the medium in which the smectite-to-illite transition takes place. Another way of saying the same thing is that the water-to-rock ratio appears to be critical in determining how transitions happen. There has been a great deal of discussion about whether I/S is one phase (Ransom and Helgeson, 1994), or two phases (Garrels, 1984), or several (Rosenberg et al., 1990). We will leave this issue to future considerations. Answers may suggest themselves when we better understand this (these) mineral(s). This applies to all the mixed-layered clay minerals. Let us leave our comparison of these models, and the question of the number of phases, with the thought that the insights from each and the disagreements among people thinking about them will encourage further work by stimulating new questions. Perhaps you will think of the question, the answer to which will resolve some of this dilemma.

Chlorite/smectite (C/S)

In the I/S sequence, the entire span of compositions from smectite to illite has been the focus of attention. Rectorite has never yet been found in association with this sequence. The reverse seems to be the case for C/S. Until recently, corrensite has received most of the attention paid to the C/S sequence, and not much attention has been paid to compositions on either side of the R1 50/50 mineral. Corrensite was named for C. W. Correns (1893-1980) by his former student at Göttingen University, Friedrich Lippmann (1956). It occurs in a great variety of geological situations: contact metamorphic zones of shales, old carbonate sequences, Lake Superior iron ores, hydrothermal alteration and weathering products of ophiolitic rocks and dolomites, and sediments containing sufficient Mg subjected to burial diagenesis and to weathering. It is most consistently associated with terrestrial and marine saline deposits younger than approximately Ordovician. The occurrence of dioctahedral chlorite/smectite, tosudite, is much more limited. Its occurrence seems to be restricted primarily to hydrothermally altered intermediate to acidic igneous rocks (Sudo et al. 1954).

   More recently, physical mixtures that are apparently discrete chlorite, either *R*1 or *R*0 C/S, and corrensite have been tentatively recognized (Moore et al., 1989; Shau et al., 1990; Schiffman and Fridlesisson, 1991; Shau and Peacor, 1992). This combination has been causing people to pull their hair, especially, as is often the case, if I/S is also present. Furthermore, there are those who contend this sequence is discontinuous, that there is no complete series from unmixed chlorite to unmixed smectite, but only chlorite with small amounts of smectite and smectite with small amounts of chlorite. Corensite seems to be a single phase that occurs alone or in physical admixture, or corrensite may be interstratified with additional chlorite, but apparently never with smectite (Shau and Peacor, 1992). Yet another complication is that

Hillier (1995) has claimed, based on modeled XRD tracings, that the $R0$ varieties of C/S minerals at each end of the series cannot be distinguished from $R1$ varieties, nor can a physical mixture of several $R1$ chlorite/smectites be distinguished from an $R0$ chlorite/corrensite. This brings to mind the cautions about modeling suggested by Oreskes et al. (1994) to the effect that modeling can never prove what happens in nature, only suggest possibilities. There will be a fair amount of head scratching before this dilemma is sorted out to everyone's satisfaction. Perhaps it will help to put the digitally recorded intensity versus 2θ data in a spreadsheet and subtract a pattern for pure chlorite, adjusting the pattern for intensities appropriate for Fe/Mg content, or subtract a corrensite pattern, or both. One redeeming feature of these messy combinations of chlorite, C/S, corrensite, and I/S is that they seem to be restricted to weathering debris from basic and ultrabasic rocks.

The contradictions noted above may be, in part, due to insufficiently sensitive XRD analytical procedures. Ethylene glycol solvated and air-dried preparations yield little information on mixtures versus mixed layering in the C/S series because of severe peak interferences. Tracings from heated preparations show promise of definitive analysis. However, we are aware that some, perhaps all, C/S rehydrates very rapidly, e.g., between the furnace and the sample holder of the X-ray machine. We will take up the problems of identifying the components of these mixtures in Chapter 8.

Recall our speculation about illite as having a different layer charge on the tetrahedral sheet that faces the outside world. Perhaps this is the case for chlorite as well. If it is, 2:1 layers with enough layer charge to hold a hydroxyl sheet in the interlayer space are stacked until there is a tetrahedral sheet with low charge, one that can expand. Then when N, the number of layers in the stack, is small enough, the mineral will behave like C/S. The trend then, as we have previously pointed out, is for N to increase with time, with increasing temperature, or with increasing tectonic stress. This too-simple view is where we begin as we try to use what we understand. It sheds no light on why the C/S sequence is discontinuous, if this is truly the case. As mentioned in the section on chlorite, de Caritat et al. (1993) and Jiang et al. (1994) concluded that chlorite geothermometry should be used cautiously and only with other corroborating methods of estimating paleotemperatures. We suggest the same approach to using a chlorite crystallinity index. Crystallinity, and less clearly geothermometry, are related to the increase of N in C/S with time and increasing temperature and tectonic stress. An additional complicating factor when using chlorite crystallinity is that the absence of interlayered 7 Å material needs to be established. Because, like the illite crystallinity index, that of chlorite is based on peak width, the presence of 7 Å layers broadens the odd-ordered peaks in the $00l$ series (see the next section). Reynolds (1988) has offered a review of the nature, occurrence, and identification of mixed-layer chlorite minerals.

Serpentine/chlorite
The character of the 1:1 layer interstratified with chlorite has not yet been agreed upon. Berthierine, kaolinite, and odinite have been suggested as candidates. Because each of these minerals represents different circumstances of formation, we will use the broader term serpentine, but the nature of this mineral, and therefore its usefulness as an interpretive tool, will not be resolved until it is identified (or perhaps it can be any of these candidates, but each for different geological circumstances).

Serpentine/chlorite is an interesting example of a mineral that is only recently beginning to be commonly reported in sedimentary rocks. Dean (1983) observed odd-order line broadening in "chlorites" that he interpreted (correctly, we believe) as evidence of interstratification of 7 and 14 Å layer types. Although in studies of chlorites synthesized in the laboratory a 7 Å phase often is reported to precede chlorite, only recently has a trioctahedral 7 Å layer been observed interstratified with chlorite in nature. Ahn and Peacor (1985) reviewed these and published lattice images, obtained by high-resolution transmission electron microscopy, of chlorite from Tertiary Gulf Coast shales that contains some 7 Å layers. Walker and Thompson (1990) found it in the Salton Sea geothermal wells. Moore and Hughes (1990) reported that it is common as grain coatings in sandstone reservoirs of Carboniferous age in the Illinois Basin, and in 91.5% of 250 samples from 9 other sandstone reservoirs from 8 petroleum provinces. Reynolds et al. (1992) gave a detailed description of serpentine/chlorite from the Tuscaloosa Formation and described a method for quantifying the proportions of serpentine layers on the basis of line-broadening studies (see section on serpentine/chlorite, Chap. 8, p. 289).

Lahann et al. (1992) identified the 1:1 layer in this mineral, from a sandstone reservoir in off-shore Norway, as odinite on the basis of texture, the associated siderite and phosphates, and the environment of deposition (see p. 139 on odinite). They detected a slight decrease in the amount of 7 Å layers with increasing depth. Ryan and Reynolds (1996) reported quite convincing evidence of a decrease of 7 Å layers with increasing depth. The proportion of the 7 Å component of this grain-coating mineral in the sandstone of the Lower Tuscaloosa Formation at shallow burial depths is 16%. It decreases monotonically to 3% at 5,470 m. (The polytype of the serpentine/chlorite in the Ryan and Reynolds study also showed a regular change with depth from IbO near the surface to Ia at depth though an interstratified series of IbO/Ia.) As so often happens in any discipline, once practitioners became aware of some previously unrecognized aspect, it is spotted more and more frequently and in more variety, e.g., Banfield et al. (1994) described specimens from a skarn in the Ertsberg mining district of Irian Jaya in which serpentine and chlorite are regularly and randomly stacked, and the patterns of ordered stacking include repeat distances of 21, 28, 35, 42, and 49 Å! (How can this be? What are these stacking arrangements?) Bailey et al. (1995) have had the mineral dozyite, a 1:1 regular interstratification of serpentine and chlorite, accepted as a valid species. It is included in Table 5.2.

Most occurrences of serpentine/chlorite in sedimentary rocks are high-Fe compositions, and for these a good guess is that the 7 Å layer is berthierine, though the chemical data are not good enough to establish this as fact. Brindley and Gillery (1953) described a kaolinite/chlorite, based on a combination of chemical and X-ray data. Hillier and Velde (1992) reported an occurrence from the subsurface of offshore Norway and interpreted their chemical data as evidence that some or all of the 7 Å layers are kaolinite. They may have a case because their relatively Al-rich mineral contains 15% 7 Å layers. Perhaps what we are calling serpentine/chlorite should be called 7 Å/chlorite; perhaps both kaolinite/chlorite and serpentine/chlorite exist in nature. Until more definitive studies emerge, we will continue to use the term serpentine/chlorite.

Based on this catalog of observations and conjectures, where are we? Again, there seems to be a difference between the behavior of minerals in sandstones compared to that in shales. Variation with depth of the amount of 7 Å layers in this mineral seems real, at least for sandstones. This leads to the reasonable conclusion that there is a progression from a 7 Å mineral, probably berthierine, but perhaps also odinite, to discrete chlorite. Now we need to learn how to apply this as an interpretive tool.

Kaolinite/expandables (K/E)

Hughes et al. (1993) reported that this mineral has been identified in soils and paleosols from Holocene to Pennsylvanian in age (and perhaps older) and in most fireclays, ball clays, and other poorly crystallized kaolinites. Apparently, it is always a product of weathering. It is a mixed-layered mineral composed of an expandable 2:1 layer and a kaolinitic 1:1 (7 Å) layer. Because the 2:1 layer in this mineral did not seem to precisely match the characteristics of smectite, they used expandable instead of smectite. Brindley et al. (1983) described 18 kaolinite/smectites with a range of proportions that make up about 75% of the Argiles Plastiques Formation in the Paris Basin, a unit long used as an industrial raw material. They concluded that these clay minerals were continental deposits, inherited from erosion of weathering profiles developed in Paleocene and Late Cretaceous time. Hughes et al. (1993) suggested that because the proportion of kaolinite in K/E seems to be related to the intensity or the duration, or both, of weathering, K/E from paleosols shows promise of helping reconstruct past climates. See their paper for a general review of this mineral.

THE ORIGIN OF CLAY MINERALS

As it has become possible to look at the surfaces of minerals on an atomic scale with TEM and AFM, a number of answers to questions become almost self-evident, e.g., the transition of muscovite into kaolinite, and kaolinite into halloysite, as shown in Fig. 5.3, p. 142. Evidence accumulates in favor of a blend of transformation and neoformation as mechanisms for the formation of clay minerals. The transformation part of this blend is covered by the term "inherited structure." We have in mind reports such as those of Banfield and Barker (1994), Jiang et al., (1995), Banfield et al., (1995), and Smith et al. (1987) in which they observed various clay minerals and hydrated oxides attached to primary minerals, the secondary minerals and the primary minerals having one, often two, and sometimes three, crystallographic directions in common. Banfield et al. referred to these secondary minerals as topotactically oriented reaction products. In the case of olivine (Smith et al., 1987) the silica tetrahedra have had to rearrange their bonding arrangements and their Si:O ratio; in the case of amphiboles reported by Banfield and Barker, x^* of the amphibole and z^* of smectite are parallel, and the y^* of both coincide. An additional conclusion of these studies is that there is a striking difference between natural weathering rates and those measured in the laboratory. The difference in rates may be that in natural settings the rates are slower because mineral surfaces are wetted episodically, whereas in laboratory experiments the primary minerals are in constant contact with liquid through which diffusion can take place (Banfield and Barker, 1994; Banfield et al., 1995). Reaction products are not restricted to a single mineral. For example, Smith et al. (1987) observed goethite, Fe-rich trioctahedral smectite (saponite), dioctahedral smectite, and halloysite forming from the weathering of an olivine, Fo_{80}. Furthermore, all these minerals except halloysite are crystallographically orientation in relation to their host, the olivine (Fig. 5.20).

Clay minerals, most notably nontronite, are actively being formed at deep-sea hydrothermal vent systems along midocean ridges in association with sulfides, carbonates, and amorphous hydrated oxides. These systems are numerous enough that they are major components in determining the chemical balance of the oceans, and although we are unaware of any estimates of the quantities of clay minerals formed in these systems, they must be a significant contribution to the total body of marine sediments. It has been generally held that clay minerals in deep-sea sediments are entirely pelagic, mainly pedogenically formed terrigenous minerals transported to the ocean basins unchanged. Clay minerals are distributed in the ocean basins roughly according to latitude with chlorite and illite more abundant in higher latitudes, smectites and kaolins more abundant in the lower latitudes. Recent studies by Clauer et al. (1990) indicate that we need to rethink the generalization that the deep-sea clay minerals are strictly detrital. Clauer et al. presented evidence that at least smectite transported to the ocean basins is diagenetically altered before it is buried. In addition, it is clear that volcanic sediments are an exception because there are numerous studies showing rapid alteration of glass and volcanic minerals, usually to smectite, but to zeolites in some cases.

In terms of the soil environment, primary minerals often are represented

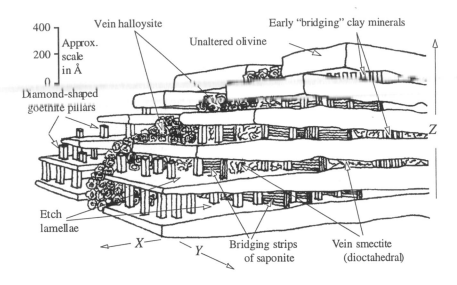

Fig. 5.20. Relation between various stages of the alteration of olivine as host and goethite, trioctahedral smectite (saponite), dioctahedral smectite, and halloysite as guest minerals. The diagram is laterally compressed (from Smith et al., 1987 with permission).

only by ghosts, their shapes the only indication that they once existed. They are reborn most often as clay minerals. Jackson (1963) stated that the endpoint of the various weathering reactions in soils is kaolinite. These reactions often don't reach their end point, but vast deposits of kaolinite in equatorial regions testify to the soundness of Jackson's assertion. Smectite is one of the most commonly formed secondary minerals, based on work such as that mentioned in the first paragraph in this section. Środoń (1980) concluded that there are two possible paths for the alteration of smectite: the one at surface temperatures and pressures is towards kaolinite, and the other path is along the geothermal gradient as smectite is buried and converted to illite. On the former path, layer charge decreases; along the latter, it increases. The path at surface conditions has two branches: one goes directly from smectite to kaolinite; the other passes through a mixed-layered stage. Hughes et al. (1993) have looked with particular care at this latter path. It is primarily confined to pedogenic processes, and its chief characteristic is the formation of K/E. This generally neglected mineral (neglected because it is easy to overlook, Chapter 8, p. 284) seems most abundant in the B horizon of soils, where aluminous-

and iron-rich 2:1 clay minerals are abundant in the parent material. K/E seems most likely to form in climates that are humid, temperate or tropical, and where silica-extracting plants are abundant. From the same parent material, K/E has been observed where the soil was drained, while in low spots with stagnant water, smectite was present. K/E has been found in soils and paleosols from recent to Pennsylvanian in age. We strongly suspect that locked in K/E and the associated minerals in paleosols are characterizing features of climate and rates of soil formation.

SUMMARY

Clay minerals formed in place represent a response to the chemical and physical conditions under which they were formed or altered, just like mineral assemblages in any petrologic study. Clay minerals usually do not occur as monomineralic deposits, a point that is partly obscured by the way in which we have presented the material in this chapter, but which is illustrated by Fig. 5.20. Detrital assemblages reflect the sources from which they have come. A major part of the interpretation of mineralogy of the clay-sized fraction of sedimentary rocks is distinguishing the diagenetic from the detrital parts of the assemblage. The clay minerals in this chapter have been considered separately for convenience. Please don't let this mislead you. Because we haven't discussed them as assemblages doesn't mean the topic isn't important; it means that it is outside our competence.

Our understanding of clay minerals is not complete. You, as well as we, need to keep working on our understanding of these materials that make up a large part of our immediate world. In spite of our incomplete understanding, we do use what we know to help solve problems, and you can too. In summary, the following are some of the more intriguing topics that you might consider pursuing:

1. Relatively small differences in structural or chemical characteristics are all that distinguish one clay mineral from another.
2. It seems as if every clay mineral can transform, one way or another, to any other clay mineral.
3. The degree to which we characterize a clay mineral and differentiate it from others is often closely related to how it has been analyzed, and to our preconceptions about the nature of these minerals.
4. It seems possible that there are different layer charges on tetrahedral sheets that face the outside world compared to those on both sides of fixed interlayer cations.
5. Perhaps what is clearest in historical perspective is that frequently, after adequate understanding has been won, both sides of a debate have contributed some insights and both have had some blind spots.

Let us leave you with this reminder before we go on to the chapters that discuss sample preparation and the interpretation of XRD results: It is important for you to understand that you now understand enough about clay minerals to begin applying them as interpretive tools, in, for example, working out the geologic history of a basin, or in the evaluation of a petroleum reservoir, or to see the relation between their properties and an engineering problem, or to see how one of their properties might be modified to make a more useful industrial or pharmaceutical product.

EXERCISE: CALCULATING STRUCTURAL FORMULAS

The following is a procedure for converting a chemical analysis of a clay mineral into a structural formula. This process will allow you to do the bookkeeping necessary to assign the source of layer charge to the tetrahedral sheets or to the octahedral sheet and to determine the total number of octahedral cations. Ross and Hendricks (1945) devised this procedure, and it has since been modified by many. This particular version is from John Hower. Water in the formula is not considered, therefore, the procedure for a 2:1 layer silicate is based on a formula unit with 11 oxygens, 8 from the tetrahedral sheets and 3 from the octahedral sheet. (See the formula at the end of this section. If you count the oxygens, there are 12. If you subtract one H_2O, you have 11 oxygens and no hydrogen in the formula unit. Using the same reasoning, the unit cell has 22 oxygens.) As a caveat, you should be aware that, even if the assumptions used in the method we suggest are valid, the procedure assumes that the analysis you are using is of a single, homogeneous phase. This is an assumption that appears OK for crystals large enough to allow direct determination of cell dimensions and density. But clay minerals are not large enough and, based on what is revealed by TEM examination, homogeneous phases are the exception rather than the rule. Clay minerals frequently contain minor impurities, perhaps adsorbed on their surfaces, perhaps interstratified with other clay minerals, and are far more often than not, structurally inhomogeneous. Warren and Ransom (1992) and Jiang et al. (1994) pointed out problems with analyses that do not recognize contamination. See Newman (1987, pp. 10ff) for a complete treatment of the topic of calculating structural formulas.

The following example (Table 5.3) is for the composition of a pink montmorillonite from an altered rhyolite tuff near Santa Rita, New Mexico, and is taken from Deere et al. (1963, Vol. 3, p. 232). Explanations for the headings of the columns are as follows (he first three should be self-evident): For equivalent weight (eq. wt.), divide the formula weight by the valence or cation charge in the oxide to obtain formula weight per unit of cation charge; e.g., for Al_2O_3, divide 101.959 by 6 to get 16.993 because there are 2 Al with 3 cation charges for each aluminum. For gram equivalence (gm. equiv.), divide weight percent by equivalent weight to obtain the atomic proportion per unit of cation charge; e.g., again, for Al_2O_3, divide 15.97 by 16.993 to get 0.940. The sum of this column is divided by 22, the valence of the 11 oxygens, to get a normalization factor, or the weight percent that will be balanced by each valence charge from the oxygens. Then divide the normalization factor, in this case 0.222, into each equivalence value to obtain the values in the sixth column, the amount of cation charge per formula unit for the cation; i.e., for the example given, there are 16.189 cation charges in the formula unit attributable to Si. This normalizes the cation charges relative to the charges on the oxygens in a formula unit containing 11 oxygens $(O_{10}(OH)_2 - 1H_2O)$. A sum of this column should give you 22 and serves as a check on your arithmetic. The value in the seventh column is obtained by dividing the value in the sixth column by the oxidation number, or valence, of the cation, e.g., 4 for Si, 3 for Al, 1 for K, etc. The value in the seventh column is the sum of the subscripts that a particular cation will have in the structural formula. Note that the values in columns 1, 3, and 4 are constant, so you may want to make a form that helps you calculate structural formulas.

Always assign cations to the tetrahedral sheet first, then to the octahedral sheet, and then to the interlayer cation positions. For our example, there are 4.047 cations of Si in the formula unit. The only place we can have Si is in the tetrahedral sheet, so it must be $Si_{4.047}O_{10}$, giving the tetrahedral sheet a +0.188 contribution toward the layer charge (0.047 x 4 = 0.188). If there had not been at least four Si cations in the unit cell, then you would have had to take enough Al to have four cations in the tetrahedral sheet. For example, if you had had 3.94 for Si and 2.05 for Al, you would assign 3.94 as the subscript for Si in the tetrahedral sheet, and then use 0.06 Al to complete the tetrahedral sheet. That would leave 1.99 Al for the octahedral sheet. We

Table 5.3. Calculation of structural formulas of layer silicates based on a formula unit of 11 oxygens (H_2O not considered)

Oxide	Wt %	Form. wt.	Eq. wt.	Charge/ gm. equiv	Cat. ch./ f. unit	No. cat./ f. unit
SiO_2	53.98	60.084	15.021	3.594	16.189	4.047
TiO_2	0.08	79.898	19.974	0.004	0.018	0.004
Al_2O_3	15.97	101.959	16.993	0.940	4.234	1.411
Fe_2O_3	0.95	159.691	26.615	0.036	0.162	0.054
FeO	0.19	71.846	35.923	0.005	0.022	0.011
MgO	4.47	40.304	20.152	0.222	1.000	0.500
CaO	2.30	56.079	28.040	0.082	0.369	0.184
Na_2O	0.13	61.979	30.990	0.004	0.018	0.018
K_2O	0.12	94.195	47.097	<u>0.002</u>	0.009	0.009
				4.889/22 = 0.222[a]		

[a]Normalized to 11 oxygens; f. unit = formula unit.

should be suspicious when there are more than 4.00 Si. It probably indicates the presence of silica-rich glass in the sample, which is not too surprising for a rhyolite tuff. This emphasizes the caveat we made in the first paragraph of this section. The remainder of the Al cations are in the octahedral sheet. As for the rest of the elements, Fe is not found as an interlayer cation and will not be found in the tetrahedral sheet if Al is present, so it is assigned to the octahedral sheet along with Mg, Ca is usually present as an interlayer cation, but you could try it both ways, one with it in the octahedral sheet and one with it as an interlayer cation, and see how the charges balance. Without Ca, the octahedral sheet has

$$(Al_{1.411},Mg_{0.5},Fe^{3+}_{0.054},Fe^{2+}_{0.011},Ti_{0.004})$$

giving the octahedral sheet a total cation charge of +5.433 to balance the -6.00 from the oxygens for the octahedral sheet, or it contributes -0.567 (5.433 - 6.00 = -0.567) to the layer charge. Now, the sum of the charges of the interlayer cations, (2 x 0.184) + 0.009 + 0.018 = 0.395, balances the layer charge (-0.567 from the octahedral sheet and +0.189 from the tetrahedral sheet = -0.38). There will be small errors due to rounding of the numbers, therefore, the balance of +0.395 from the interlayer cations against the -0.38 of the 2:1 layer is satisfactory. The final structural formula is

$$(Ca_{0.184},Na_{0.018},K_{0.009})(Al_{1.411},Mg_{0.5},Fe^{3+}_{0.054}{}^{3}Fe^{2+}_{0.011}Ti_{0.004})-$$
$$(Si_{4.047}O_{10})(OH)_2$$

Could you work one on your own now? See Newman (1987) for other clay mineral analyses you can try.

EXERCISE: MAKING STRUCTURAL MODELS OF LAYER SILICATES

Photocopy Figs. 5.21 and 5.22 onto card stock. Keep in mind that photocopying often changes the size of the image by a few percent. You may use this to your advantage. The difference in sizes can represent polyhedra with different amounts of substitution of other ions that have changed the dimensions of the polyhedra. This change in dimension will model the problem of fitting sheets together (see Figs. 4.4 and 4.5 and related text, page 108ff). To fold for gluing, place a ruler along the lines to be folded and score them with a ballpoint pen. Then fold in so that the lines are inside the polyhedron when you've finished. White glue is probably the best. To hold one polyhedron to another and to give the joints some small amount of articulation, cut 3- to 4-cm pieces of pipe cleaner with wire snips. Snip off just enough of the corner of a polyhedron so the piece of pipe cleaner can be inserted. Then it can be joined to another polyhedron, bent to the right angle, but it doesn't need to be glued until you're satisfied you have the structure the way you want it.

Toothpicks may be used to hold layers apart to represent interlayer space. Cork or Styrofoam balls may be used to represent interlayer cations. Polyhedra can be sprayed with enamel before or after assembly. Be imaginative and inventive.

Fig. 5.21. Coordination octahedra

Fig. 5.22. Coordination tetrahedra

REFERENCES

Ahn, J. H., and Peacor, D. R. (1986) Transmission and analytical electron microscopy of the smectite-to-illite transition: *Clays and Clay Minerals* **34**, 165-80.

Altaner, S. P. (1986) Comparison of rates of smectite illitization with rates of K-feldspar dissolution: *Clays and Clay Minerals* **34**, 608-11.

Altaner, S. P. (1989) Calculation of K diffusional rates in bentonite hedoi *Geochim. Cosmochim. Acta* **53**, 923-31.

Altaner, S. P., and Bethke, C. M. (1988) Interlayer order in illite/smectite: *Amer. Minerl.* **73**, 766-74.

Altaner, S. P., Weiss, C. A., and Kirkpatrick, R. J. (1988) Evaluation of structural models of mixed-layer illite/smectite: ^{29}Si NMR evidence: *Nature* **331**, 699-702.

Altaner, S. P., and Ylagan, R. F. (1993) Interlayer-by-interlayer dissolution: A new mechanism for smectite illitization: *Program with Abstracts*, 30th Annual Mtg, Clay Minerals Society, San Diego, California, p. 88.

Anjos, S. M. C. (1986) Absence of clay diagenesis in Cretaceous-Tertiary marine shales, Campos Basin, Brazil: *Clays and Clay Minerals* **34**, 424-34.

April, R. H., Hluchy, M. M., and Newton, R. M. (1986) The nature of vermiculite in Adirondack soils and till: *Clays and Clay Minerals* **34**, 549-56.

Aronson, J. L., and Douthitt, C. B. (1986) K/Ar systematics of an acid-treated illite/smectite: implications for evaluating age and crystal structure: *Clays and Clay Minerals* **34**, 473-82.

Awwiller, D. N. (1994) Geochronology and mass transfer in Gulf Coast mudrocks (south-central Texas, U.S.A.): Rb-Sr, Sm-Nd, and REE systematics: *Chem. Geology* **116**, 61-71.

Bailey, S. W. (1966) The status of clay mineral structures: *Proceedings*, 14th National Conf. on Clays and Clay Minerals, Pergamon Press, New York, 1-23.

Bailey, S. W. (1967) Crystal structures of layer silicates—I: in Layer Silicates: *Short Course Lecture Notes*, American Geological Institute, Washington, SB-1A to SB-17C.

Bailey, S. W. (1980a) Summary and recommendations of AIPEA nomenclature committee: *Clays and Clay Minerals* **28**, 73-8.

Bailey, S. W. (1980b) Structures of layer silicates: in Brindley, G.W., and Brown, G., editors, *Crystal Structures of Clay Minerals and Their X-Ray Identification*, Monograph No. **5**, Mineralogical Society, London, 1-123.

Bailey, S. W. (1982) Nomenclature for regular interstratifications: *Amer. Minerl.* **67**, 394-98.

Bailey, S. W. (1984) Classification and structures of the micas: in Bailey, S. W., editor, *Micas*, Vol. **13** *in* Reviews in Mineralogy, Mineralogical Society of America, Washington, D.C., 1-12.

Bailey, S. W. (1988a) Introduction; Polytypism of 1:1 layer silicates: in Bailey, S. W., editor, *Hydrous Phyllosilicates (exclusive of micas)*, Vol. **19** in Reviews in Mineralogy, Mineralogical Society of America, Washington, D.C., 1-8.

Bailey, S. W. (1988b) Structures and compositions of other trioctahedral 1:1 phyllosilicates: in Bailey, S. W., editor, *Hydrous Phyllosilicates (exclusive of micas)*: Vol. **19** in Reviews in Mineralogy, Mineralogical Society of America, Washington, D.C., 169-88.

Bailey, S. W. (1988c) Chlorites: Structures and crystal chemistry: in Bailey, S. W., editor, *Hydrous Phyllosilicates (exclusive of micas)*, Vol. **19** in Reviews in Mineralogy, Mineralogical Society of America, Washington, D.C., 347-403.

Bailey, S. W. (1988d) Odinite, a new dioctahedral-trioctahedral Fe^{3+}-rich clay mineral: *Clay Minerals* **23**, 237-47.

Bailey, S. W. (1989) Halloysite—a critical assessment: in Farmer, V. C., and Tardy, Y., editors, *Sciences Géologiques Mém.* **86**, Vol. II, Proceedings 9th International Clay Conference, Strasbourg, 89-98.

Bailey, S. W., Brindley, G. W., Kodama, H., and Martin, R. T. (1979) Comment: Report of the Clay Minerals Society Nomenclature Committee for 1977 and 1978: *Clays and Clay Minerals* **27**, 238-39.

Bailey, S. W., Banfield, J. F., Barker, W. W., and Katchan, G. (1995) Dozyite, a 1:1 regular interstratification of serpentine and chlorite: *Amer. Minerl.* **80**, 65-77.

Banfield, J. F., Jones, B. F., and Veblen, D. R. (1991) An AEM-TEM study of weathering and diagenesis, Abert Lake, Oregon: I. Weathering reactions in the volcanics: *Geochim. Cosmochim. Acta* **55**, 2781-93.

Banfield, J. F., Bailey, S. W., and Barker, W. W. (1994) Polysomatism, polytypism, defect microstructures, and reaction mechanisms in regularly and randomly interstratified serpentine and chlorite: *Contr. Min. Petrol.* **117**, 137-50.

Banfield, J. F., and Barker, W. W. (1994) Direct observation of reactant-product interfaces formed in natural weathering of exsolved, defective amphibole to smectite: Evidence for episodic, isovolumetric reactions involving structural inheritance: *Geochim. Cosmochim. Acta* **58**, 1419-29.

Banfield, J. F., Ferruzzi, G. G., Casey, W. H., and Westrich, H. R. (1995) HRTEM study comparing naturally and experimentally weathered pyroxenoids: *Geochim. Cosmochim. Acta* **59**, 19-31.

Bates, R. L., and Jackson, J. A. (1980) *Glossary of Geology*: 2nd ed., American Geological Institute, Falls Church, Va., 751 pp.

Berner, R. A. (1971) *Principles of Chemical Sedimentation*: McGraw-Hill, New York, 240 pp.

Bethke, C. M., and Altaner, S. P. (1986) Layer-by-layer mechanism of smectite illitization and application to a new rate law: *Clays and Clay Minerals* **34**, 136-45.

Bethke, C. M., Vergo, N., and Altaner, S. P. (1986) Pathways of smectite illitization: *Clays and Clay Minerals* **34**, 125-35.

Bodine, M. W., Jr., and Standaert, R. R. (1977) Chlorite and illite compositions from Upper Silurian rock salts, Retsof, New York: *Clays and Clay Minerals* **25**, 57-71.

Bohor, B. F., and Triplehorn, D. M. (1993) Tonsteins: Altered volcanic-ash layers in coal-bearing sequences: *Special Paper* **285**, Geological Soc. Amer., Boulder, Colorado, 44 pp.

Boles, J. R., and Franks, S. G. (1979) Clay diagenesis in Wilcox Sandstones of southwest Texas: *J. Sed. Petrol.* **49**, 55-70.

Bowen, N. L. (1928) *The Evolution of Igneous Rocks* : Princeton Univ. Press, Princeton, N.J., 334 pp.

Bradley, W. F., Grim, R. E., and Clark, G. F. (1937) A study of the behavior of montmorillonite on wetting: *Z. Kristallogr.* **97**, 260-70.

Brindley, G. W. (1980) Order-disorder in clay mineral structures: in Brindley, G. W., and Brown, G., editors, *Crystal Structures of Clay Minerals and Their X-Ray Identification*, Monograph No. 5, Mineralogical Society, London, 125-195.

Brindley, G. W. (1981) X-ray identification (with ancillary techniques) of clay minerals: in Longstaffe, F. J., editor, *Short Course in Clays for the Resource Geologist*: Mineralogical Association of Canada, Toronto, 22-38.

Brindley, G. W. (1982) Chemical composition of berthierines— a review: *Clays and Clay Minerals* **30**, 153-55.

Brindley, G. W., Suzuki, T., and Thiry, M. (1983) Interstratified kaolinite/smectites from the Paris Basin; correlations of layer proportions, chemical compositions and other data: *Bull. Mineral.* **106**, 403-10.

Bruce, C. H. (1984) Smectite dehydration—its relation to structural development and hydrocarbon accumulation in northern Gulf of Mexico Basin: *AAPG Bull.* **68**, 673-83.

Burst, J. F. (1959) Post diagenetic clay mineral-environmental relationships in the Gulf Coast Eocene in clays and clay minerals: *Clays and Clay Minerals* **6**, 327-41.

Calle, C. de la, and Suquet, H. (1988) Vermiculite: in Bailey, S. W., editor, *Hydrous Phyllosilicates (exclusive of micas)*, Vol. **19** in Reviews *in* Mineralogy, Mineralogical Society of America, Washington, D.C., 455-496.

Carman, J. H. (1974) Synthetic sodium phlogopite and its two hydrates: stabilities, properties, and mineralogic implications: *Amer. Minerl.* **59**, 261-73.

Cathelineau, M., and Nieva, D. (1985) A chlorite solid solution geothermometer. The Los Azufres (Mexico) geothermal system: *Contrib. Mineral. Petrol.* **91**, 235-44.

Chang, H. K., Mackenzie, F. T., and Schoonmaker, J. (1986) Comparisons between the diagenesis of dioctahedral and trioctahedral smectite, Brazilian offshore basins: *Clays and Clay Minerals* **34**, 407-23.

Cho, M., and Fawcett, J. J. (1986) A kinetic study of clinochlore and its high temperature equivalent forsterite-cordierite-spinel at 2 kbar water pressure: *Amer. Minerl.* **71**, 68-77.

Clauer, N., O'Neil, J. R., Bonnot-Courtois, C., and Holtzapffel, T. (1990) Morphological, chemical, and isotopic evidence for an early diagenetic evolution of detrital smectite in marine sediments: *Clays and Clay Minerals* **38**, 33-46.

Colton, V. A. (1986) Hydration states of smectite in NaCl brines at elevated pressures and temperatures: *Clays and Clay Minerals* **34**, 385-89.

Cotton, F. A., and Wilkinson, G. (1972) *Advanced Inorganic Chemistry*: Wiley, New York, 1145 pp.

Cradwick, P. D. G., Farmer, V. C., Russell, J. D., Masson, C. R., Wada, K., and Yoshinaga, N. (1972) Imogolite, a hydrated aluminum silicate of tubular structure: *Nature Phys. Sci.* **240**, 187-89.

Daniels, E. J. and Altaner, S. P. (1990) Clay mineral authigenesis in coal and shale from the Anthracite region, Pennsylvania: *Amer. Minerl.* **75**, 825-39.

Dean, R. S. (1983) Authigenic trioctahedral clay minerals coating Clearwater Formation sand grains at Cold Lake, Alberta, Canada: *Programs with Abstracts*, 20th Annual Meeting, The Clay Minerals Society, Buffalo, New York, 1983, 79.

de Caritat, P., Hutcheon, I., and Walshe, J. L. (1993) Chlorite geothermometry: A review: *Clays and Clay Minerals* **41**, 219-39.

Deer, W. A., Howie, R. A., and Zussman, T. (1962) *Rock Forming Minerals*, Vol. 3, *Sheet Silicates*: Wiley, New York, 270 pp.

Deer, W. A., Howie, R. A., and Zussman, J., (1963) *Rock Forming Minerals*, Vol. 4, *Framework Silicates*: Wiley, New York, 435 pp.

Douglas, L. A. (1989) Vermiculites: in Dixon, J. B., and Weed, S. B., editors, *Minerals in Soil Environments*, Second Edition, Soil Science Society of America, 635-74.

Drits, V. A. (1987) Mixed-layer minerals: diffraction methods and structural features: in Schultz, L. G., van Olphen, H., and Mumpton, F. A., editors, *Proceedings*, International Clay Conference, 1985, The Clay Minerals Society, Bloomington, Ind., 33-45.

Drits, V. A., and Tchoubar, C. (1990) *X-Ray Diffraction by Disordered Lamellar Structures*: Springer-Verlag, Berlin, 371 pp.

Droste, J. B. (1956) Alteration of clay minerals by weathering in Wisconsin tills: *Geol. Soc. Am. Bull.* **67**, 911-18.

Duba, D., and Williams-Jones, A. E. (1983) The application of illite crystallinity, organic matter reflectance, and isotopic techniques to mineral exploration: A case study in southwestern Gaspé, Quebec: *Econ. Geol.* **78**, 1350-63.

Eberl, D. D., Jones, B. F., and Khoury, H. N. (1982) Mixed-layer kerolite/stevensite from the Amargosa Desert, Nevada: *Clays and Clay Minerals* **30**, 321-26.

Eberl, D. D., Srodoń, J., Lee, M., Nadeau, P. H., and Northrop, H. R. (1987) Sericite from the Silverton caldera, Colorado: correlation among structure, composition, origin, and particle thickness: *Amer. Minerl.* **72**, 914-34.

Eberl, D. D., and Velde, B. (1989) Beyond the Kubler index: *Clay Minerals* **24**, 571-77.

Eberl, D. D., Srodoń, J., Kralik, M., Taylor, B. E., and Peterman, Z. E. (1990) Ostwald ripening of clays and metamorphic minerals: *Science* **248**, 474-77.

Eggleton, R. A. (1986) The relation between crystal structure and silicate weathering rates: in Colman, S. M., and Dethier, D. P., editors, *Rates of Chemical Weathering of Rocks and Minerals*, Academic Press, New York, 21-40.

Ehrenberg, S. N., Aagaard, P., Wilson, M. J., Fraser, A. R., and Duthie, D. M. L. (1993) Depth-dependent transformation of kaolinite to dickite in sandstones of the Norwegian Continental Shelf: *Clay Minerals* **28**, 325-52.

Evans, B. W., and Guggenheim, S. (1988) Talc, pyrophyllite, and related minerals: in Bailey, S. W., editor, *Hydrous Phyllosilicates (exclusive of micas)*, Vol. **19** in Reviews in Mineralogy, Mineralogical Society of America, Washington, D.C., 225-94

Fanning, D. S., Keramidas, V. Z., El-Desoky, M. A. (1989) Micas: in Dixon, J. B., and Weed, S. B., editors, *Minerals in the Soil Environment*, 2nd Edition, Soil Science Society of America, Madison, Wisconsin, 551-634.

Freed, R. L., and Peacor, D. R. (1992) Diagenesis and the formation of authigenic illite-rich I/S crystals in Gulf Coast shales: TEM study of clay separates: *J. Sed. Pet.* **62**, 220-34.

Fripiat, J. J., and Cruz-Cumplido, M. I. (1974) Clays as catalysts for natural processes: in Donath, F. A., editor, *Annual Review of Earth and Planetary Sciences*: Annual Review Inc., Palo Alto, Ca., 239-56.

Frye, J. C., Glass, H. D., and Willman, H. B. (1968) Mineral Zonation of Woodfordian Loesses of Illinois. Ill. State Geol. Surv. Circ. **427**, 44 pp.

Garrels, R. M. (1984) Montmorillonite/illite stability diagrams: *Clays and Clay Minerals* **32**, 161-66.

Gast, R. G. (1977) Surface and colloid chemistry: in Dixon, J. B., and Weed, S. B., editors, *Minerals in the Soil Environment*: Soil Science Society of America, 27-73.

Giese, R. F., Jr. (1973) Interlayer bonding in kaolinite, dickite and nacrite: *Clays and Clay Minerals* **21**, 145-49.

Glass, H. D., Hughes, R. E., Moore, D. M., and Reynolds, R. C., Jr. (1992) Expandable clay minerals: Are those in soils, sediments, and bentonites different? in *Agronomy Abstracts*, 1992 Annual Meeting, American Society of Agronomy joint meeting with The Clay Minerals Society, Minneapolis, Minnesota, 371.

Grathoff, G. H., Moore, D. M., and Hay, R. L. (1994) Illite mineralogy of the Lower Paleozoic in the Illinois Basin: in *Program and Abstracts* of the 31st Annual Meeting of the Clay Minerals Society, University of Saskatchewan, Saskatoon, August 13-18, 36.

Grathoff, G. H., and Moore, D. M. (1996) Illite polytype quantification using WILDFIRE© calculated XRD patterns: *Clays and Clay Minerals* **44**, in press.

Grim, R. E. (1968) *Clay Mineralogy*, 2nd ed.: McGraw-Hill, New York, 596 pp.

Grim, R. E., Bray, R. H., and Bradley, W. F. (1937) The mica in argillaceous sediments: *Amer. Minerl.* **22**, 813-29.

Güven, N. (1971) The crystal structure of 2*M*-phengite and 2*M*-muscovite: *Zeit. Krist.* **134**, 196-212.

Harter, R. D. (1977) Reactions of minerals with organic compounds in soils: in Dixon, J. B., and Weed, S. B., editors, *Minerals in the Soil Environment*: Soil Science Society of America, 709-39.

Hay, R. L., Hughes, R. E., Kyser, T. K., Glass, H. D., and Liu, J. (1995) Magnesium-rich clays of the Meerschaum Mines in the Amboseli Basin, Tanzania and Kenya: *Clays and Clay Minerals* **43**, 455-66.

Hayes, J. B. (1970) Polytypism of chlorite in sedimentary rocks: *Clays and Clay Minerals* **18**, 285-306.

Hillier, S. (1995) Mafic phyllosilicates in low-grade metabasites. Characterization using deconvolution analysis—Discussion: *Clay Minerals* **30**, 67-73.

Hillier, S., and Velde, B. (1991) Octahedral occupancy and the chemical composition of diagenetic (low-temperature) chlorites: *Clay Minerals* **26**, 149-68.

Hillier, S., and Velde, B. (1992) Chlorite interstratification with a 7 Å mineral: An example from offshore Norway and possible implications for the interpretation of the composition of diagenetic chlorites: *Clay Minerals* **27**, 475-86.

Hluchy, M. M. (1988) *The Chemistry or Clay Minerals Associated with Evaporites in New York and Utah*: unpub. Ph.D. thesis, Dartmouth College, 160 pp.

Horton, D. G. (1985) Mixed-layer illite/smectite as a paleotemperature indicator in the Amethyst vein system, Creede, Colorado, USA: *Contrib. Mineral. Petrol.* **91**, 171-79.

Hower, John (1981) Shale diagenesis: in Longstaffe, F. J., editor, *Clays and the Resource Geologist*: Mineralogical Association of Canada, Toronto, 60-80.

Hower, John, Eslinger, E. V., Hower, M. E., and Perry, E. A. (1976) Mechanism of burial metamorphism of argillaceous sediments: *Geol. Soc. Amer. Bull.* **87**, 725-37.

Hower, John, and Mowatt, T. C. (1966) The mineralogy of illite and mixed-layer illite/montmorillonite: *Amer. Minerl.* **51**, 825-54.

Huang, W-L., Longo, J. D., and Pevear, D. R. (1993) An experimentally derived kinetic model for smectite-to-illite conversion and its use as a geothermometer: *Clays and Clay Minerals* **41**, 162-77.

Huff, W. D., Bergstrom, S. M., and Kolata, D. R. (1992) Gigantic Ordovician ash fall in North America and Europe: biological, tectonomagmatic, and event-stratigraphic significance: *Geology* **20**, 875-78.

Hughes, RE., DeMaris, PJ., White, WA., and Cowin, DK. (1987) Origin of clay minerals in Pennsylvanian Strata of the Illinois Basin: in Schultz, L. G., van Olphen, H., and Mumpton, F. A., editors, *Proceedings* , International Clay Conference, Denver, 1985, The Clay Minerals Society, Bloomington, Ind., 97-104.

Hughes, R. E., Moore, D. M., Berres, T. E., and Farnsworth, K. B. (1990) Berthierine pipestones of Native Americans in the mid-continent: *Program with Abstracts*, 27th Annual Mtg, Clay Minerals Society, Columbus, MO, 64.

Hughes, R. E., Moore, D. M., and Reynolds, R. C., Jr., (1993) The nature, detection, and occurrence, and origin of kaolinite/smectite: in Murray, H. H., Bundy, W. M., and Harvey, C. C., editors, *Kaolin Genesis and Utilization*, Spec. Publication No. **1**, Clay Minerals Society, Boulder, CO, 291-323.

Hughes, R. E., Moore, D. M., and Glass, H. D. (1994) Qualitative and quantitative analysis of clay minerals in soils: *in* Amonette, J.E., and Zelazny, L. W., editors, *Quantitative Methods in Soil Mineralogy*, SSSA Miscell. Pub., 330-59.

Ildefonse, Ph., Kirkpatrick, R. J., Montez, B., Calas, G., Flank, A. M., and Lagarde, P. (1994) ^{27}Al MAS NMR and Aluminum X-ray absorption near edge structure study of imogolite and allophanes: *Clays and Clay Minerals* **42**, 276-87.

Inoue, A., Kohyama, N., Kitagawa, R., and Watanabe, T. (1987) Chemical and morphological evidence for the conversion of smectite to illite: *Clays and Clay Minerals* **35**, 111-20.

Inoue, A., Velde, B., Meunier, A., and Touchard, G. (1987) Mechanism of illite formation during smectite-to-illite conversion in a hydrothermal system: *Amer. Minerl.* **73**, 1325-34.

Ireland, B. J., Curtis, C. D., and Whiteman, J. A. (1983) Compositional variation within some glauconites and illites and implications for their stability and origins: *Sedimentology* **30**, 769-86.

Jadgozinski, H. (1949) Eindimensionale Fehlordnung in Kristallen und ihr Einfluss auf die Röntgeninterferenzen. I. Berechnung des Fehlordnungsgrades aus der Röntgenintensitäten: *Acta Crystallogr.* **2**, 201-07.

Jakobsen, H. J., Nielsen, N. C., and Lindgreen, H. (1995) Sequences of charged sheets in rectorite: *Amer. Minerl.* **80**, 247-53.

Jarhren, J. S. (1991) Evidence of Ostwald ripening related to recrystallization of diagenetic chlorites from reservoir rocks offshore Norway: *Clay Minerals* **26**, 169-78.

Jiang, W-T., and Peacor, D. R. (1994) Prograde transitions of corrensite and chlorite in low-grade pelitic rocks from the Gaspé Peninsula, Quebec: *Clays and Clay Minerals* **42**, 497-517.

Jiang, W-T., Peacor, D. R., and Buseck, P. R. (1994) Chlorite geothermometry? –Contamination and apparent octahedral vacancies: *Clays and Clay Minerals* **42**, 593-605.

Juster, T. C., Brown, P. E., and Bailey, S. W. (1987) NH_4-bearing illite in very low grade metamorphic rocks associated with coal, northeastern Pennsylvania: *Amer. Minerl.* **72**, 555-65.

Keller, W. D. (1958) Glauconitic mica in the Morrison Formation in Colorado: in Swineford, A., editor, *Proceedings*, 12th National Conf. on Clays and Clay Minerals, Urbana, Ill., National Academy of Science-National Research Council, Washington, D.C., 120-28.

Kübler, B. (1964) Les argiles, indicateurs de métamorphishme: *Rev. Inst. Franc. Petrole* **19**, 1093-112.

Lagaly, G. (1992) Layer charge determination by alkylammonium ions: in Mermut, A. R., editor, *Layer Charge Characteristics of 2:1 Silicate Clay Minerals*: CMS Workshop Lectures, Vol. **6**, The Clay Mineral Society, Boulder, Colorado, 1-46.

Laird,J. (1988) Chlorites: Metamorphic petrology: in Bailey, S. W., editor, *Hydrous Phyllosilicates (exclusive of micas)*, Vol. **19** in Reviews in Mineralogy, Mineralogical Society of America, Washington, D.C., 405-53.

Land, L. S., and Milliken, K. L. (1981) Feldspar diagenesis in the Frio Formation, Brazoria County, Texas Gulf Coast: *Geology* **9**, 314-18.

Lee, J. H., Ahn, J. H., and Peacor, D. R. (1985) Textures in layered silicates. Progressive changes through and low-temperature metamorphism: *J. Sed. Petrol.* **55**, 532-40.

Lippmann, F. (1956) Clay minerals from the Röt member of the Triassic near Göttingen, Germany: *J. Sed. Petrol.* **26**, 125-39.

Lippmann, F. (1982) The thermodynamic status of clay minerals: in van Olphen, H., and Veniale, F., editors, *International Clay Conference*, Developments in Sedimentology **35**, Elsevier Scientific Publishing Co., Amsterdam, 475-85.

Lynch, L., and Reynolds, R. C., Jr. (1985) The stoichiometry of the smectite-illite reaction: *Program with Abstracts*, 21st Annual Mtg, Clay Minerals Society, Baton Rouge, LA, 84.

MacEwan, D. M. M., and Wilson, M. J. (1980) Interlayer and intercalation complexes of clay minerals: in Brindley, G.W., and Brown, G., editors, *Crystal Structures of Clay Minerals and Their X-Ray Identification*: Monograph No. **5**, Mineralogical Society, London, 197-248.

Malla, P. B., and Douglas, L. A. (1987) Identification of expanding layer silicates: Layer charge vs. expansion properties: in *Proceedings*, International Clay Conference, Denver, 1985, The Clay Minerals Society, Bloomington, IN, 277-83.

Marshall, C. E. (1977) *The Physical Chemistry and Mineralogy of Soils*, Vol. II: *Soils in Place*: Wiley, New York, 313 pp.

Matthews, J. C., Velde, B., and Johansen, H. (1994) Significance of K-Ar ages of authigenic illitic clay minerals in sandstones and shales from the North Sea: *Clay Mineral.* **29**, 379-89.

Mooney, R. W., Keenan, A. C., and Wood, L. A. (1952) Absorption of water vapour by montmorillonite: *J. Amer. Chem. Soc.* **74**, 1367-74.

Moore, D. M. (1982) Shallow burial diagenesis of the Purington Shale: *Program with Abstracts*, 19th Annual Mtg, Clay Minerals Society, Hilo, HI, 73.

Moore, D. M., and Hower, John (1986) Ordered interstratification of dehydrated and hydrated Na-smectite: *Clays and Clay Minerals* **34**, 379-84.

Moore, D. M., Ahmed, J., and Grathoff, G. (1989) Mineralogy of the Eocene Ghazij Shale, Western Indus Basin, Pakistan: in *Program with Abstracts*, 9th International Clay Conference, Strasbourg, France, 266.

Moore, D. M., and Hughes, R. E. (1990) The clay mineralogy of two Mississippian sandstone reservoirs in the Illinois Basin: in *Program with Abstracts*, 27th Annual Mtg, Clay Minerals Society, Columbia, MO, 90.

Moore, D. M., and Hughes, R. E. (1991) Characteristics of chlorite interlayered with a 7 Å mineral as found in sandstone reservoirs: in *Program with Abstracts*, 28th Annual Mtg, Clay Minerals Society, Houston, TX, 115.

Mortland, M. M. (1970) Clay-organic complexes and interactions: in Brady, N.C., editor, *Advances in Agronomy*, Academic Press, New York, 75-117.

Morton, J. P. (1985) Rb-Sr evidence for punctuated illite/smectite diagenesis in the Oligocene Frio Formation, Texas Gulf Coast: *Geol. Soc. Am. Bull.* **96**, 114-22.

Murray, H. H. (1988) Kaolin minerals: Their genesis and occurrences: in Bailey, S. W., editor, *Hydrous Phyllosilicates (exclusive of micas)*, Vol. **19** in Reviews in Mineralogy, Mineralogical Society of America, Washington, D.C., 67-89.

Nadeau, P. H., and Reynolds, R. C., Jr. (1981) Volcanic components in pelitic sediments: *Nature* **294**, 72-4.

Nadeau, P. H., Wilson, M. J., McHardy, W. J., and Tait, J. M. (1984) Interstratified clays as fundamental particles: *Science* **225**, 923-25.

Nagelschmidt, G. (1936) The structure of montmorillonite: *Z. Kristallogr.* **93**, 481-87.

Newman, A. C. D., and Brown, G. (1987) The chemical constitution of clays: in Newman, A. C. D., editor, *Chemistry of Clays and Clay Minerals*: Monograph No. **6**, Mineralogical Society, London, 1-128.

Odin, G. S., editor (1988) *Green Marine Clays*: Elsevier, Amsterdam, 445 pp.

Odom, I. E. (1984) Glauconite and celadonite minerals: in Bailey, S. W., editor, *Micas*, Vol. **13** in Reviews in Mineralogy, Mineralogical Society of America, Washington, D. C., 545-72.

Ohr, M., Halliday, A. N., and Peacor, D. R. (1991) Sr and Nd isotopic evidence for punctuated clay diagenesis, Texas Gulf Coast: *Earth and Planet. Sci. Letters* **105**, 110-26.

Oreskes, N., Shrader-Frechette, K., and Belitz, K. (1994) Verification, validation, and confirmation of numerical models in the Earth Sciences: *Science* **263**, 641-46.

Parham, W. E. (1964) Lateral clay mineral variations in certain Pennsylvanian underclays: in Bradley, W. F., editor, *Proceedings*, 12th National Conf. on Clays and Clay Minerals, Atlanta, Ga., 1963, Pergamon Press, New York, 581-602.

Perry, E. A., and Hower, J. (1970) Burial diagenesis of Gulf Coast pelitic sediments: *Clays and Clay Minerals* **18**, 165-77.

Pevear, D. R. (1992) Illite age analysis, a new tool for basin thermal history analysis: in Kharaka, Y. K., and Maest, A. S., editors, *Water-Rock Interaction*: A. A. Balkema, Rotterdam, 1251-54.

Pinnavaia, T. J. (1983) Intercalated clay catalysts: *Science* **220**, 365-71.

Pollastro, R. M. (1985) Mineralogical and morphological evidence for the formation of illite at the expense of illite/smectite: *Clays and Clay Minerals* **33**, 265-74.

Powers, M. C. (1967) Fluid release mechanisms in compacting marine mudrocks and their importance in oil exploration: *AAPG Bull.* **51**, 1240-54.

Pytte, A. M., and Reynolds, R. C. (1989) The kinetics of the smectite to illite reaction in contact metamorphic shales: in Naeser, N.D., and McCulloch, T.H., editors, *The Thermal History of Sedimentary Basins* , Springer-Verlag, New York, 133-40.

Ransom, B., and Helgeson, H. C. (1994) A chemical and thermodynamic model of aluminous dioctahedral 2:1 clay minerals in diagenetic processes: Regular solution representation of interlayer dehydration in smectite: *Amer. Jour. Sci.* **294**, 449-84.

Retallack, G. J. (1983) A paleopedological approach to the interpretation of terrestrial sedimentary rocks: The mid-Tertiary fossil soils of Badlands National Park, South Dakota: *Geol. Soc. Amer. Bull.* **94**, 823-40.

Reynolds, R. C., Jr. (1980) Interstratified clay minerals: in Brindley, G. W., and Brown, G., editors, *Crystal Structures of Clay Minerals and Their X-Ray Identification:* Monograph No. **5**, Mineralogical Society, London, 249-303.

Reynolds, R. C., Jr. (1985) *NEWMOD©* *a Computer Program for the Calculation of One-Dimensional Diffraction Patterns of Mixed-Layered Clays*: R. C. Reynolds, 8 Brook Rd., Hanover, NH.

Reynolds, R. C., Jr. (1988) Mixed layer chlorite minerals: in Bailey, S. W., editor, *Hydrous Phyllosilicates (exclusive of micas)*: Vol. **19**, Reviews in Mineralogy, Mineralogical Society of America, Washington, D.C., 601-29.

Reynolds, RC., Jr., and Hower, J. (1970) The nature of interlayering in mixed-layer illite-montmorillonite: *Clays and Clay Minerals* **18**, 25-36.

Reynolds, R. C., Jr., DiStefano, M. P., and Lahann, R. W. (1992) Randomly interstratified serpentine/chlorite: Its detection and quantification by powder X-ray diffraction methods: *Clays and Clay Minerals* **40**, 262-67.

Rimmer, S. M., and Eberl, D. D. (1982) Origin of an underclay as revealed by vertical variations in mineralogy and chemistry: *Clays and Clay Minerals* **30**, 422-30.

Robertson, I. D. M., and Eggleton, R. A. (1991) Weathering of granitic muscovite to kaolinite and halloysite and of plagioclase-derived kaolinite to halloysite: *Clays and Clay Minerals* **39**, 113-26.

Rosenberg, P. E., Kittrick, J. A., and Aja, S. U. (1990) Mixed-layer illite/smectite: A multiphase model: *Amer. Minerl.* **75**, 1182-85.

Ross, C. S., and Hendricks, S. D. (1945) Minerals of the montmorillonite group: U.S. Geol. Survey Professional Paper, **205-B**, 23-79.

Ryan, P. C. and Reynolds, R. C., Jr (1996) The origin and diagenesis of grain-coating serpentine/chlorite, Tuscaloosa Formation: *Amer. Minerl.* **81**, 213-25.

Ruiz Cruz, M. D. (1994) Diagenetic development of clay and related minerals in deep water sandstones (S. Spain): Evidence of lithological control: *Clay Mineral.* **29**, 93-104.

Sawhney, B. L. (1960) Aluminum interlayers in clay minerals: *Transactions*, 7th International Congress Soil Science, Vol. 4, 476-81.

Schiffman, P., and Fridlesisson, G. O. (1991) The smectite-chlorite transition in drillhole NJ-15, Nesjavellir geothermal field, Iceland: XRD, BSE and electron microprobe investigations: *Jour. Metamorphic Geol.* **9**, 679-96.

Shau, Y-H., Peacor, D. R., and Essene, E. J. (1990) Corrensite and mixed-layer chlorite/corrensite in metabasalt from northern Taiwan: TEM/AEM, EPMA, XRD, and optical studies: *Contrib. Mineral. Petrol.* **105**, 123-42.

Shau, Y-H., and Peacor, D. R. (1992) Phyllosilicates in hydrothermally altered basalts from DSDP Hole 504B, Leg 83—a TEM and AEM study: *Contrib. Mineral. Petrol.* **112**, 119-33.

Singer, A. (1989) Palygorskite and sepiolite group minerals: in Dixon, J. B., and Weed, S. B., editors, *Minerals in the Soil Environment*, 2nd Edition, Soil Science Society of America, Madison, Wisconsin, 829-72.

Smith, K. L., Milnes, A. R., and Eggleton, R. A. (1987) Weathering of basalt: formation of iddingsite: *Clays and Clay Minerals* **35**, 418-28.

Smith, J. V., and Yoder, H. S. (1956) Experimental and theoretical studies of mica polymorphs: *Mineralog. Mag.* **31**, 209-35.

Smith, R. W. (1929) *Sedimentary Kaolins of the Coastal Plain of Georgia*: Bull. 44, Geological Survey of Georgia, Atlanta, Georgia, 482 pp.

Środoń, J. (1984) X-ray identification of illitic materials: *Clays and Clay Minerals* **32**, 337-49.

Środoń, J., and Eberl, D. D. (1984) Illite: in Bailey, S. W., editor, *Micas*, Vol. **13** in Reviews in Mineralogy, Mineralogical Society of America, Washington, D. C., 495-544.

Środoń, J., Morgan, D. J., Eslinger, E. V., Eberl, DD., and Karlinger, M. R. (1986) Chemistry of illite/smectite and end-member illite: *Clays and Clay Minerals* **34**, 368-78.

Środoń, J., Elass, F., McHardy, W. J., and Morgan, D. J. (1992) Chemistry of illite-smectite inferred from TEM measurements of fundamental particles: *Clay Minerals* **27**, 137-58.

Sterne, E. J., Reynolds, R. C., Jr., and Zantop, H. (1982) Natural ammonium illites from black shales hosting a stratiform base metal deposit, Delong Mountains, northern Alaska: *Clays and Clay Minerals* **30**, 161-66.

Suquet, H., and Pézerat, H. (1988) Comments of the classification of trioctahedral 2:1 phyllosilicates: *Clays and Clay Minerals* **36**, 184-86.

Talibudeen, O., and Goulding, K. W. T. (1983a) Apparent charge heterogeneity in kaolins in relation to their 2:1 phyllosilicate content: *Clays and Clay Minerals* **31**, 137-42.

Talibudeen, O., and Goulding, K. W. T. (1983b) Charge heterogeneity in the smectites: *Clays and Clay Minerals* **31**, 37-42.

Tamura, T. (1958) Identification of clay minerals from acid soils: *J. Soil Sci.* **2**, 141-47.

Thompson, G. R., and Hower, John (1973) An explanation for the low radiometric ages from glauconite: *Geochim. Cosmochim. Acta* **37**, 1473-91.

Thompson, G. R., and Hower, John (1975) The mineralogy of glauconite: *Clays and Clay Minerals* **23**, 289-300.

Van Olphen, H. (1977) *An Introduction to Clay Colloid Chemistry*, 2nd ed.: Wiley, New York, 301 pp.

Veblen, D. R. (1991) Polysomatism and polysomatic series: A review and applications: *Amer. Minerl.* **76**, 801-26.

Veblen, D. R., Guthrie, G. D., Jr., Livi, K. J. T., and Reynolds, R. C., Jr. (1990) High-resolution transmission electron microscopy and electron diffraction of mixed-layer illite/smectite: Experimental results: *Clays and Clay Minerals* **38**, 1-13.

Velde, B. (1987) Petrologic phase equilibria in natural clay systems: in Newman, A. C. D., editor, *Chemistry of Clays and Clay Minerals*: Monograph No. **6**, Mineralogical Society, London, 423-458.

Velde, B., Suzuki, T., and Nicot, E. (1986) Pressure-temperature-composition of illite/smectite mixed-layer minerals: Niger Delta mudstones and other examples: *Clays and Clay Minerals* **34**, 435-41.

Vergo, N. (1984) *Wallrock Alteration at the Bulldog Mountain Mine, Creede Mining District, Colorado*: unpub. MS thesis, Univ. of Illinois-Urbana, 88 pp.

Wada, K. (1989) Allophane and imogolite: in Dixon, J. B., and Weed, S. B., editors, *Minerals in the Soil Environment*, 2nd Edition, Soil Science Society of America, 1051-87.

Walker, G. F. (1958) Reactions of expanding lattice minerals with glycerol and ethylene glycol: *Clay Miner. Bull.* **3**, 302-13.

Walker, J. R. (1987) *Structural and Compositional Aspects of Low-Grade Metamorphic Chlorite*: Ph.D. Dissertation, Dartmouth College, Hanover, N. H.

Walker, J. R. (1993) Chlorite polytype geothermometry: *Clays and Clay Minerals* **41**, 260-67.

Walker, J. R., and Thompson, G. R. (1990) Structural variations in illite and chlorite in a diagenetic sequence from the Imperial Valley, California: *Clays and Clay Minerals* **38**, 315-21.

Warren, E. A., and Ransom, B. (1992) The influence of analytical error upon the interpretation of chemical variations in clay minerals: *Clay Minerals* **27**, 193-209.

Weaver, C. E., and Beck, K. C. (1971) Clay water diagenesis during burial: how mud becomes gneiss: *Geol. Soc. Amer., Spec. Paper* **134**, 96 pp.

Weaver, C. E., and Brockstra, B. R. (1984) Illite-mica: in Weaver, C. E., and associates, *Shale Slate Metamorphism in the Southern Applachians*, Elsevier, Amsterdam, 67-199.

Weaver, C. E., and Pollard, L. D. (1973) *The Chemistry of Clay Minerals*: Elsevier, New York, 213 pp.

Weiss, A. (1969) Organic derivatives of clay minerals, zeolites, and related minerals: in Eglinton, G., and Murphy, M. T. J., editors, *Organic Geochemistry*: Springer-Verlag, New York, 737-81.

Whitehouse, U. G., Jeffery, L. M., and Debbrecht, J. D. (1960) Differential settling tendencies of clay minerals in saline waters: in *Proceedings*, 7th National Conf. on Clays and Clay Minerals, Pergamon Press, New York, 1-79.

Whitney, G., and Northrop, R. (1988) Experimental investigation of the smectite to illite reaction: Dual reaction mechanisms and oxygen-isotope systematics: *Amer. Minerl.* **73**, 77-90.

Wicks, F. J., and O'Hanley, D. S. (1988) Serpentine minerals: structures and petrology: in Bailey, S. W., editor, *Hydrous Phyllosilicates (exclusive of micas)*, Vol. **19** in Reviews in Mineralogy, Mineralogical Society of America, Washington, D.C., 91-167.

Willman, H. B., Glass, H. D., and Frye, J. C. (1989) *Glaciation and Origin of the Geest in the Driftless Area of Northwestern Illinois*: Circular **535**, Illinois State Geological Survey, Champaign, IL, 44 pp.

Wnuk, C., and Pfefferkorn, H. W. (1987) A Pennsylvanian age terrestrial storm deposit: using plant fossils to characterize the history and process of sediment accumulation: *J. Sed. Pet.* **57**, 212-21.

Yamada, H., Nakazawa, H., and Hashizume, H. (1994) Formation of smectite crystals at high pressures and temperatures: *Clays and Clay Minerals* **42**, 674-78.

Yau, Y.-C., Peacor, D. R., and McDowell, S. D. (1987) Smectite-to-illite reactions in Salton Sea shales: a transmission and analytical electron microscopy study: *J. Sed. Pet.* **57**, 335-42.

Young, T. P., and Taylor, W. E. G., editors (1989) *Phanerozoic Ironstones*: Special Publication **46**, London, Geological Society, 251 pp.

Zhou, T., and Phillips, G. N. (1994) Sudoite in the Archean Witwatersrand Basin: *Contrib. Mineral. Petrol.* **116**, 352-59.

Zoltai, T., and Stout, J.H. (1984) *Mineralogy: Concepts and Principles*: Burgess, Minneapolis, Minn., 505 pp.

Chapter 6
Sample Preparation Techniques for Clay Minerals

There is no *single* way to prepare materials for X-ray diffraction analysis. Preparation techniques depend on (1) supplies and equipment available; (2) purposes of analysis, e.g., whether qualitative or quantitative identification is the goal; (3) the material itself; and, probably most important, (4) your sense of organization and habits, your goals, your understanding of the principles of X-ray diffraction, and your ingenuity. You must design and execute suitable procedures because they are so dependent on the characteristics of the sample and what you want from the sample. When someone said "research is 90% tedium," sample preparation may have been what he or she had in mind. We discuss several methods of preparation, keeping in mind that we may be addressing neophytes and that the laboratory facilities available may be limited. We have borrowed heavily from Brown and Brindley (1980), Jackson (1969), and Bish and Reynolds (1989).

This chapter may seem pedestrian—a bit like giving instruction on typing or driving or other subjects that might seem unscientific. But unless you make good samples that produce clean, intense diffraction patterns, all the world's theory and creativity cannot be usefully applied to the interpretation of XRD results.

Our discussion is directed toward the treatment of common types of clay minerals in sedimentary rocks. We show no flowcharts because, although they are nice devices for summary, they usually are so complicated that they are intelligible only to those who do not need the information in the first place.

The most important point to be made here is that *there must be a single method of preparation for a given set of samples if X-ray diffraction data are to be compared within the set,* particularly for quantitative analysis. However, treatment of different sets of samples may vary up to the point at which they are exposed to the X-ray beam, because each set may present characteristics and preparation problems of its own. For example, carbonates present different problems than tills, and shales require a different treatment than sandstones.

Procedures for preparing samples for X-ray diffraction fall into two groups: one in which perfectly random orientation of the grains is the goal, and one in which perfect orientation of the clay mineral flakes parallel to the substrate is the goal. Neither goal will be met, but we try.

EVALUATING THE SAMPLE

We assume you have at hand a representative sample, one you have taken carefully at the outcrop with specific questions in mind. The first step is a preliminary assessment. Split out 1 to 2 g, crush it in a mortar, and grind it as a paste with water or propanol. Dry the sample and load into a sample holder as a randomly oriented powder (see subsequent discussion on preparation of random powder mounts). Note that we crush by impact (as gently as will do the job) rather than by grinding, and we accomplish further reduction in grain size by wet grinding. Aggressive dry grinding can cause changes of phase, and in extreme cases, can lead to strains on the crystal structure that cause XRD line broadening or even the production of X-ray amorphous material. The X-ray diffraction tracing from a randomly ordered powder will give you a rough idea of the proportion of clay and nonclay minerals. All clay minerals diffract from the 020 and 110 spacing. These spacings are very nearly the same dimension for all of them. The intensity of this combined peak (they diffract at approximately the same Bragg angle) is used by many workers to estimate the amount of clay minerals relative to the nonclay minerals. This preliminary tracing will also be the basis for designing subsequent steps.

Certainly the nonclay minerals will be of interest, but most of them, when present, mask some of the 00l reflections of the clay minerals. In addition, the weak 00l clay mineral intensities produced by such a sample means that you will not be able to do much identification of specific clay minerals with this preparation.

For many studies, you will know enough about the samples to eliminate the random powder XRD reconnaissance and proceed directly to a separation of the clay minerals.

The most common nonclay minerals are quartz, feldspars, carbonates, gypsum, pyrite, zeolites, and iron oxides (either amorphous or crystalline). Organic matter also can be present. Most of these minerals can be separated from the clay minerals by extracting a fine enough particle-size fraction, leaving the nonclay minerals in the coarser residue, but not always.

Good orientation of the clay minerals requires that nonplaty minerals such as quartz be removed, for their equant crystal shapes destroy the preferred orientation we are seeking. In addition, the clay minerals must be dispersed into individual colloidal particles in the suspension before you prepare the sample, because in the flocculated condition they produce submicroscopic polymineralic aggregates within which the orientation is poor to random. All this preparation must be accomplished with a minimum of chemical and physical damage to the clay minerals present. The rule is *do as little as possible to the sample before presenting it to the X-ray beam.* Use chemical treatment only as a last resort, and use as little physical treatment as possible because there is always the danger of changing, in some unanticipated way, these fragile phyllosilicates with large, reactive surfaces.

In what follows, we have tried not to present suggestions as recipes, but as general guidelines. The suggestions, those that appear as recipes and otherwise, must be adapted to your problem, your lab, and your skills.

DISAGGREGATING THE ROCK

Rock samples that require different treatments can be categorized as: (1) Friable sandstones and moderately to poorly lithified shales; (2) silica-cemented sandstones, flint clays, and well-lithified (siliceous) shales; (3) limestones and dolomites; (4) gypsum-anhydrite rocks; and (5) unconsolidated materials such as till, soil, and modern sediments. The first serious mistake, and one that is all too commonly made, is to pulverize the sample with a grinder, shatter-box, or ball mill. This reduces the coarse-grained, nonclay minerals to the clay-size range from which they can never be separated. The samples must be crushed, not ground, with some device such as a large iron mortar and pestle. The purpose of this operation is to increase the specific surface area of the grains so that the following procedures are more effective. Stop when the grains average a few millimeters and don't worry about the inevitable presence of some larger grains. Ten grams of starting material is just about right for most clastic rocks, with perhaps 20 g for carbonates and evaporites.

Separating Clay Minerals from Clastic Rocks

Suppose that the sample is some sort of sandstone or shale. The next step is a preliminary disaggregation in an industrial-grade Waring® blender, perhaps after an overnight soak. The home kitchen-grade device will fail so quickly that it is not cost-effective. For a mix of 200 mL of distilled water and ~10 g of crushed rock, give it 2 to 3 min at full power. Quickly decant the obvious fine fraction and transfer it to a plastic container for ultrasonic treatment. If the rock is hard or silica-cemented or both, the blender treatment is ineffective and should be eliminated.

Ultrasonic disaggregation is a crucial contributor to good preparations. The best instruments are the horn-type devices that produce 100 or more acoustical watts at the transducer tip. If you use one of these devices, be certain that suspensions are always irradiated in plastic containers because the horn tip will quickly punch a hole in a glass cup or beaker if it contacts it. In addition, glass becomes crazed by exposure to the ultrasonic energy, with the result that a beaker will suddenly crumble, usually when you are treating a valuable sample for which only small amounts are available. Wear ear plugs or enclose the instrument in a sound-dampening box or both. Irradiate the sample for a few minutes (1 min at 300 W is nominal), allow it to settle for a minute or so, and decant the fines to a centrifuge cup. Much longer irradiation times are not advisable because the ultrasound causes some exfoliation of clay mineral crystallites and, you will notice, quickly heats the sample. If you are dealing with well-cemented clastic rocks, it may take hours in the ultrasonic instrument to separate enough material for a good preparation. The way to accomplish this separation is to irradiate for 3-5 min, decant off the fines, add more water, irradiate again, etc. This procedure ensures that liberated crystallites spend only short periods exposed to the ultrasound. In the end, you will finish with a very large volume of water that contains only small amounts of clay, but, as we will see, there are easy ways to concentrate such a dilute suspension.

Separating Clay Minerals from Carbonate Rocks

Carbonate rocks, or clastic rocks with significant carbonate contents, need to have the calcite or dolomite removed. Removal is accomplished by using a method tested by Ostrum (1961). He found that clay minerals can be extracted from carbonate rocks without affecting them if acetic acid ≤0.3 molar is used, and if the process is closely watched so that the sample is rinsed as soon as all the carbonate has been dissolved. The carbonate itself serves nicely as a buffer if closely monitored. As long as the sample is effervescing, carbonate is being dissolved and that controls pH. Heating will speed up the reaction and may be necessary to dissolve dolomite. Watch the sample carefully for at least $1/2$ h to make certain that the increased reaction rate due to heating does not cause sufficient effervescence to result in sample loss. Such loss can be a mess to clean up. For samples with large amounts of calcite, have a pan of water cooled with ice in case the reaction becomes too violent. If it does, you may place the beaker in the pan of cold water to cool it. Then proceed with the digestion at a lower temperature. Jackson (1969) recommended heating the crushed sample and using a sodium acetate-acetic acid solution buffer at pH 5.

In spite of the results reported by Ostrum (1961), mixed-layered clay minerals, particularly the ordered ones commonly found in soils, may be damaged by high temperature and low pH. Although we can find no written mention, palygorskite and sepiolite are quite acid soluble. So, be cautious. Try to find a way to test the clay minerals before and after exposure to acid and heat.

Separating Clay Minerals from Sulfate Rocks

Clay minerals can be extracted from gypsum-anhydrite rocks by dissolving the sulfates in the sodium salt of ethylenediaminetetraacetic acid (EDTA), an alternative method for carbonate rocks because it works about as well as acid for the removal of calcite and dolomite but is somewhat slower. (To extract gypsum or anhydrite from argillaceous rocks, water may do the trick. They dissolve in amounts of about 0.2 g per 100 cc of water.) The following procedure is summarized from Bodine and Fernalld (1973).

Per one liter of solution, dissolve 74.45 g of reagent-grade disodium EDTA in 800 to 900 mL of water and adjust the pH to about 11 by carefully adding sodium hydroxide pellets. Dilute to 1 L to produce a 0.2 M EDTA solution. Add 20 g of crushed rock to about 600 mL of reagent and boil for 4 h. Centrifuge and discard the supernatant. Wash the insoluble residue repeatedly by centrifugation until dispersion. The high pH of the reagent may, for some samples, prohibit flocculation and thus make it impossible to separate the insoluble residue from the EDTA. If calcium is added to promote flocculation, calcium sulfate will precipitate, and if the pH is lowered, the acid salt of EDTA will precipitate and mix with the insoluble residue. In this situation, the best procedure is to separate and wash the insoluble residue in its dispersed condition by ultracentrifugation if available. Bodine and Fernalld (1973) have demonstrated that this treatment has no deleterious effects on the simple clay minerals, but research is needed to ascertain whether or not it is entirely safe for mixed-layered clay minerals.

Separating Clay Minerals from Unconsolidated Materials

Unconsolidated materials are the easiest to work with and often require only dispersion by ultrasound or the Waring® blender and removal of salt by centrifugation. Some samples may contain carbonates that must be removed (see preceding section), or else the solubility will keep the Ca ion activity in solution high enough to inhibit full dispersion of some of the clay minerals. Organic matter and iron oxides can present real problems, and if they occur together with highly disordered mixed-layered clay minerals, you will have to face the fact that there is no ideal way to proceed. Treatments to eliminate organic matter and iron oxides may alter the mixed-layered clay minerals. Yet if these substances are not removed, the diffraction patterns may be so poor that you will not be able to make acceptable interpretations. Our tendency in such situations is to use smear mounts on porous tiles and run the samples while they are moist, as with peelers, as discussed below. (If you suspect smectite is

---

Box 6.1.   Glacial Deposits of the North American Interior

Resolution of the broad features of the stratigraphy of the glacial deposits covering the interior of the North American continent and distinguishing one till unit from another rest on a database of the clay mineralogy of approximately 150,000 samples, run in duplicate, and prepared by first soaking overnight, then disaggregating with a milk shake mixer, another period of standing for flocculation, a decanting, adding more water and a few grains of Calgon®, and then a pouring off those that seemed too concentrated. Then samples were stirred, allowed to settle for 15 min, and using an eyedropper, a portion was drawn from the top half cm and deposited on a glass slide that had been cleaned with a bit of spit. The slides dried on the lab bench, and were then placed in an ethylene glycol atmosphere to be run after two days. The intensities of the peaks are adjusted according to a set of empirically derived factors and tabulated. There is the art in all this of the good cook or the butcher who picks up exactly the amount of hamburger you request. Herbert Glass of the Illinois State Geological Survey created this database, and continues to add to it. He is a cook's cook. The details of his method were presented by Hallberg et al. (1978).

present in the sample, glycol may be added to the paste.) The intrinsically poor diffraction pattern is made acceptable by using step-scanning procedures with long count-time intervals or by recording the pattern at slow goniometer and strip-chart speeds in conjunction with long time constants (see Chapter 2, p. 49).

CHEMICAL PRETREATMENTS

Removal of Iron Oxides

If you are using a Cu (atomic number 29) target X-ray tube, fluorescent X-rays from Fe (atomic number 26) are a problem because they produce a high background that can mask peaks. Iron oxides also cement particles together and thus inhibit dispersion of the clay mineral particles. You can overcome the problem of fluorescence from Fe by placing a crystal monochromator in the beam path just in front of the detector. The monochromator diffracts the diffracted beam through the spacings of a crystal, commonly graphite, so that only the 1.54 Å CuKα radiation is diffracted at an angle that allows it to enter the detector. The problem can also be solved by using an Fe tube, but the monochromator is less expensive, solves the problem equally well, and eliminates the trouble of changing and aligning tubes.

Fe oxides can be removed chemically. The most commonly used treatment is the citrate-bicarbonate-dithionite (CBD) method (Jackson, 1969, p. 44). Jackson notes that the method he recommends also removes calcite and phosphates. This method can change the X-ray diffraction response of mixed-layered clay minerals, and it is not recommended unless it is absolutely necessary to remove very large amounts of Fe oxides.

Removal of Organic Materials

Organic matter can produce broad X-ray diffraction peaks, increase the background, and inhibit dispersal of other minerals if present in significant amounts (a few percent). The organic matter can be removed chemically. The solutions used are strong oxidizing agents, so you must be alert to possible changes in the clay minerals. Commercial bleach (Chlorox™, Purex™, etc.) is NaOCl (sodium hypochlorite) and seems quicker, cheaper, and safer (for your health) than the frequently recommended hydrogen peroxide.

The following procedure is recommended. Treat with 10 to 20 mL of NaOCl that has been adjusted to pH 9.5 with HCl just before treatment. Heat the mixture in a boiling-water bath about 15 min. Centrifuge at 800 rpm for 5 min and decant and discard the supernatant. Repeat the procedure until organic material is sufficiently removed, as evidenced by a change in sample color to white, gray, or red. The procedure can oxidize ferrous iron in octahedral sites, causing a change in silicate layer charge and altered clay mineral diffraction characteristics. Do not use this method unless you must, and then do so with the full realization that mixed-layered clay mineral identifications are suspect.

A sample that is predominantly organic matter, such as coal, requires that the organic material be removed by low-temperature ashing. This technique, described by Gluskoter (1965), treats the sample in an atmosphere of electronically excited oxygen at a temperature at or slightly above 100°C. The procedure oxidizes the organic material in the sample. Possible effects on clay minerals have not been documented.

Saturating the Clay Minerals with Different Cations

Clay minerals adsorb anions and cations and hold them in an exchangeable state. The X-ray diffraction characteristics of the air-dried and the ethylene glycol-solvated states of smectites depend on the type of cation that is held in the exchange sites. Saturation with Mg and solvation with glycerol is standard for tests for differentiating vermiculite from smectite (see Chapter 5, p.160).

Techniques for exchanging cations are relatively simple. Treat the clay minerals with a 1 M solution of the chloride of the univalent cation of choice or a 0.1 M solution of the divalent cation of choice. An exchange reaction will occur; cation *A*, adsorbed on the clay mineral, will be replaced by cation *B* from the solution containing excess *B* cations. The clay mineral will become saturated with *B* cations if it is separated from the solution 3 to 5 times, replacing the liquid each time with a fresh solution containing an excess of *B* cations. For the last step, wash the clay minerals with deionized water and then with a 50/50 ethanol/water mixture[1] until essentially all chloride ions are removed, for the absence of chloride ions means that the cations from the salt solution have also been removed. There is a simple test for this. Prepare a small amount (10 mL) of AgNO$_3$ solution and keep it in a light-proof dropper bottle. (To make the bottle light-proof, etch the bottle and dip it into the liquid material used for making plastic handle grips on pliers.) A drop or two of AgNO$_3$ will cause the precipitation of AgCl if there is even a very small amount of Cl anion present (the solubility product of AgCl = 1.8 x 10^{-10}). If all Cl anions, as cation indicators, cannot be removed this way, final washings should be done by dialysis. To do this, put the clay mineral suspension in semipermeable dialysis tubing (available from hospital or biological lab suppliers) and immerse in a large volume of warm, deionized water. Gentle stirring is helpful, and the water should be changed about four to five times the first day or until no more Cl is detected.

Cation saturation can also be accomplished very simply if you use the Millipore® transfer method for sample preparation (see p. 215).

---

[1]The ethanol-water mixture is used to minimize hydrogen ion substitution for the other exchangeable cations, i.e., to stop hydrolysis.

PARTICLE-SIZE SEPARATION

At this point we have a suspension in which, ideally, the particles are single crystals. One of the important reasons for washing the suspension free of salt is that if there is enough dissolved salt in the suspension, it will cause flocculation. Suspensions must be washed free of salt by centrifugation before particle-size separations are performed. Balance the centrifuge cups and spin them for a few minutes at 2,000 rpm. If the suspension has flocculated, the supernatant liquid after centrifugation will be crystal clear. Decant the water and discard it. Redisperse the suspension in another 200 mL of distilled water by means of ultrasound and repeat the centrifugation. After three or four of these washings, the supernatant will show some turbidity that may be extreme or only a faint opalescence. This condition indicates full or incipient dispersion, and that process should be completed by the addition of a suitable dispersing agent.

Dispersing agents all have one feature in common—they produce a buffered pH from neutral to high. The most effective ones also have phosphate ions that promote dispersion, presumably by the adsorption of the phosphate ions on clay edges where they reverse normally positive edge charge and thus help prohibit flocculation. Sodium pyrophosphate is our choice, although you may want to try other sodium phosphates if you encounter particularly recalcitrant sample. You don't want to add too much dispersing agent, or you can actually promote flocculation because of an increase in ionic strength. About 10^{-3} to 10^{-4} M is suitable, and for 200 mL of suspension, that is about equal to 20 to 30 mg of the reagent. Take about half a "pinch" of the powder, and you will see that this is just about right. Such an approximate measure is good enough, and you need not go to the trouble of weighing the powder that is added to each suspension. Add the reagent and disperse by ultrasound. Let it stand a few minutes and gently stir the surface with a rod or spatula. Notice that the streamlines stand out, just as they do when you stir aluminum paint (that is, if you've ever stirred aluminum paint). This effect is caused by the orientation of the clay mineral crystals in the stream lines with reflection of light from their cleavage faces, and signifies that dispersion is good and that the suspension is suitable for particle-size separation. Seldom will dispersions be good enough to see this effect, but it is a goal.

With the sample well dispersed, you may now split the sample into a number of size fractions. For different samples, you will have to move to smaller sizes to get below the grain size of quartz and feldspar. Other reasons for making a series of size fractions will become clear in the discussions of polytypes and the dating of various size fractions in Chapter 10; the smaller the size fraction, the more it is dominated by polytypes of diagenetically formed clay minerals.

Before we move on to the next subject, we make a few comments about difficult samples. Some materials resist dispersion, even after repeated washing by centrifuge and the addition of normal peptizing (dispersing) agents. If this happens, try adding 20 to 30 mg (per 200 mL) of sodium carbonate; it often works wonders. We have been able to disperse numerous samples of palygorskite and some materials rich in opaline silica only by this procedure. If you're faced with dispersing a flint clay, Bohor and Triplehorn (1993, p. 6) suggested a three to four week soak in water and dimethyl sulfoxide (DMSO).

The next step is the separation of the clay-size fraction, which we take here to be the < 2 μm equivalent spherical diameter (< 2 μm e.s.d.). Particle-size separations are based on Stokes's law, and it applies strictly to spherical particles, which platy clay minerals are not. So when we say that a clay mineral crystallite is 1 μm, we mean that it settles at the same velocity as a 1 μm sphere of equal density. Intuitively, you know that a leaf will fall through air more slowly than a ball of the same volume, so you won't be surprised that the maximum diameters of "1 μm" clay crystallites are a good deal more than 1 μm.

Below a particle size of about 20 μm, particles settling in a fluid approximately obey Stokes's law, which is a numerical expression that describes a particle being pulled by gravity but whose fall is resisted by a viscous fluid. The balance between these two forces results in a terminal velocity V_T (i.e., no longer accelerating) that is inversely proportional to the viscosity of the liquid η and proportional to the force of gravity g (in cm/sec^2). It is also directly proportional to the difference in density between the particle and the liquid ($d_p - d_l$) and the particle diameter squared, D^2, in square centimeters (i.e., the size of the surface resisting movement through the fluid). The equation for Stokes's law is

$$V_T = g(d_p - d_l)D^2/18\eta \tag{6.1}$$

Stokes's law can be put into a more useful form by using the relation velocity = distance/time ($V = h/t$), and the height h of the cylinder in which we have a dispersion as the distance in the relation $V = h/t$. Hence,

$$t = 18\eta h/g(d_p-d_l)D^2 \tag{6.2}$$

Equation (6.2) will allow you to figure the distance h a particle falls (in centimeters) during a given time interval t (in sec). Particles settling through a fluid in a centrifuge obey the same law except that the settling force is increased

Table 6.1. Settling times for gravity sedimentation of particles in water at 20°C[a]

Particle diameter (μm)	h	min	sec
50	—	—	22
20	—	2	20
5	—	37	30
2	3	50	

Data apply to a settling distance of 5 cm and a mineral density of 2.65 (Jackson, 1969).
[a]The viscosity of water is a function of temperature.

Table 6.2. Settling times for a specific centrifuge for sedimentation of particles[a]

Particle diameter (μm)	sp. g. mineral	Centrifuge speed (RPM)	time (min)
5	2.65	300	3.3
2	2.65	750	3.3
0.2	2.50	2400	35.4

[a]Data apply to 20°C, a distance of 15 cm from the centrifuge axis to the liquid meniscus, a 10-cm suspension depth, and 1 cm of sediment at the bottom of the centrifuge tube (Jackson, 1969).

as a function of the speed and radius of the centrifuge.

Tables 6.1 and 6.2 give solutions of Stokes's law for normal gravity and for centrifuge sedimentation (Jackson, 1969). Table 6.1 gives the settling times for a standing cylinder, and Table 6.2 gives times for one centrifuge.

The data in Table 6.1 are easily modified for other conditions. Just remember that if you double the settling distance you must double the time [see Eq. (6.2)]. Notice that a lower specific gravity (sp. g.) is used for 0.2 μm clay particles (Table 6.2). This reduction in sp. g. is necessary to take into account the bound water at the mineral surface because very fine particles have very high specific surface areas and their absorbed water is no longer a negligible portion of their volume. Particle-size separations should be made as soon as dispersion is achieved, because some clay minerals flocculate slowly even though they were once well dispersed.

Normal gravity settling in tubes is not recommended because it takes too long. Centrifugation is the best method. If you have a 15-cm machine, spin for 3.3 min at 750 rpm (Table 6.2) and decant the supernatant liquid into a separate container. The supernatant is the yield of the process. All the particles in it are < 2 μm e.s.d., but the material in the bottom of the centrifuge cup is not entirely >2 μm. It contains a good deal of the < 2 μm suspension, so if sample size is limited and there is not yet enough clay in the yield or if you wish to measure the amount of the < 2 μm fraction, redisperse the sediment from the cup by ultrasound, centrifuge again, and add the supernatant to the yield from the first separation. Three separations are about all that are practical because that constitutes almost all the < 2 μm material in the suspension.

Minerals can be separated based on their Fe contents. Dispersions of the entire <2 μm portion of a sample or separate size fractions can be passed through a system with a reservoir, a tube filled with stainless steel wool passing through a Frantz Isodynamic Magnetic Separator™, and a pump. Clay minerals that contain iron can be separated from those that do not. It is not uncommon to separate an Fe-poor chlorite from an Fe-rich one. The details of this technique have been given by Tellier et al. (1988).

PREPARING THE ORIENTED CLAY MINERAL AGGREGATE

We now have a dispersed suspension of clay-size material. Let us proceed to the preparation of the oriented clay aggregate. The first step often required is the concentration of this material to a level that it is suitable for the various sample preparation procedures.

The easiest way to make the suspension more concentrated is to collect the clay by ultracentrifugation and then redisperse it in a small volume of water. Any of the available angle-head centrifuges capable of 20,000 rpm are suitable for this purpose. If such a centrifuge is not on hand, or if the volume of the suspension is large (>1 L), then the clay must be flocculated, collected by gravity settling, and washed to redispersion in a small volume of water. Make the suspension 0.1M with respect to $CaCl_2$ and do something else for a while. Calcium is a good ion to use because it will saturate the ion-exchange positions in smectites, producing a glycol solvated type that will give unambiguous diffraction characteristics. The floccules will settle after a few hours and the supernatant can be removed by means of a vacuum hose attached to a tap-water-driven aspirator. The concentrated suspension must be centrifuged, redispersed in water, centrifuged, etc., until dispersion is attained. Try to achieve a sediment concentration of 60 mg of clay per milliliter of liquid in the final suspension.

X-ray diffraction samples must be smooth, flat, long enough, thick enough, and should be mineralogically homogeneous throughout their depth or thickness. These characteristics are crucial for quantitative representation (Hughes et al., 1994), and acceptable limits of deviation from the ideal for each of them are discussed in Chapter 9. There is a bewildering array of sample preparation methods in use throughout the many laboratories that deal principally with clay science. We recommend that you become proficient with four: (1) the so-called glass slide method; (2) the smear method; (3) the Millipore™ filter transfer method (Drever, 1973); and (4) the centrifuged porous plate (Kinter and Diamond, 1956). Table 6.3 summarizes, in order of ease of mastery, the strengths, weaknesses, and application for each method.

The Glass Slide Method

The glass slide method is best described as old faithful. It probably is the most commonly used routine method. Its only advantage, however, is its ease of application. Orientation is only fair, the aggregate usually is particle-size segregated with the finest material on the top, and the clay films are usually too thin for accurate diffraction intensities at moderate to high diffraction angles. Place a glass microscope slide (2.7 by 4.6 cm) in an oven at 90°C. It is a good idea to use a porous ceramic plate under the slide to catch any spills. Use an eye dropper to add the suspension so that the liquid covers the entire surface of the slide. You will be surprised at how much you can add without overflow. Four

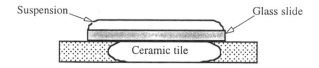

Fig. 6.1. Surface tension holds the suspension on a clean glass slide. The tile catches spills.

mL is easily possible on a standard slide (Fig. 6.1). Drying usually takes about 1 h at 90°C.

Some readers may object to the high drying temperatures recommended here. Poorly crystallized clay minerals, such as those found in soils, can be damaged by such high temperatures. Studies of hydrated halloysite, for example, require samples that have been dried at room temperature, or even to be run while wet. Most sedimentary rock clay minerals, however, are unaffected (as measured by X-ray methods) by temperatures near 100°C, and such temperatures greatly diminish sample preparation time. All the methods described above produce samples that can be dried at room temperature, if you require that option.

Glass substrates are not very useful for heat treatments because they soften unacceptably or warp at temperatures that are much above 300°C. Fused silica glass can be used, but it is expensive.

The Smear Mount Method

For identifying the constituents in a bulk sample and roughly estimating the quantities, the smear technique is a good compromise between time and skill required and the quality of the results. Done properly, phases should be represented in their true proportions, i.e., without segregation. Quarter the sample for a representative portion. For the quickest preparation, place about half of the end of a microspatula (a few mm^3) of sample in a medium (~ 100 mm diameter) mortar and thoroughly grind until the material is smeared on the mortar. Brush the powder onto a labeled slide and add one or two drops of a

Table 6.3 Features of different sample preparations

Method	Advantage	Disadvantage	Level of skill required	Application
Glass slide	Quick	All	Low	Qualitative analysis
Smear slide	Quick, moderately homogenous	Most	Moderate	Clay and non-clay minerals
Filter transfer	Homogeneous aggregate	Fair intensities	Moderate	Quantitative representation
Porous plate	Best intensities	Inhomogeneous aggregate	High	Crystal structure studies

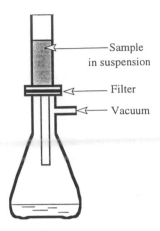

Fig. 6.2. Filtration apparatus.

dispersant solution to the powder, mixing with microspatula until a butter-like paste is formed. Sample material also can be collected from a dispersion by centrifugation. In this case, pour off the supernatant, mix the material in the bottom of the centrifuge tube, again to a butter-like paste. Smear by spreading the paste uniformly over the slide with the microspatula subparallel to the slide. Practice may be required to obtain a thin, even coating on the slide. You may find it helpful to hold the slide by attaching it to a vacuum hose while smearing. Preparation with a McCrone™ mill, i.e., micronizing the sample, as described below in the discussion about particle size and diffraction effects, will improve the precision of this technique. Virtually any size fraction can be smeared in this way. Some laboratories use the <16 µm size fraction as a way to "catch" all the clay minerals and also have a sampling of the nonclay minerals. However, samples with abundant kaolinite are not always as well sampled by this technique because kaolinite often occurs in vermicules up to 100 µm.

The Millipore® Filter Transfer Method

The filter transfer method produces only fair crystallite orientation, but the clay surface presented to the beam is likely to be representative of the proportions of the different minerals present. Therefore it is the recommended method for quantitative analysis of the size fraction being examined.

The method requires a vacuum filter apparatus, such as the one provided by the Millipore™ Corporation. The apparatus consists of a side-necked vacuum flask and a funnel reservoir clamped to a flat porous glass base. The filter separates the two pieces of the device (Fig. 6.2). The Millipore™ apparatus is the best we have seen because the junction between the permeable glass filter and the enclosing glass funnel has been ground to a flush joint. Other filter apparatus have a distinct ridge at this point that will be impressed into any sample prepared from such a device. You don't want any device that has a raised pattern on the glass filter plug unless you like sample surfaces that resemble waffles. We use the Gelman™ GA-6, 0.45 µm pore, 47 mm diameter Metricel™ filter, although others are doubtless equally useful. Insert the filter, add the suspension, and apply the vacuum.

We presume at this point that the application is quantitative analysis, so no particle-size segregation is acceptable. In other words, you cannot extend the filtration period beyond about 3 min or else the surface of the filter cake will be enriched in fine particles. Use the correction method for measuring μ^* described in Chapter 9 if, after this period, there is insufficient material on the filter to provide infinite thickness (the usual case). An alternative method is to stir the suspension in the funnel; stir just enough so that settling velocities are overcome. This keeps the suspension homogeneous. Then suction time can be as long as necessary to get a layer that is thick enough. To finish, be certain that liquid is still present in the funnel and then turn off the vacuum and bring the vacuum flask up to room pressure. This sequence is necessary because if air is drawn through the filter cake, it probably will not adhere properly to the glass substrate. Drain the excess suspension from the filter funnel, remove the wet filter, invert it, and carefully lay it face down on a glass substrate. Do this just as you would apply a decal onto an automobile windshield (Fig. 6.3). Start at one end and lower the filter surface sequentially in a smooth motion. Then invert the sample and examine the clay film through the glass. If there are any air pockets, which show as reflective (silver) areas, the sample is useless because it will fail to adhere to the glass at those points. You will have to try again and do better this time. In the original method, Drever (1973) recommends pressing the sample onto the glass by means of a roller. We do not use this procedure because it may produce a non-Gaussian particle orientation, which in turn produces an unknowable Lorentz factor.

Drying the sample-filter-glass slide combination correctly is crucial to a successful preparation. We place it in a oven at 50°C and check it frequently. It usually requires about 3 or 4 min to reach the critical point at which the filter must be stripped off. That point is reached when the filter surface shows opaque and translucent streaks. If you strip the filter too soon, the clay film may come off with the filter. If you wait so long that the filter surface is opaque and white, the filter is so brittle that you won't be able to remove it. But if you catch the filter with the correct moisture content and place the tip of your tongue precisely on the point of your upper, right canine tooth, you will have a well-oriented, smooth, and uniformly thick clay film centered on the glass slide.

Some may question the need to invert the clay film by transference from the filter to a different substrate, for it is this step that requires the most skill and causes most of the sample preparation failures. But it is the crucial one for good quantitative procedures because other methods (except for the smear method) invariably lead to particle-size segregations within the sample. Think of it this way. At first, flow rates through the filter are high because the filter has not yet been clogged with the sample. So long as the vertical movement of the liquid is fast compared with the settling rate of the coarsest particles, there can be no significant particle-size segregation. This condition applies for the sample

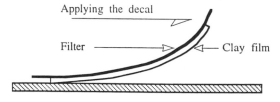

Fig. 6.3. Application of a wet clay film onto a glass substrate.

portion nearest to the filter face. When the filter is inverted, the most particle-size-representative portion of the sample is presented to the X-ray beam, and the least representative (enriched in the finest sizes) becomes the bottom. Due to the logarithmic absorption of the beam in the sample, the surface of the inverted clay mineral film produces a disproportionately large proportion of the diffraction intensity, with the result that errors due to particle-size segregation are minimized.

As mentioned before, cation exchange or saturation can be accomplished relatively simply with the Millipore™ method. When the sample has reached sufficient thickness, remove the vacuum, add a few milliliters of the exchange solution, draw it through the clay cake, and follow it with a few milliliters of distilled water. This procedure leaves the clay minerals homoionic and sufficiently salt-free for X-ray diffraction analysis. Cation saturation can be eliminated if the sample is known to be free of expandable clay minerals.

The Centrifuged Porous Plate Method

The porous plate method produces thick aggregates that have very high degrees of preferred orientation and therefore produce excellent diffraction patterns. Indeed, the best-prepared tiles produce integrated diffraction intensities comparable to those from single crystals. We have studied examples of illite/smectite and illite that give measurable intensities for 00l peaks out to the limit of 130° 2θ, corresponding to the illite 00,12. Unfortunately, the method suffers from particle-size segregation effects and is therefore useless for quantitative analysis. It is too skill intensive and too time consuming for routine qualitative studies, but it is the one indicated for crystal structure studies or the detailed characterization of pure clay minerals. It requires unglazed ceramic tile that has been cut into rectangles approximately 2.5 by 4.5 cm to serve as sample substrates.

You or your machine shop must prepare a sample holder assembly similar to the one in Fig. 6.4. Aluminum is a suitable fabrication material, although others are acceptable. The porous plate is faced with rubber gaskets above and below. Insert it into the assembly and tighten the screws to provide a liquid-tight seal for the whole apparatus. Put a pair of these assemblies into centrifuge cups on opposite pans of a laboratory balance, and add clay suspensions to each cup until they balance. Screw a threaded rod into the tapped hole in the top of each assembly to facilitate transfer to the centrifuge cup. Centrifuge the samples for about 10 min at 2,000 rpm. Remove the sample assemblies, decant any supernatant liquid, disassemble the apparatus, and dry the porous plate preparations in an oven below 100°C. Higher temperatures are undesirable because boiling of the pore fluids will destroy the uniformity of the aggregate. Many things can go wrong during this procedure, and we suggest that you practice on a suitable standard clay before you apply it to a valuable sample that may be in short supply.

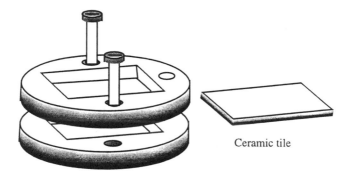

Ceramic tile

Aluminum tile holder assembly

Fig. 6.4. Aluminum assembly for holding porous plates.

A strong advantage of this method lies in the character of the porous tile substrate. After the sample is prepared, dried, and analyzed, it can be carefully replaced in the centrifuge plate assembly, and the exchangeable cation can be changed by means of the simple expedient of centrifuging a small amount of a chloride solution through the clay film and porous plate. In addition, the ceramic tile is unaffected by high temperatures, so the substrate is an ideal one for the various heat treatments that can be essential for correct identification of some clay minerals, and clay films on ceramic tiles do not lose glycol as readily as films on glass slides do because the porous tile acts as a reservoir for glycol retention.

Dealing with Curlers or Peelers

The most vexing problem in sample preparation is the occasional encounter with a "peeler." Some materials, no matter how they are treated, shrink on drying with the result that the sample curls, breaks up, and separates from the substrate. This behavior is usually due to (1) the presence of angular grains, such as quartz, that destroy the preferred orientation of the clay mineral crystallites; (2) partial flocculation of the clay minerals during sample preparation, which also destroys preferred orientation; (3) a large particle-size range with the finest material on the surface of the clay film, leading to differential shrinkage during drying; and (4) the presence of gelatinous, hydrated colloidal material, such as some hydrated iron hydroxides and organic matter in soils. Samples that produce peelers often cannot be dealt with in any fashion that will produce good 00*l* diffraction patterns. The best that you can do is collect the material by ultracentrifugation as a first step. Then drain off the supernatant water. Using a spatula, smear the clay paste onto a dry unglazed ceramic tile. Capillary forces will draw the pore fluids into the plate quickly, leaving a moist but not wet clay film. Run this sample on the diffractometer before it dries and curls. If ethylene glycol solvation is required, stir into the ple is smeared onto the tile. This procedure will produce poor diffraction patterns, but they are better than nothing. For some reason, some expandable clay minerals do not respond normally to glycol solvation performed by this method, so be wary when you interpret the diffraction pattern.

MAKING THE RANDOM POWDER MOUNT

In the preceding sections we have described and discussed procedures for maximizing the preferred orientation of the flake morphologies of clay minerals. For three-dimensional diffraction studies you will need to see all (*hkl*) reflections of the mineral or minerals present, with the correct relative intensities of their reflections, which requires perfectly random orientation of the particles of the sample. However, the platy character of the clay minerals makes random orientation difficult to achieve. Most crystals break or cleave more readily along some planes than others. When packed as a powder, orientation of individual grains tends to be governed by the juxtaposition of faces formed by these preferential breakages, and this process works against the production of a randomly oriented aggregate. Sample preparations that have no preferred orientation are crucial for quantitative analyses of mixtures of clay and nonclay minerals and for interpretations of diffraction patterns of clay mineral polytypes (Chap. 10). Various procedures for achieving random orientation of the sample powder, some elaborate, some simple, have been suggested. Spray drying holds great promise, but to date a suitable commercial apparatus for this has not become widely available (Smith et al., 1979). Brown and Brindley (1980, p. 310), Brindley (1980, pp. 426-27), and Bish and Reynolds (1989) review a number of other methods. A back-loading procedure was described in an earlier edition of this book, but recently we have had better results with side-loading.

One of the most important steps is to start with a powder of small and uniform particle size, 5 to 10 μm. Perhaps the easiest way to grasp the importance of particle size is to think about particle statistics. Picture two particles mounted on a slide that you're going to present to the X-ray beam. As the diffractometer moves through the arc in search of signals, the two particles may be oriented so that they will diffract from one, or even two, of their many spacings. However, two particles cannot, under any circumstances, provide a diffraction signal for all the spacings they contain. If you increase the number of particles to 100, your chances of getting diffraction from most of the spacings is much improved. However, the chances that the spacings will yield diffraction signals in proportion to the real distribution of spacings are pretty slim. For any sort of quantitative determination, precise, i.e., repeatable, relative intensities are very important. This sort of relative intensity will come only from the averaging of diffraction from millions to billions of individual crystal fragments. If you had a sample of quartz particles packed into a holder and its volume was 25 mm^3, and the particles were 10 μm cubes, how many particles would you have? How does about 25 million sound? And what if you were to double the size to 20 μm, or halve the size of the particles to 5 μm, how many would you have?

Klug and Alexander (1974, pp. 365-68) offered an example of the importance of particle size. Using powdered quartz, they did ten replicate analyses on each of four particle-size ranges and found these mean percentage deviations in peak intensity: ±18.2% for 15-20 μm powder; ±10.1% for 5-50 μm powder; ±2.1% for 5-15 μm powder; and ±1.2% for <5 μm powder. Hand grinding does not yield particles much less than 40 μm. You can see that using the material that passes through a 325-mesh screen, i.e., <44 μm, as is so often recommended, is going to give you poor precision. Small particles, and we recommend trying to get the narrowest possible size range centered at 5 μm, give you two other distinct advantages in addition to improved precision: (1) They decrease or eliminate the preferential absorption of the X-ray beam by minerals with heavy elements in them, a phenomenon called microabsorption (see Bish and Reynolds, 1989, p. 82); and (2) for whole rock samples, the nonclay minerals with good cleavage are far less likely to preferentially orient. You will need to use some kind of grinding device and grind your sample in water or alcohol. We like the McCrone™ mill.

The relative intensities of the set of peaks from a random mount for a given mineral are used in the Joint Committee on Powder Diffraction Standards (JCPDS) *Powder Diffraction File* system of identification, so if you wish to compare relative intensities for a series of peaks from one mineral, you will need a randomly oriented aggregate. Relative intensity is one of two criteria used to classify all diffraction patterns, the other being the size of the spacings. The thousands (probably 35,000 by now) of cards in the JCPDS file are subdivided into Hanawalt groups according to a range of spacing sizes responsible for the most intense peak. Within each Hanawalt group, compounds are arranged according to the size of the spacing responsible for the second most intense peak. This system was established by two papers, Hanawalt and Rinn (1936) and Hanawalt, Rinn, and Frevel (1938). Both of these historically important papers have been reprinted in the journal *Powder Diffraction*, Vol. 1, No, 1, March 1986. The Committee and the International Centre for Diffraction Data, 1601 Park Lane, Swarthmore, PA 19081-2389, publish in book form the minerals selected from all the compounds they catalog. The 1993 edition, *Mineral Powder Diffraction File Databook, Sets 1-42*, catalogs 3800 data cards representing 3200 minerals. An appendix has X-ray diffraction tracings of typical clay minerals because recognition of the clay minerals is often a "gestalt" process, i.e., the pattern just looks, in its entirety, like such-and-such clay mineral.

Deviation from perfect randomness can be tolerated for some problems. For others, it cannot. We recommend two methods, one for every day problems (side-loading of the sample holder) and one for those instances when as little deviation as possible must be achieved. The latter requires freeze drying and side loading.

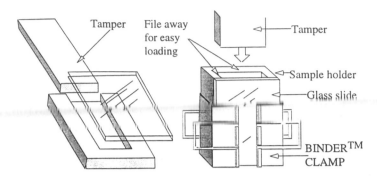

Fig. 6.5. Sample holder and accessories for side-loading method for a random powder mount.

Everyday random powder packs

For the every-day method, an example of the simple type of device used in most labs for side-loading is shown in Fig. 6.5. It is an advantage to have the side of the glass slide frosted, or rough on a microscopic scale, next to the powder. This helps prevent particles from orienting against the plane of the glass. The glass is held to the sample holder by a pair of Binder™ clips. These serve as legs so the assemble can stand upright. Different labs use different funnel-like devices to guide the powder into the holder. We have one machined from an aluminum cylinder that has a rectangular notch cut into the bottom that just fits over the top of the assembly shown in Fig. 6.5, and has a funnel machined into the top. It takes judgment and some practice to apply the proper amount of packing or tamping of the powder to fit it snugly between the glass plate and the bottom of the sample holder for enough adhesion that it won't fall out. If packed too hard, orientation will be introduced. Or, the tamping pressure may store some strain in the aggregate, so when the glass slide is removed, the surface of the powder may bulge upward, making it impossible to align a flat sample surface with the goniometer axis. If the powder is not packed enough, it may fall out of the holder. This method is appropriate for whole samples powdered to an appropriate grain size (as discussed above), or for size fractions of a sample.

Freeze-dried random powder packs

For circumstances in which deviation from perfect randomness must be minimized, freeze-dried clay suspensions that are side-loaded into specially designed sample holders (Fig. 6.5) have given us our best results. Freeze-drying equipment is readily available commercially. You will need a shell freezer which rotates rapidly in a freezing bath and a cylindrical vessel that contains the dispersed sample. This step in the process lines a cylindrical vessel with a concentric layer of ice that has the high specific surface area necessary for efficient evaporation in the freeze drier. A good place to start is with 200 mL of suspension containing 200 to 500 mg of sample. The frozen suspension is then attached to the freeze drier and pumped to dryness, which takes overnight. This may seem like a long time, but you can dry perhaps six samples at once, and drying in an oven is not a quick operation either.

The freeze-dried powder has an extremely low density and cannot be handled unless it is somewhat compacted. We attach a rubber stopper to a glass rod and whip the powder much like you whip eggs to scramble them or make an omelette. If your lab is in a dry climate, or if you work in winter when the relative humidity is low, an alpha-emitting probe is useful to combat the static electricity that makes the powder very mobile. (**N.B.** Remember our caution from Box 2.2, p. 29, that alpha particles do about 20 times as much damage to tissue as X-ray photons.) We use the same sample-holding assemblage as for the everyday random powder pack, as shown in Fig. 6.5. The procedure rarely produces a sample that is dense enough to produce good diffraction tracings without packing, so careful tamping is more important for freeze-dried samples than for those simply powdered. The carefully packed sample may be run in a chamber in which the atmosphere can be controlled. We stream tank N_2 through the chamber to prevent heated samples from rehydrating. (Chambers can be made relatively easily. We have seen pieces of 6 in. diameter Plexiglas™ tube, clear Nalgene™ buckets cut in half, etc., work effectively as chambers.)

The area of the sample well in the holder that will be exposed to the X-ray beam depends on the slit system you use and the lowest diffraction angles anticipated. The lowest-angle clay mineral *hkl* peaks are at about 19° 2θ (CuKα), and for a one-degree divergence slit, the sample length should be about 2.5 cm for typical (20 cm) goniometer radii; this is shorter than that necessary for diffraction tracings that include the first-order peaks of the clay minerals. The depth should be 1 mm or greater for infinite sample thickness (CuKα) at diffraction angles up to 70°, which are the highest you will probably record. A full sample holder will require between 150 and 300 mg of powder, depending on how tightly it is compacted.

This all takes some practice, but we have found that it often yields aggregates that, judging by agreement between experimental and calculated diffraction patterns, are almost perfectly randomly oriented.

There are a few pluses and minuses. The sample suspensions must be washed free of salt because all the dissolved ions in the 200 mL of solution will be concentrated as crystals in a few hundred milligrams of powder. The method works poorly on Na-saturated expandable minerals, so Ca or Mg saturation should be routine. Pure Wyoming bentonite has so far resisted all our attempts to make a random powder--it always forms a leathery film of oriented crystals. But there is a way to deal with that is better than nothing. Cut the film into strips with scissors, roll them into tiny tubes, and put them in epoxy. The hardened epoxy surface can be ground flat to make a good sample surface. Of course, the diffraction pattern will contain the very broad maxima due to the epoxy, but that is easily dealt with by running a pattern from the pure epoxy, and then by means of a simple computer program, subtracting that from the clay mineral/epoxy data file.

An important plus is that the freeze-dried powder is very easily dispersed into suspension. Add water, a quick zap with the ultrasonic probe, and you have a stable dispersion suitable for making slides of oriented aggregates. Freeze-dried powders are in a good form for storage in your sample or standard library because they are easily studied as random powders or oriented aggregates, and need no further preparation for other studies such as chemical and isotope analyses.

ETHYLENE GLYCOL SOLVATION

Most clay mineral samples should be analyzed in an air-dried condition, an ethylene glycol-solvated condition, and after enough heating to collapse any expandable layers. If you are experienced and already know something about the mineralogy of the samples in a suite, you may elect to eliminate one or the other of these analyses, but that election should only rarely eliminate the ethylene glycol analysis because, otherwise, you risk a serious misidentification. The diagnostic adsorption of ethylene glycol by smectite was discovered by Bradley (1945).

The best general method of ethylene glycol solvation is to expose the sample to the vapor of the reagent for at least 8 h at 60°C. Use a large desiccator or a Pyrex casserole dish with lid, add 100 to 200 mL of ethylene glycol, insert some sort of a platform similar to the ones used in desiccators, and put the setup in an oven at 60°C. You will have to dedicate one oven permanently for this purpose. Place sample mounts face up on the platform and do not allow them to contact the liquid glycol. Label the sample slides with a diamond marker, because ethylene glycol is a good solvent and may remove identification marks that have been made with a felt-tipped pen. An ordinary pencil works well for ceramic tiles, if those are what you are using. After solvation, analyze the samples immediately. If the clay film is on a glass substrate, you have about 1 h to complete the analysis, because, for longer times, the glycol will evaporate away sufficiently to affect the expansion of clay minerals. Good procedure requires that you scan the low-angle region first, run the complete diffraction pattern, and repeat the low-angle scan. If the two low-angle scans are identical, you are assured that there has been no significant change in the solvation state during the analysis, and there frequently is. An alternative is to rig some sort of environmental chamber around the sample and place an open dish in the chamber with a few mm of glycol and a wick in it, or pass glycol-saturated gas through the chamber while running the glycolated sample (as mentioned in the section on freeze drying). Such chambers also are required when running samples that have been heated to keep them from rehydrating. In handling large numbers of samples, you might use a large desiccator with a few cm of glycol in the bottom kept at room temperature. Samples should stay in such a desiccator for at least two days.

Novich and Martin (1983) suggested two other methods: (1) while the sample is dispersed, mix glycol in the dispersion; and (2) saturate a laboratory tissue with ethylene glycol and leave the sample mounts face down on the tissue for about 8 h. For clay minerals that solvate with difficulty, this procedure may be carried out in a closed chamber at 60°C (Whitney and Northrop, 1987).

Ethylene glycol solvation is difficult to accomplish with random powders if they contain large amounts of smectite, for the smectite swells enough to displace the powdered mineral surface above the plane of the surface of the sample holder. The best method is the vapor technique described before. Simply place the random powder mount in an ethylene glycol atmosphere at 60°C for 12 h. The surface may swell above the surface of the aluminum plate and require pressing with a spatula. Unfortunately, such pressing will introduce some orientation of the clay minerals at the surface, but you will have a preparation that is sufficiently randomly oriented to serve well for the identification of the various clay mineral polytypes or to provide good 060 intensities.

There are several other special sample preparation techniques for reducing background, for enhancing peak intensities, and for dealing with very small amounts of sample. These are thoroughly covered by Bish and Reynolds (1989, pp. 93ff).

FINAL NOTE

Środoń and Eberl (1980) made a plea in which we concur. When you write a report or submit a paper for publication, include the X-ray diffraction tracings in the form in which they came off of the diffractometer, not after they have been smoothed by a draftsperson. Such traces are data that can be reinterpreted in the future as more is learned about peak shapes, relative intensities, and other, as yet unrecognized features. Środoń and Eberl have some other good suggestions, too, that you may want to look at.

REFERENCES

Bish, D. L., and Reynolds, R. C., Jr. (1989) Sample preparation for X-ray diffraction: in Bish, D. L., and Post, J. E., editors, *Modern Powder Diffraction*, Reviews in Mineralogy **20**, Mineralogical Society of America, Washington, D.C., 73-99.

Bodine, M. W., Jr., and Fernalld, T. H. (1973) EDTA dissolution of gypsum, anhydrite, and Ca-Mg carbonates: *J. Sed Pet.* **43**, 1152-56.

Bohor, B. F., and Triplehorn, D. M. (1993) *Tonsteins: Altered Volcanic-Ash Layers in Coal-Bearing Sequences*: Geological Society of America, Special Paper 285, Geological Society of America, Inc., Boulder, Colorado, 44 pp.

Bradley, W. F. (1945) Molecular associations between montmorillonite and some polyfunctional organic liquids: *J. Amer. Chem. Soc.* **67**, 975-81.

Brindley, G. W. (1980) Quantitative X-ray mineral analysis of clays: in *Crystal Structures of Clay Minerals and Their X-Ray Identification*, Brown, G., Brindley, G. W., editors, Monograph No. 5, Mineralogical Society, London, pp. 411-38.

Brown, G. and Brindley, G. W. (1980) X-ray diffraction procedures for clay mineral identification: in *Crystal Structures of Clay Minerals and Their X-Ray Identification*, Brown, G., Brindley, G. W., editors, Monograph No. 5, Mineralogical Society, London, pp. 305-59.

Drever, J. I. (1973) The preparation of oriented clay mineral specimens for X-ray diffraction analysis by a filter-membrane peel technique: *Amer. Minerl.* **58**, 553-54.

Gluskoter, H. J. (1965) Electronic low-temperature ashing of bituminous coal: *Fuel* **44**, 285-91.

Hallberg, G. R., Lucas, J. R., and Goodman, C. M. 1978. Semiquantitative analysis of clay mineralogy. Part I: in Standard Procedures for Evaluation of Quaternary Materials in Iowa. 5-21. Iowa Geol. Survey, Tech. Information Series No. 8.

Hughes, R. E., Moore, D. M., and Glass, H. D. (1994) Qualitative and quantitative analysis of clay minerals in soils: in Amonette, J. E., and Zelazny, L. W., editors, *Quantitative Methods in Soil Mineralogy*: SSSA Miscellaneous Publications, Soil Science Society of America, Madison, Wis., 330-59.

Jackson, M. L. (1969) *Soil Chemical Analysis-Advanced Course*: 2nd Ed., published by the author, Madison, Wis., 895 pp.

Kinter, E. G., and Diamond, S. (1956) A new method for preparation and treatment of oriented-aggregate specimens of soil clays for X-ray diffraction analysis: *Soil Sci.* **81**, 111-20.

Klug, H. P., and Alexander,L. E. (1974) *X-Ray Diffraction Procedures*, 2nd ed.: Wiley, New York, 966 pp.

Novich, B. E. and Martin, R. T. (1983) Solvation methods for expandable layers: *Clays and Clay Minerals* **31**, 235-38.

Ostrum, M. E. (1961) Separation of clay minerals from carbonate rocks by using acid: *J. Sed. Petrol.* **31**, 123-29.

Smith, S. T., Synder, R. L., and Brownell, W. E. (1979) Minimization of preferred orientation in powders by spray drying: *Adv. X-Ray Analy.* **22**, 77-87.

Środoń, J., and Eberl, D. D. (1980) The presentation of X-ray data for clay minerals: *Clay Minerals*, **15**, 317-20.

Tellier, K. E., Hluchy, M. M., Walker, J. E., and Reynolds, R. C., Jr. (1988) Application of high gradient magnetic separation (HGMS) to structural and compositional studies of clay mineral mixtures: *J. Sed. Petrol.* **58**, 761-63.

Whitney, G., and Northrop, H. R. (1987) Diagenesis and fluid flow in the San Juan Basin, New Mexico—regional zonation in the mineralogy and stable isotope composition of clay minerals in sandstone: *Amer. J. Sci.* **287**, 353-82.

Chapter 7
Identification of Clay Minerals and Associated Minerals

We have considered the structures and the chemical nature of clay minerals as well as their interaction with X-rays. Now we will focus on the process of gathering and using X-ray diffraction data for the identification and analysis of clay minerals, but we must be careful not to narrow our focus too much for three reasons. First, not all minerals in the clay-sized fraction are clay minerals. Second, there is almost always more to a rock sample than the clay-sized fraction. Third, X-ray diffraction, though extremely powerful, often must be supplemented by information from other analytical techniques.

There are two levels of analysis, qualitative and quantitative, and, depending on the time available and the nature of a project, there are different levels of qualitative analysis. But if an analysis is going to be called quantitative, there is only one level and that is the very best that you can do, and even that will often not be good enough. It seems safe to assume that this text provides sufficient background for anyone who wants to do simple qualitative analysis of geologic materials consisting of several components. On the other hand, if quantitative analysis is achievable at all, it requires a great deal of experience, patience, luck, and skill. At present, quantitative analysis may be more of an art than a science.

The qualitative identification procedure begins by searching for a mineral that will explain the strongest peak or peaks, then confirming the choice by finding the positions of weaker peaks for the same mineral. Once a set of peaks is confirmed as belonging to a mineral, these peaks are eliminated from consideration. From the remaining peaks, again search for a mineral that will explain the strongest remaining peak or peaks and then confirm this by looking for its peaks of lesser intensity. Repeat until all peaks are identified. For example, though quartz is not a clay mineral, it is very often present in the clay-size fraction. Its strongest peak occurs at 26.65° 2θ for CuKα. If this peak is present, then automatically check the 20.85° 2θ position, where the second-most-intense peak occurs. By using Table 7.8B you can locate the other peaks of quartz, all which then may be removed from consideration as you concentrate on the remaining peaks. Quartz can be more of a friend than a foe. Its peak positions are, for our purposes, invariant because the quartz structure tolerates no significant atomic substitutions. Thus the quartz pattern is a built-in internal standard against which you can estimate the accuracy and precision of peak positions for the other phases present. Precise measurements of d are required to measure the composition of mixed-layered minerals, and for the very best work you must add such a standard to your sample if quartz is not present. Errors in peak position arise mostly from the sample position errors mentioned in Chapters 2 (p. 44) and 9 (p. 300), and you often cannot minimize them to acceptable levels. The use of an internal standard is the best, and often the only, solution.

We encourage you to collect a set of X-ray diffraction tracings of the minerals you commonly encounter. Comparing tracings of unknowns against those of known minerals is an effective, quick, and accurate identification procedure. For example, Frondel (1962, p. 34) has a foldout of a diffraction tracing and table of spacings of quartz that you will find useful. Make your own at the conditions at which you normally run your machine.

CLAY MINERAL IDENTIFICATION—GENERAL PRINCIPLES

Clay minerals are identified by using X-ray diffraction patterns (diffractograms or diffraction tracings) of oriented aggregates that enhance the basal, or 00l, reflections. hkl peaks are not very diagnostic because the structures of most clay minerals are very similar in their X and Y directions. It is the atomic pattern along Z that is the most different from mineral to mineral. The JCPDS powder diffraction cards that are so useful for the identification of other minerals are useless or worse for clay minerals because they usually have been prepared from random orientations of the mineral powder; often what is listed as the strongest reflection for a clay mineral or mica will not even be visible on a pattern prepared from an oriented aggregate. Some of the newer cards, due to G. W. Brindley, are exceptions, and do indeed show the 00l diffraction profile for oriented aggregates. In addition, in the JCPDS 1993 edition of the *Mineral Powder Diffraction File Databook*, Appendix A shows 00l XRD tracings of common clay minerals. In this book, all diffraction pattern references for clay minerals apply to the 00l series unless otherwise specified, and CuKα radiation is assumed for all tables and calculated and experimental diffraction patterns.

The intent here is to provide a practical basis for the identification of the "simple" or noninterstratified clay minerals. The interstratified minerals are worth a chapter of their own, so methods for their identification are deferred to Chapter 8.

Clay mineral diffraction patterns contain a good deal of character. This character is manifested by the peak's position, intensity, shape, and breadth. Peak position is determined by the Bragg law (Chapter 3), which is written as $n\lambda = 2d\sin\theta$. If the analysis is one dimensional, l may be substituted for n and the equation rearranged to give $l\lambda/2d = \sin\theta$. Now we have two constants, $d = d(001)$ and λ, and if θ is small, the angle may be substituted for its sine and we have a working result of $\theta = l \times$ (constant). This equation means that at small diffraction angles the various members of the 00l series are equidistant.

What are "small diffraction angles"? Well, most of the important clay peaks are at 2θ values of 40° or less; therefore, θ is 20° or less, and that value is sufficiently small to fit the foregoing argument to a pretty good approximation. Figure 7.1 shows the separation of the 00*l* series for chlorite, and you can see that the peaks are evenly spaced. In practice a ruler is unnecessary—simply holding your fingers at the proper fixed separation is good enough to identify the different reflections that belong to the same clay mineral.

Peak intensity was treated in Chapter 3. We simply remind you that relative 00*l* intensities are controlled by chemical composition and the positions of atoms in the unit cell, some characteristics of the sample, and the Soller slits on the goniometer.

Peak breadth for the 00*l* reflections from clay minerals is inversely proportional to the mean dimension (in Å) normal to the diffracting planes in an optically coherent domain, i.e., the thinner the crystallites (= domain), the broader the peak. Recall that we discussed reasons for peak breadth in the text associated with Fig. 3.13, p. 85. By breadth, we mean the width of the diffraction maximum or peak at half its height above background (*B* in Fig. 3.12). Well-crystallized minerals such as quartz, for which domains are thousands of Ångstroms, produce sharp lines whose breadths depend only on the optical distortions inherent in the diffraction apparatus. But as domain size becomes smaller, noticeable line broadening occurs and is easily evident in diffraction patterns of clay minerals for which the optical domains are a few hundred Ångstroms or less in thickness. *The importance of line breadth for*

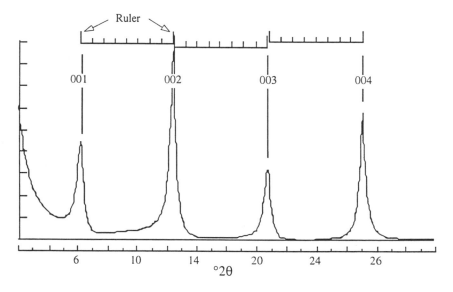

Fig. 7.1. Uniform separation of members of the 00*l* diffraction series for chlorite.

qualitative analysis is that it tells us at a glance which reflections are due to clay minerals and which might be assigned to other minerals, such as quartz, calcite, zeolites, pyrite, etc. In addition, all members of the 00*l* series from a given species have the same breadth (at low 2θ), so if your pattern contains broad (> 0.2 to 0.3°) lines, some of which are broader than others, the sample contains more than one clay mineral type. (We will deal with exceptions to this in the next chapter.) Figure 7.2 shows a pattern of a sample that contains thick crystallites of illite (sharp peaks), thin crystallites of kaolinite (broad peaks), and quartz (very sharp peaks). With no other knowledge, you should be able to conclude that the sample contains at least two clay minerals and one nonclay mineral. The rule of equidistant 00*l* peaks shows that the broad peaks belong to one series and the sharper peaks to another. You will, after you gain a bit of experience, also guess that the illite 003 reflection is superimposed on an intense peak of a nonclay mineral. This conclusion can be verified by a high-resolution scan plotted at an expanded horizontal scale that will produce, for the reflection near 26.6° 2θ, a shape that can only be interpreted as the superposition of a sharp peak on a much broader one.

   Identification of clay minerals can be accomplished by careful consideration of peak positions and intensities, which are compared to published values in the literature. But with a little experience, you will not identify clay minerals this way. After all, the identification of clay minerals is a simple procedure, much simpler than the qualitative analyses of bulk rock powders that can contain many minerals, each of which has a complicated pattern that produces many interferences on the peaks of other minerals

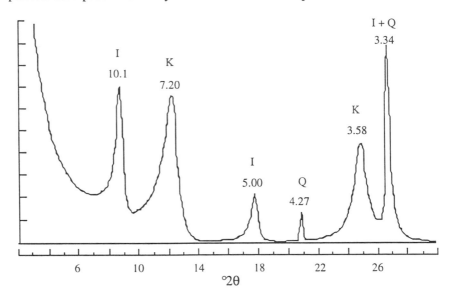

Fig. 7.2. A mixture of illite, kaolinite, and quartz.

present. The diffractionist will quickly grow familiar with the "image," that is, the diffraction pattern, of a given clay mineral species and will recognize it in a diffractogram in much the same way that the face of a friend can be instantly recognized in a crowd. If a mixture is too complicated for that, the best way to proceed is to ask, "Is illite present?" Then, "Is vermiculite present?" "Is kaolinite present?" "Is chlorite present?" "Is smectite present?" If these questions can be answered, the analysis is complete, for with the exceptions of palygorskite, sepiolite, and the mixed-layered clay minerals, the interrogation has considered all the clay minerals likely to be found in sedimentary rocks, and if halloysite is included, the list covers soils as well. Some may regard this subjective procedure as "cheating." But why do things the hard way? If you are a geologist and have collected the rocks yourself, or at least communicated with the person who did, then you know that the sample is a limestone or a shale or some other sedimentary rock type. Limestones probably do not contain olivine, nor is a sandstone likely to be rich in cuprite. Qualitative analysis contains elements of creative art. It is not entirely a science. One element is involved with interpreting the shapes, intensities, and positions of peaks, and another is the consideration of the geological setting of the sample when interpreting the X-ray diffraction pattern.

For some samples, peaks will remain after others have been assigned to identified minerals. Then a second level of inquiry is needed to identify unfamiliar (or exotic) minerals. At that point, reference must be made to published data on the possibilities. In this category lie talc, pyrophyllite, paragonite, and the nonclay minerals that make up the aluminum and iron hydroxides and oxy-hydroxides, which may produce diffraction peaks that are broad like those of the clay minerals. Zeolites can be troublesome because they have reflections in the same low-2θ range as the clay minerals, but they can usually be identified because their peaks are sharp and thus do not resemble clay mineral reflections. If you are reduced to identifying a completely unknown phase, first check the phases that match the peaks at the lowest diffraction angles. These are diagnostic because there are few phases with peaks in the very low-angle range. Check this out by examining the JCPDS index for inorganic phases. Not many minerals have reflections for which $d > 10$ Å, but many substances have peaks of $d = 2$ Å or so. Chen's (1977) booklet listing the key peaks of minerals by 2θ and d can be helpful for pursuing the identity of unknowns.

These methods provide only a first step in qualitative analysis, because no consideration has been given to the speciation of the clay groups identified. For example, the word *smectite* covers a number of different smectites, each of which is given a name and has a defined range of composition (as in Table 4.3, p. 131). But for many applications a crude analysis is all that is required, and that can often be accomplished simply by inspection of the diffraction pattern.

Many identifications will be compromised by peak interferences that preclude or at least complicate the simple approach we have described. Then the material must be subjected to one or more chemical treatments (see Chapter 5) and reexamined by X-ray diffraction methods.

Let us now discuss the diffraction patterns of the various clay minerals, point out the common interferences, and suggest ways to deal with them. The diffraction patterns shown here have been calculated by computer methods and record correct relative intensities for each pattern. Absolute intensities have been adjusted to provide the scale most favorable for viewing the diagnostic features of each pattern. The relative intensities are correct only if the sample is thick enough and long enough (see Chapter 9). A 1° beam (divergence) slit is assumed in conjunction with a sample length of 4.6 cm, and this configuration causes weakening of peaks in the 5° to 6° 2θ region by about 50%. These conditions were selected because the 1° slit and such sample lengths are routine in most laboratories. We need not worry about the lowangle intensity distortion because it isn't very important for qualitative analysis, and the low-angle peaks are not suitable for quantitative analysis for reasons that are discussed in Chapter 9. For all calculations, a Lorentz-polarization factor has been applied that is realistic for typical, well-oriented samples, analyzed using a diffractometer equipped with only one Soller slit (Reynolds, 1986). If your instrument uses two Soller slits, the intensities at high 2θ will be diminished somewhat, compared with the patterns shown here.

The X-ray diffraction patterns in Chapters 3 through 9 have been calculated by the computer program NEWMOD© (Reynolds, 1985), details of which are given in the Appendix. The program handles mixed-layered and simple clay minerals and takes into account diffractometer characteristics and setup. It provides adjustable chemical and structural parameters. With a few exceptions, we have elected to use calculated diffractograms because of the difficulties inherent in accumulating clay mineral standards that (1) have excellent diffraction characteristics, (2) are free from other mineral components that produce diffraction peak interferences, (3) have well-documented compositions and structures, and (4) are sufficiently diverse to illustrate the many examples required. Our approach is to show you many complete 00*l* diffraction patterns to provide the kind of experience necessary for the development of good analytical instincts; we don't think that tables of spacings and intensities accomplish that end. Finally, as you will see in Chapter 9, calculated diffraction patterns provide the best means of standardization for quantitative analysis and will be used for that purpose throughout our discussion. A disadvantage of calculated patterns may be that they "look too good" and don't provide you with an appreciation of the real world. A collection of your own diffractograms, therefore, is still an effective device for preliminary identification.

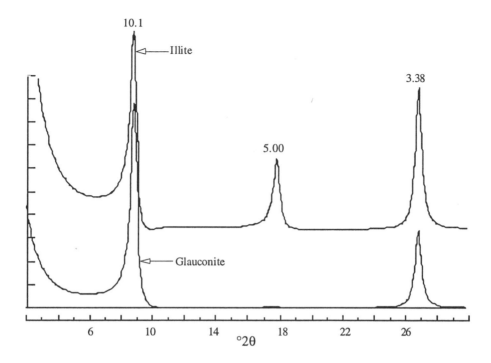

Fig. 7.3. Illite and glauconite.

Illite and Glauconite

Figure 7.3 shows patterns for pure illite and pure glauconite. The profiles are unaffected by ethylene glycol solvation and heating to 550°C and are difficult to confuse with those of any other clay mineral. Glauconite has a higher 001/003 intensity ratio than illite, but the main difference is the very weak or nonexistent glauconite 002 reflection whose weakness is caused by heavy scattering from octahedral iron. (See Fig. 3.10, p. 75 and related discussion. Can you adapt the problem associated with Fig. 3.10 to explain the difference in the 002 intensities in Fig. 7.3?) Diffractograms of the dioctahedral glauconite are difficult to distinguish from those of trioctahedral phlogopite and biotite unless the 060 reflection (see Table 7.4) can be measured to differentiate trioctahedral from dioctahedral minerals, but phlogopite-biotite minerals are rarely encountered in <2 μm size fractions of sedimentary rocks.

Chlorite and Kaolinite

Chlorite and kaolinite have very different structures and geological occurrences, but they are discussed together here because of the difficulties they present in mutual mixtures. Chlorite has a basal series of diffraction peaks superimpose or nearly superimpose on the members of the kaolinite 00*l* series. High-Fe chlorites have weak odd-order reflections, so weak that the 001 peak is easily obscured or not noticed, so the distinction between chlorite and kaolinite is most difficult when Fe-rich chlorites are involved. Figure 7.4 illustrates the problem.

There are several methods for identifying these minerals in mixtures. Most kaolinites have the 002 peak at 24.9° 2θ, and common chlorites have their 004 reflection at ~25.1° 2θ. The lines will be sharp if the crystallites are thick, and if they are, you can see resolution or partial resolution of the two. For some samples, evidence for both phases is suggested by a greater peak breadth at 25° 2θ than at 12.5° 2θ. If large concentrations of both phases are present, the best method of identification is to look for the kaolinite 003 and the chlorite 003 peaks. Neither of these weak reflections is interfered with by reflections from the other mineral. The intensities of the 00*l* series can also be of help here. The intensity ratio for the kaolinite 002/003 is about 10. If the measured ratio is much greater than this, a significant contribution from the chlorite 004 is indicated for the peak at 25° 2θ. Positive identification of chlorite is provided by peaks at 6.2 and 18.8° 2θ, but these peaks are often weak and may not be detectable if the chlorite concentration is low or if its Fe content is high.

A difficulty arises when only weak reflections are present at 12.5 and 25° 2θ. Such a sample could contain only chlorite, only kaolinite, or a mixture of the two. In this situation, the sample may be treated chemically or heated and then reexamined. Heating chlorite to 550°C for 1 h causes dehydroxylation of the hydroxide sheet with attendant changes in the diffraction pattern. The intensity of the 001 reflection usually increases greatly and shifts to about 6.3 to 6.4° 2θ, and the 002, 003, and 004 reflections are much weakened (but not eliminated). At this temperature, kaolinite becomes amorphous to X-rays and its diffraction pattern disappears. This test may suggest that chlorite is present or that kaolinite is present and chlorite is absent. It will not give final and complete results for a sample that contains both kaolinite and chlorite. An additional treatment is required for dealing with such mixtures.

An aliquot of the sample should be boiled for 2 h in 1 N HCl. This step dissolves most chlorites, and any residual peaks at 12.5 and 25° 2θ indicate the presence of kaolinite. Mg-rich chlorites may be unaffected by this treatment, however, and though not common in sedimentary rocks, their possible presence always leaves some doubt about the results of an acid digestion.

There is a definitive method for identifying mixtures of kaolinite and chlorite, and only the difficulty of application makes it nonroutine. Kaolinite can be expanded by certain reagents that form strong hydrogen bonds (MacEwan and Wilson, 1980, p. 239). The expanded kaolinite produces a changed and recognizable diffraction pattern that is not interfered with by

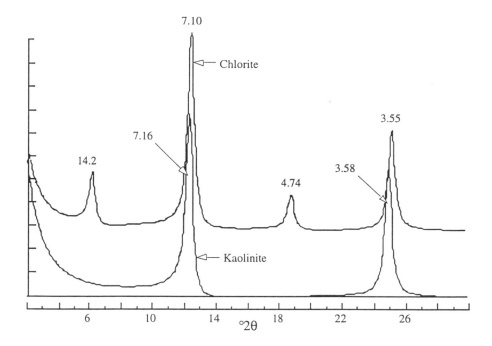

Fig. 7.4. Kaolinite and high-Fe chlorite.

chlorite. The presence of either or both phases can be determined with certainty. Several procedures have been proposed, but the easiest to use is to place an oriented aggregate in an atmosphere of formamide, much as you place samples with smectite in an ethylene glycol atmosphere, only it takes about two weeks or more to get full expansion to 10.4 Å. Halloysite expands much faster than kaolinite in an atmosphere of formamide and the procedure can be used to distinguish collapsed halloysite from kaolinite (Churchman et al., 1984). A more involved method that intercalates kaolinite with DMSO to give $d(001) = 11.2$ Å has been devised by Calvert (1984). Bohor and Triplehorn (1993) reviewed intercalation processes in relation to the dispersion of kaolinitic flint clays. Studies of samples before and after any of these treatments allow estimates of the diffraction contributions of chlorite, halloysite, kaolinite, and any mixed-layered clay minerals that produce possible interferences in the 12° 2θ region. More about halloysite in the section on *sepiolite, palygorskite, and halloysite*.

The relative intensities of the chlorite 00*l* series can be used to determine the total heavy metal content of the mineral as well as the distribution of the heavy metals between the silicate and hydroxide octahedral sites. Heavy metals in chlorite usually consist of Co, Cr, Fe, Mn, and Ni. Of these, Fe is by far the most important. In our discussion, we assume that the only heavy metal present is Fe. Mg and Al are the light metals in chlorite octahedral sites (except for Li in the rare chlorite cookeite). Mg is assumed here because the difference in scattering power between Mg and Al is negligible for our purposes. The distribution of Fe with respect to Mg in the two layers is called the *symmetry*, and if the symmetry is zero, then equal proportions of Fe and Mg exist in the two sites. The use of the 00l intensities for these determinations requires that correct conditions are maintained for sample-length-beam-slit relations and for sample thickness, as discussed in Chapter 9.

The procedure described next is modified from Brown and Brindley (1980). It allows the determination of the total Fe content of chlorite, designated Y, and a measure of the symmetry of Fe substitution, defined as D and equal to the number of Fe atoms in the octahedral sheet of the silicate layer minus the number of Fe atoms in the hydroxide sheet. Both Y and D are based on a chlorite formula that contains a total of six metal atoms in both sites, and whose formula can be represented by $(Mg,Al)_{6-Y}Fe_Y(Si,Al)_4O_{10}(OH)_8$. The maximum and minimum values for D are 3 and -3, respectively, and Y varies between 0 and 6. Brindley's method uses several 00l reflections, but we recommend only the 002, 003, 004, and 005, partly for the sake of simplicity, but also to eliminate errors in the 001 intensity that are caused by sample-length problems and uncertainties in the Lorentz factor.

Table 7.1 shows the intensity ratios $I(003)/I(005)$ and corresponding values of D. It is easy to measure this intensity ratio and to estimate D because there are no interferences from common minerals on these two reflections. Brown and Brindley's values have been calculated under the assumption of random orientation of the chlorite crystallites in the powder; consequently, you do not want very highly oriented sample preparations here. Very high 00l intensities indicate preferred orientation, and when that occurs the safest procedure is to base your intensity measurements on an unoriented

Table 7.1. $I(003)/I(005)$ and the symmetry of Fe distribution D

D	$I(003)/I(005)$
3.0	0.242
2.5	0.43
2.0	0.68
1.5	1.09
1.0	1.67
0.5	2.54
0.0	3.83
-0.5	5.76
-1.0	8.74
-1.5	13.3
-2.0	21.2
-2.5	34.1
-3.0	54.1

Values from Brown and Brindley (1980).

powder that has been packed into the back or side-loaded into a cavity-type specimen holder (see Chapter 6). Unfortunately, the odd-order reflections are weak and may not be detectable from randomly oriented powders unless high concentrations of chlorite are present.

The determination of Y, the total number of Fe atoms, may be a bit difficult to apply in practice if there are kaolinite or serpentine interferences on the chlorite 002 and 004 peaks; otherwise the procedure is straightforward. First, using the following equation (Brown and Brindley, 1980), correct the value of $I(003)$ for any asymmetry to give $I(003)'$. The equation is given with the relevant constants.

$$I(003)' = I(003) \frac{I(003)\,(114)^2}{(114 - 12.1D)^2} \tag{7.1}$$

A measure of Y is then obtained according to the values shown in Table 7.2.

The application of these methods can yield precise results if you pay attention to details, and if there are no interferences on the chlorite reflections. Figure 7.5 shows calculated patterns for chlorites with different amounts of total Fe. The numerical methods are definitive, but we believe that the images are still important to the student for quick evaluations and for building up the critical experience that leads to an intuitive appreciation of the compositions involved.

The diffraction patterns of Fig. 7.5 illustrate the effects of total Fe on the intensities of the chlorite 001, 002, 003, and 004 reflections. The examples treat 2, 1, and 0 Fe atoms in each of the two octahedral sites: the silicate (Sil) and hydroxide (Hyd) sheets. The D value (symmetry) is zero. The example shows nicely how the odd-order peaks weaken, with respect to the even-order reflections, as the concentration of symmetrically distributed Fe increases.

Figure 7.6 illustrates the effects of Fe symmetry in chlorites. All calculated traces depict compositions with 1 Fe per Si_4O_{10}. Examples are shown for the Fe sited in the interlayer hydroxide sheet, in the silicate layer, and evenly distributed between the two layers. The large effect of Fe

Table 7.2. Estimation of the number of Fe atoms in six octahedral sites (Y)

Y	$[I(002) + I(004)] / I(003)'$
0	2.38
1	3.54
2	5.0
3	6.7
4	8.6
5	10.8
6	13.4

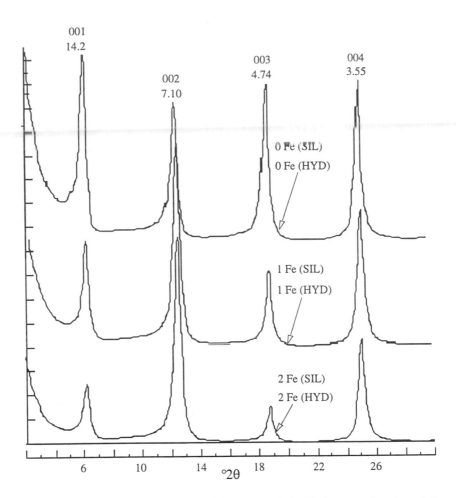

Fig. 7.5. Even/odd peak intensities in symmetrical chlorites as a function of Fe content. Numbers of Fe atoms refer to three octahedral sites per Si_4O_{10}.

symmetry on the intensity ratio of the 001 to 003 reflections is noteworthy. As a practical matter, you should remember that chlorites have symmetrical Fe substitution if their diffraction patterns show approximately equal intensities for the 001 and 003 peaks (Fig. 7.5).

The significant effect that Fe substitution has on diffracted intensity is due to the great difference in scattering power between Fe and Al or Mg (Chapter 3, p. 79). All this is perfectly consistent with the principles already outlined; still, it is nice to see the results in graphic form. Let us summarize the effects of Fe substitution in chlorite.

1. Increasing Fe or other heavy metal concentrations causes a weakening of the 001, 003, and 005 reflections relative to the 002 and 004 reflections.

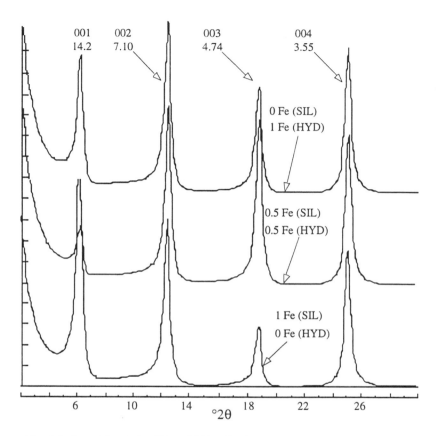

Fig. 7.6. Intensities of the 001 and 003 reflections of chlorites with identical total Fe contents, but with different values for Fe symmetry. Numbers of Fe atoms refer to each of the two octahedral sites that contain three atoms each.

2. Asymmetry of substitution that involves a relative enrichment of Fe in the silicate layer increases the intensities of the 001 and 005 reflections relative to the 003 reflections. For extreme cases, the diffraction patterns of such minerals resemble those of air-dried Mg-vermiculites.

3. Enrichment of Fe in the hydroxide sheet produces effects that are the reverse of those described in the second item. The 001 and 005 peaks are weak compared to the 003 reflection. This produces a strange-looking diffraction pattern that cannot be confused with that of any other clay mineral or mixture of clay minerals.

Vermiculite
Figure 7.7 shows the Mg-vermiculite diffraction pattern. It looks pretty much the same whether it is analyzed dry or after glycol solvation, assuming that it

is the Mg-saturated form that is common for most vermiculites. Na-saturated vermiculites give a strong peak in the 7° 2θ region, so if you are in doubt, Mg-saturate the sample and reanalyze it. The accepted definition of vermiculite is operational and requires that Mg-vermiculite retain a d(001) of 14.5 Å (like the one in Fig. 7.7) after *glycerol* (not ethylene glycol) solvation, whereas smectite produces a first-order peak at about 17.7 Å (5.0° 2θ) after such treatment. Samples will be encountered that produce spacings intermediate between 14.5 and 17.7 Å with glycerol and that collapse to 10 Å after K saturation in the air-dried condition (Walker, 1958). These may be expandable clay minerals whose layer charges are intermediate between vermiculite and smectite, or they may be mixed-layered structures. At present, such minerals are poorly understood.

The vermiculite 001 reflection is intense, allowing the detection of very small amounts. Problems arise when small amounts of vermiculite are mixed with chlorite. The chlorite and vermiculite 00*l* reflections interfere, producing a pattern that looks like a chlorite with asymmetrical Fe substitution—in this case, a chlorite containing more Fe in the silicate layer than in the hydroxide sheet. The difficulty is easily resolved. K-saturate the sample or heat to 300°C

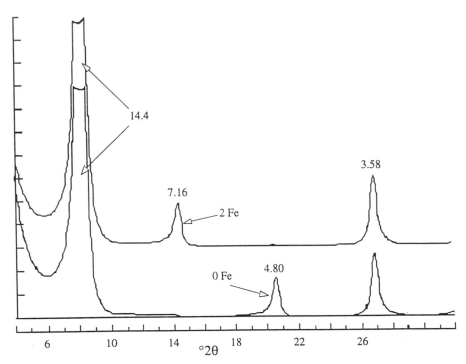

Fig. 7.7. Air-dried trioctahedral Mg-vermiculite. Fe values refer to three octahedral sites per Si_4O_{10}.

for an hour, and run a new diffraction pattern. These treatments collapse vermiculite to 10 Å, remove the 00l interferences, and change the vermiculite diffraction pattern to one similar to those of glauconite and biotite (the 002 reflection is very weak; see Fig. 7.3). Dioctahedral vermiculites can also be identified by these methods because the heat-treated mica-like pattern more closely resembles that of illite. The best procedure to distinguish trioctahedral from dioctahedral varieties is, however, to use the position of 060 reflection.

Figure 7.7 illustrates the effect of Fe for Mg substitution on the intensities of the vermiculite 00l series. The intensity ratio of the 002/003 is a good measure of Fe content. Minerals that have 1 Fe per three sites show approximately equal intensities for these two reflections, and low-Fe dioctahedral varieties have diffraction patterns much like the 0 Fe composition of Fig. 7.7.

Smectite

Smectite is easily identified by comparing diffraction patterns of air-dried and ethylene glycol-solvated preparations. The glycol-treated preparation gives a very strong 001 reflection at about 5.2° 2θ (16.9 Å), which, in the air-dried condition, shifts to about 6° (15 Å) if the clay is saturated with a divalent ion and equilibrated with air at room temperature and moderate humidity. Confirmation of the identification, if necessary, is accomplished by K-saturation and drying at 300°C. This treatment collapses smectite to 10 Å, producing a diffraction pattern similar to that of illite.

The major difficulty in smectite identification lies in judging whether the mineral is mixed-layered with illite. Small amounts of illite interstratification show up as a shift in the 003 peak to higher diffraction angles. Unfortunately, all glycol-solvated smectites do not produce a $d(001)$ of 16.9 Å, and a thinner glycol spacing also displaces the 003 reflection (Środoń, 1980). To establish the presence of a different glycol thickness, you will have to measure carefully the 00l positions and test for a consistent $d(001)$ by means of Bragg's law. If the diffraction pattern is rational (see Chapter 8), peak displacement cannot be due to interstratification. The 002 and higher-order reflections are relatively weak and may not be detectable on the diffraction pattern unless the sample contains significant amounts of smectite. If the higher-order reflections are not detectable, some useful information can be obtained from the shape of the 001 reflection (ethylene glycol-solvated), which will be broad, with a high low-angle shoulder if interstratified illite is present. But we defer further discussion of such details to the section on mixed-layered clay minerals. Figure 7.8 shows diffraction patterns of a common smectite, montmorillonite, in the air-dried (two-water-layer) and ethylene glycol-solvated states. Note that, like vermiculite, the 001 is very intense, allowing detection of amounts as small as a few percent.

The smectite 00l intensities, like those of chlorite, can be used to gain information on Fe substitution or, more accurately, on the scattering from the

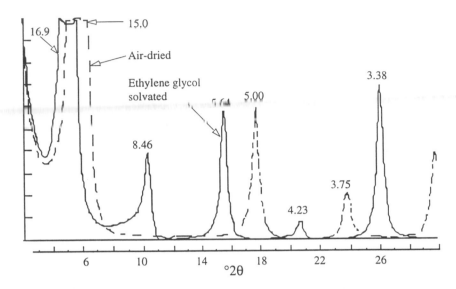

Fig. 7.8. Montmorillonite in the air-dried and ethylene glycol-solvated states.

octahedral sheet. Figure 7.9 shows patterns for three minerals with different amounts of Fe per two octahedral sites, montmorillonite with 0.1, Mg-saponite with none, and nontronite with 1.7. The intensity ratio of the 002 to the 003 increases sharply as the total number of electrons in the octahedral sites increases; this intensity ratio is useful for differentiating dioctahedral from trioctahedral smectites, though moderate Fe concentrations in a dioctahedral variety will produce diffraction patterns that are indistinguishable from Mg-rich trioctahedral varieties. The dilemma can be resolved by measuring the spacing of the 060, which differentiates dioctahedral from trioctahedral species provided that the sample does not contain other clay minerals that interfere with this reflection.

Montmorillonite can be distinguished from nontronite, saponite, and beidellite by its irreversible collapse after Li saturation and heat treatment. Speculation is that Li ions migrate into the octahedral sheet and neutralize the layer charge if the charge is due to octahedral substitution. The elimination of charge converts montmorillonite to a pyrophyllite-like mineral that does not expand upon treatment with water, glycerol, or ethylene glycol. The procedure is known as the Greene-Kelly test (Greene-Kelly, 1952, 1953). Modifications of the original procedure have been made that improve the performance of the test (Byström-Brusewitz, 1975), and they are included in the following description. Li-saturate a clay sample and remove excess LiCl reagent by multiple washings with water followed by silver nitrate tests to ascertain the absence of chloride. Mount the sample on an *opaque* fused silica slide (not glass or clear silica) and heat to 300°C for 12 h, saturate with *glycerol* and analyze immediately on the diffractometer. Rigid standardization

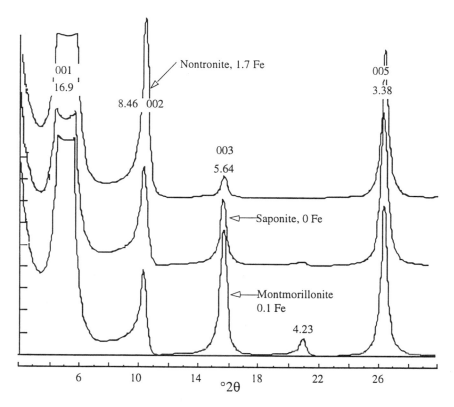

Fig. 7.9. Peak intensity as a function of octahedral scattering (Fe) in ethylene glycol-solvated smectites. Fe values refer to two or three octahedral sites per Si_4O_{10}.

of the procedures is essential for repeatable and significant results. We have found this method tricky to use, and ambiguous results can be obtained if attention is not paid to details.

After such treatment, montmorillonite is recognized by its collapsed state, which produces a first-order spacing of about 9.6 Å. The other smectites should expand to give the characteristic (glycerol-smectite) 17.7 Å reflection at 4.9 to 5.0° 2θ. Complex diffraction patterns may result, and if they do, the investigator will be hard pressed to classify them as evidence of interstratification, on the one hand, or failure of the treatment, on the other.

Sepiolite, Palygorskite, and Halloysite

Sepiolite, palygorskite, and halloysite are fibrous and cannot be oriented so that 00*l* reflections dominate the diffraction pattern. Instead, many lines from the *hkl* series are present, but because most lines are relatively weak they are difficult to detect unless the sample is rich in one of these minerals. Table 7.3 shows the relevant data for each, but bear in mind that a good deal of variation in spacings and intensities may occur from one sample to the next.

Table 7.3. X-ray data for sepiolite and palygorskite

Sepiolite			Palygorskite		
d	I	2θ	d	I	2θ
12.8	100	6.9*	10.4	100	8.5*
7.6	4	11.6	6.4	16	13.8*
5.1	8	17.4	5.4	11	16.4+
4.4	35	20.2	4.46	17	19.9
3.77	20b	23.6*	3.65	10	24.4*
3.35	30vb	26.6	3.18	16	28.1*

b = broad, vb = very broad
Reflections marked with an asterisk are common to several different specimens of these two minerals and are free from interferences by other minerals common in the clay-size fraction. The palygorskite data are a composite from Bailey (1980). The sepiolite values are taken from JCPDS powder diffraction card 29-1492.

The diffraction patterns are unaffected by mild heat treatment (< 200°C) and ethylene glycol solvation, and the minerals are most easily identified on the basis of the strong reflections at low 2θ.

In most samples from sedimentary rocks, sepiolite and palygorskite will be minor constituents and only the low-angle peaks will be detectable—take care not to confuse them with mixed-layered clay minerals or vermiculite or smectite. These and other peaks are near the diffraction angles for many of the mixed-layered and expandable clay minerals. Additional sample treatments such as heating and ethylene glycol solvation may be necessary to differentiate mixed-layered and expandable species from either of these two minerals. In the older literature, sepiolite and palygorskite may have been misidentified as mixed-layered clay minerals, particularly as illite/chlorite.

Halloysite is a kaolin mineral whose habit is most commonly fibrous instead of platy. Crystals cannot be oriented in basal parallel fashion. The 00*l* peaks are relatively weak, and nonbasal reflections are strong at about 20 and 35° 2θ. The latter are asymmetrical with slowly diminishing intensity toward high diffraction angles. They are two-dimensional diffraction bands indicative of turbostratic stacking along *Z* and resemble similar reflections from smectites. The 00*l* reflections are usually broader than those from kaolinite, though this criterion is by no means definitive. Dehydrated halloysite is a good guess for a mineral that gives broad, kaolinite-like 00*l* and strong asymmetric *hk* reflections despite efforts to achieve good preferred orientation.

060 REFLECTIONS

X-ray diffraction studies of randomly oriented powders provide useful adjuncts to the identification of the clay minerals and to refinements in qualitative classification. The 060 reflections allow the distinction between

Table 7.4. Values of $d(060)$ and 2θ for micas and clay minerals

Mineral	$d(060)$	2θ
Kaolinite	1.490	62.31
Montmorillonite	1.492-1.504	62.22-61.67
Illite (Muscovite)	1.499	61.90
Glauconite	1.511	61.35
Saponite	1.520	60.95
Nontronite	1.521	60.91
Hectorite	1.530	60.51
Serpentines	1.531-1.538	60.47-60.16
Biotite	1.538	60.16
Chlorites	1.538-1.549	60.16-59.69
Sepiolite	1.540-1.550	60.07-59.65
Vermiculite	1.541	60.03
Berthierine	1.555	59.44
Palygorskite	1.56	59.23

All data are from Bailey (1980) except for the smectites, which are taken from Brindley (1980).

dioctahedral and trioctahedral types because the b cell dimension is sensitive to the size of the cations and to site occupancy in the octahedral sheet and is unaffected by the monoclinic angle β. The peaks are weak, but can be satisfactorily resolved from the background by step-scan procedures that use long count times or by chart recording at slow goniometer and chart speeds (0.5° /min and 0.5 cm/min), in conjunction with a long time constant (4 sec) and a small scale factor (1,000 cps full-scale). Table 7.4 shows nominal values of $d(060)$ and the corresponding diffraction angles.

The $d(060)$ values vary somewhat for a given mineral species because they depend on the composition of the octahedral sheet, the amount of Al in tetrahedral coordination, and the degree of tetrahedral tilt. Nevertheless, you can see that dioctahedral and trioctahedral types are clearly separable, with the exceptions of saponite and nontronite, which have nearly identical values for $d(060)$.

Be careful when identifying trioctahedral species because of the possible presence of a quartz peak at $d = 1.542$ Å. If your sample contains quartz, look for another quartz reflection at $d = 1.82$ Å. This reflection has about the same intensity as the peak at $d = 1.542$ Å. If the peak at $d = 1.82$ Å is present, allow for the interference of quartz if you choose to interpret a peak in the 1.54 Å region as evidence for a trioctahedral clay mineral in the sample.

THE USE Of *hkl* REFLECTIONS FOR THE DETERMINATION OF POLYTYPES

Determination of polytypes is a further level of refinement in qualitative analysis. You will need a thick random powder specimen (Chapter 6) and a diffractogram with high peak-to-background resolution. The problems always center around peak interferences. The various clay mineral *hkl* reflections interfere with each other (e.g., micas and chlorites), and severe interferences can be caused by quartz and feldspars. Polytype determinations have been made infrequently because only exceptional samples are sufficiently monomineralic to allow measurements of the required diffraction peaks or the techniques for isolating the mineral you are interested in are too time consuming and complex. The most frequent polytype determinations are made for the chlorites, illite, glauconite, and the kaolin group.

Any of the polytypes can be disordered to any degree from a few rotational or stacking faults whose effects can hardly be noticed in the diffraction patterns to disorder complete enough so that only the 00*l* and the *hkl* reflections with $k = 3n$ are present. Disorder is best illustrated with calculated three-dimensional diffraction patterns that are selected to demonstrate the effects of different kinds of disorder present in different amounts. This discussion is deferred to Chapter 10. The diagnostic tables given here apply <u>only</u> to ordered species, except for the clay-size chlorites that almost invariably show only sharp 00*l* and $k = 3n$ lines on powder diffraction patterns (Bailey, 1980).

Chlorite Polytypes
The chlorite polytypes that can be easily determined from random powder studies are *Ia*, *Ib* (β=90°), *Ib* (β=97°), and *IIb*. Table 7.5 shows the most useful peaks for their identification.

Table 7.5. Diagnostic reflections for the determination of chlorite polytypes by random powder methods

Type *Ia*			Type *Ib*(β=90°)			Type *Ib*(β=97°)			Type *IIb*		
d	I	2θ	d	I	2θ	d	I	2θ	d	I	2θ
2.65	30	33.8	2.69	20	33.3	2.68	25	33.4	2.66	15	33.7
2.59	15	34.6	2.65	15	33.8	2.60	15	34.5	2.59	60	34.6
2.39	60	37.6	2.51	100	35.8	2.55	5	35.2	2.55	50	35.2
2.27	10	39.7	2.34	10	38.5	2.47	70	36.4	2.45	50	36.7
2.07	5	43.7	2.15	40	42.0	2.40	5	37.5	2.39	50	37.6
2.01	30	45.1	1.96	5	46.3	2.30	5	39.2	2.26	40	39.9
						2.11	20	42.9	2.07	10	43.7
						2.01	10	45.1	2.01	60	45.1

From Bailey (1980).

Table 7.6. Diagnostic reflections for the determination of kaolin polytypes by random powder methods

Kaolinite			Dickite			Nacrite			Metahalloysite		
d	I	2θ	d	I	2θ	d	I	2θ	d	I	2θ
3.84	45	23.15	3.26	10	27.34	3.44*	40	25.9	4.45	100	20.0
3.12*	55	28.6	3.10	10	28.83	3.09*	30	28.9	2.57	40	34.9
2.75	35	32.56	2.94	10	30.43	2.93	10	30.54	2.22	5	40.6
2.34	90	38.47	2.80	10	32.02	2.41*	100	37.3	1.69	20	54.45
2.29	80	39.34	2.32	95	38.75	2.26*	10	39.9			
2.18	30	41.34	2.21	15	40.81	2.09*	20	43.3			
1.99*	50	45.6	1.97	40	45.97	1.92*	45	47.4			
1.84	40	49.57									

The asterisk indicates averages of two or more unresolved peaks. From Bailey (1980).

The Kaolin Polytypes

Table 7.6 shows data that have been selected to provide the most useful lines for the identification of the kaolin polytypes kaolinite, dickite, nacrite, and metahalloysite.

The Micas, Illite, and Glauconite

The mica-like clay minerals in sedimentary rocks provide examples of the *1M*, *3T*, and *2M₁* polytypes. The *1M* structure is found in two modifications characterized by different patterns of octahedral cation ordering, the *trans*-vacant or *tv* (centrosymmetric) and the *cis*-vacant or *cv* (noncentrosymmetric) varieties. Think of the *tv* types as the traditional ones and the *cv* as the new kid on the block who may actually turn out to be abundant now that Drits and his colleagues have shown us how to identify it (Drits et al., 1984; Drits and Tchoubar, 1990). These structures have been briefly discussed in Chapter 4.

All these polytypes have a pair of strong diagnostic peaks that lie on either

Table 7.7. Diagnostic reflections for the mica polytypes

tv 1M			cv 1M			3T			2M₁		
d	I	2θ	d	I	2θ	d	I	2θ	d	I	2θ
4.35	15	20.4	3.88	40	22.9	3.87	35	23.0	4.29	10	20.7
4.12	10	21.6	3.58	30	24.9	3.60	30	24.7	4.09	10	21.7
3.66	50	24.3	3.12	50	28.6	3.11	30	28.7	3.88	30	22.9
3.07	50	29.1	2.86	55	31.3	2.88	40	31.1	3.72	30	23.9
2.93	10	30.5	2.68	20	33.4	2.68	10	33.4	3.49	30	25.5
									3.20	30	27.9
									2.98	35	30.0
									2.86	30	31.3
									2.79	25	32.1

Data from Bailey (1980) except for the *cv 1M* pattern, which was calculated from unit cell parameters and atomic coordinates given by Drits et al. (1984).

side of the illite 003 peak. Techniques for enhancing these peaks are discussed in Chapter 6.

The data for the *cv 1M* and *3T* structures are so much alike that it takes a good pattern to tell the difference. Some of the key reflections differ between the two by 0.15 to 0.2° 2θ, and these differences are the only way to distinguish them. To do this requires a high resolution diffraction pattern from structures with little or no disorder. To date, only two ordered *cv 1M* occurrences have been identified, so we must assume that at present we do know much about the relative abundances of these two structures. All references in the older literature to clay-size *3T* polytypes are suspect pending further study.

Problems with polytype identification are caused by interferences from the diffraction patterns of common nonclay minerals such as quartz and feldspars. These interferences can be overwhelming because the key three-dimensional polytype lines are weak unless (1) rather involved sample preparation methods are used (Chapter 6) to assure random orientation and (2) sufficiently long count times are used to produce good peak to background resolution—count times that cause typical diffraction runs to take 6 to 8 h. In addition, sedimentary rocks may contain a mixture of polytypes that probably represent diagenetic and detrital components.

For these reasons, few studies have been made of clay mineral polytypes in sedimentary rocks, compared to the voluminous literature on basal diffraction patterns. But don't be scared off. Modern instrumentation has improved and continues to improve in many ways that provide solutions to some of the experimental problems. Be willing to work a little harder and pay more attention to sample preparation and instrumental details. Settle for a small number of a samples per study that have been exhaustively characterized instead of a large diffuse database that has only statistical significance. Any field that is poorly investigated is a fruitful one for important and perhaps spectacular findings.

NONCLAY MINERALS

Clay minerals are seldom found as monomineralic material. Nonclay minerals are almost always present, often in amounts so small that only their most intense peaks can be seen. What follows is more of a tabulation than a discussion, but it provides a summary of the diffraction characteristics of the most common nonclay minerals in the clay-size fractions of sedimentary rocks. We have pared the list to diagnostic peaks with significant intensity within the 2θ range normally scanned for clay minerals. All *d* values have been rounded to two decimal places. The 2θ angles are listed to one decimal place if *d* in the sources is given to two decimal places; if *d* is given to three places, 2θ is listed at two. You may need to go to the JCPDS powder diffraction file (abbreviated as PDF) or to other authoritative sources for confirmation of an identification made with these tables. Again, we urge you to make a collection of patterns of the purest examples of minerals you will deal with most often.

A few final comments. How many peaks are necessary to identify a mineral? A rule of thumb that has been around for a long time is that you need three peaks, assuming, of course, that none of them is coincident with reflections from another mineral known to be present. Is this statistically valid? We don't know. But don't try to sell a mineral identification on a single peak. Five or more would be nice. A specification of three seems to be about right. How good should the agreement be between a mineral's relative intensities from the card and from an experimental pattern? The agreement should be very good indeed if the mineral has no cleavage such as quartz, garnet, pyrite, etc.; fair agreement (20 to 30%?) if the mineral has a few good cleavage directions such as the carbonates and some sulfates, worse is there is one predominant cleavage such as in many of the feldspars, and very poor (off by maybe a factor of 10) if the mineral has a pronounced habit (fibrous) or one excellent cleavage direction such as we find in the micas and clay minerals. Minerals with relatively wide ranges of solid solution such as the feldspars and carbonates also have peaks whose positions vary. These comments assume that you have prepared an oriented aggregate for the study of the clay minerals. If a random powder mount is used, such as that described in Chapter 6, discrepancies between the intensities from experimental patterns and the cards can be reduced for all minerals to discrepancies of perhaps 10% or less, but only if the chemical composition of your experimental sample is identical to the one described by the cards. The different scattering powers of the different ions are a factor in controlling diffraction intensities. The worst effect is due to Fe, because it has many more electrons than any of the other common cations in silicates, and in most rock-forming sulfates and carbonates. Substitutions involving Al, Si, Mg, and Na have little effect on intensity; Ca for Mg or K for Na produce significant intensity changes, but nothing like the large effects of Fe substitution for Mg in dolomite.

Remember that these minerals produce sharper peaks than the clay minerals, and this distinction is an important diagnostic criterion.

Silica Minerals

Low, or α-quartz, is by far the most common of the silica minerals in sedimentary rocks. Most clay-sized fractions contain at least a trace of it, and, as discussed previously, its diffraction lines can be used as an internal standard for the accurate and precise measurement of d values. However, you should not see much more than a trace of quartz in the < 2 μm fractions of most rocks, for if you do it usually indicates that something is wrong with the sample preparation procedures. Large amounts of quartz in the < 2 μm fraction, for properly prepared samples, indicate the presence of either glacial rock flour or siliceous microfossil types, such as radiolarians and diatoms whose original skeletal material is usually opaline silica, especially in younger sedimentary rocks.

Varieties of opaline silica (which are transformed by diagenesis to a-quartz) form a diagenetic transformation series that starts with amorphous opal (opal A), progresses through opal CT to opal C, and ends with chert (low quartz). For details on the diffraction characteristics of opaline silica, see Jones and Segnit (1971) and newer work described below.

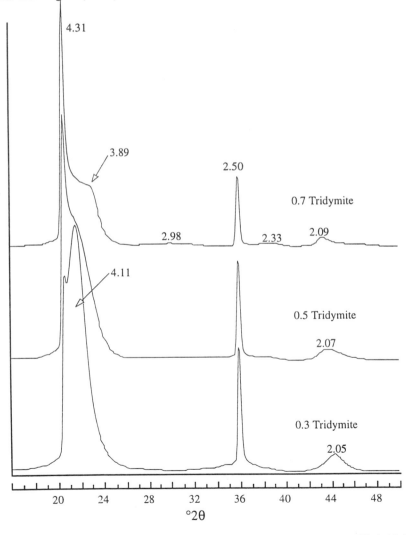

Fig. 7.10. Calculated diffraction patterns for randomly interstratified high-tridymite/high-cristobalite. See text for details.

Table 7.8A. Diffraction data for the silica minerals

Lo-Cristobalite[a]			Hi-Cristobalite[b]			Lo-Tridymite[c]			Hi-Tridymite[d]		
d	I	2θ	d	I	2θ	d	I	2θ	d	I	2θ
4.04	100	22.0	4.15	100	21.4	4.33	90	20.52	4.37	100	20.3
3.14	8	28.4	2.92	5	30.6	4.11	100	21.64	4.12	61	21.6
2.84	9	31.49	2.53	80	35.5	3.87	20	23.00	3.86	57	23.0
2.49	13	36.15	2.17	10	41.6	3.82	50	23.30	3.00	16	29.8
2.47	4	36.45	2.07	30	43.8	2.98	25	30.04	2.52	13	35.6
2.12	2	42.69	1.99	5	45.6	2.78	8	32.25	2.33	7	38.6
1.93	4	47.11	1.80	5	50.87	2.50	16	35.92	2.11	5	42.9
1.87	4	48.69	1.69	5	54.3	2.49	14	36.07			
			1.64	60	56.04	2.31	16	39.03			

[a]PDF 39-1425. [b]PDF 4-359. [c]PDF 18-1170. [d]See Brown (1980).

Figure 7.10 shows calculated three-dimensional diffraction patterns for randomly interstratified tridymite/cristobalite, or opal CT. These were computed by a version of WILDFIRE® modified in collaboration with G. D. Guthrie and D. L. Bish. Full details are given in Guthrie et al. (1995). The peak positions follow mixed-layered principles as described in Chapter 8. The high temperature structures of tridymite and cristobalite were used in the model structures, but if you compare these peak positions (Figure 7.10) with the pure end members (Table 7.8A) and allow for the effects of mixed-layering, you will see that the reflections on the calculated patterns are displaced somewhat toward higher diffraction angles. The best results for matches of experimental with calculated diffraction patterns required a reduction in the unit cell dimensions of the nominal silica layers that make up tridymite and cristobalite, and this alteration caused shifts of most reflections toward higher 2θ. The validity of this unit cell adjustment, with attendant shortening of Si-O bond lengths, and its possible significance is a subject that is still under study.

Estimates of the proportions of tridymite/cristobalite in opal CT are best made by (1) matching experimental/calculated diffraction profile shapes

Table 7.8B. Diffraction data for α (low) quartz; from PDF Card 33-1161

d	I*	2θ	d	I	2θ
4.27	22	20.8	1.979	4	45.83
3.342	100	26.67	1.818	14	50.18
2.457	8	36.57	1.672	4	54.91
2.282	8	39.49	1.659	2	55.38
2.237	4	40.32	1.541	9	60.00
2.128	6	42.50	1.453	1	64.04

*These values replace those from PDF card 5-490, the data that were taken with a Geiger counter.

between 20 and 24° 2θ, and (2) using the position of the peak near 44°. Guthrie et al. (1995) report evidence of nearest neighbor ordering in some opal CT specimens, and this is manifested by a noticeable but unresolved peak on the high 2θ side of the strong reflection near $d = 4.11$ Å.

We offer a longer list of data for α-quartz because you will find it a useful internal standard everywhere but at low angles. Quartz d values are very constant from specimen to specimen, and its value as a standard justifies listing these spacings to more significant places than for the other minerals described here. The 1.54 Å spacing, for example, is a standard against which the 060 of the clay minerals can be measured when you are distinguishing between dioctahedral and trioctahedral forms. A complete diffraction tracing and tabulation of spacings can be found in Frondel (1962, p. 34).

Feldspar

After you have established the presence of feldspar, you should ask, is it K-feldspar or plagioclase feldspar? Then, given good enough diffraction tracings, there may be a chance you can answer one or both of the following questions: What is the composition of the plagioclase? Is the K-feldspar triclinic or monoclinic or a mixture of the two? However, if you're going to get very involved with the feldspars, you had better change slit sizes to ones smaller than those you have been using for clay minerals and get ready for a career change. People have been wrestling with a reliable way to identify and measure the quantities of the varieties of feldspars for as long as there has been the recognition that the feldspars are a complex family of minerals. So, if you go much beyond the simple approximations offered here, expect to find disagreements and complications. For detailed information on feldspars see Ribbe (1983) or Smith and Brown (1988).

The low symmetry of feldspars means that they have complex diffraction patterns. Complicating the problem is that, unless they have been diagenetically homogenized, a variety of feldspars reflecting variation form source areas should be expected in sedimentary rocks. Almost as common as quartz in soils and fine-grained sedimentary rocks, they offer a confusing array of peaks. When trying to determine the polytype of illite, they are a bother. Table 7.9 consists of data selected from PDF cards 22-687, 19-926, 19-1227, 20-554, 9-466, 19-1184, 20-572, 10-393, 12-301, and 20-20, from Borg and Smith (1969a, 1969b), and from Herbert D. Glass, Illinois Geological Survey, Champaign, Illinois (personal communication, 1991). Both K-feldspar and plagioclase feldspars have variable compositions, and both are found as ordered and disordered varieties. In addition, many of the peak positions shift with composition changes. Therefore, we have rounded d off to the nearest 0.01 Å and the ° 2θ positions to the nearest 0.05°. Several of the peaks represent diffraction from more than one hkl plane. If you need to know the hkl index of a plane, you will find it on the PDF cards listed above. The most intense peak for each of the four varieties of feldspars shown in

Table 7.9 is underlined.

When looking at XRD traces from oriented aggregates, about the best you'll be able to do is to use the peak at ~27.5° 2θ to indicate that some variety of K-feldspar is present, and the peak at ~28° 2θ to indicate that some variety of plagioclase is present. For XRD traces of random powders, the peaks of the K-feldspars that are intense enough to show when feldspar is present in small amounts, that are not found in the plagioclase series, and that occur in the 2θ range normally used by clay mineralogists, are, in addition to the one mentioned above for oriented aggregates, two smaller ones at 21.10 and 25.65° 2θ for microcline, and 21.05 and 25.75° 2θ for orthoclase. The two peaks at the lowest angles can be useful only if they are resolved from the quartz peak at 20.8° 2θ. You probably won't be able to distinguish microcline from orthoclase because 2θ positions vary with composition and degree of ordering. The most intense line of orthoclase and sanidine, which is at 26.75° 2θ, is often of no use because it is interfered with by the strongest quartz line at 26.65° 2θ. High sanidine-like feldspars as a diagenetic mineral are not uncommon in sedimentary rocks. When it is the only feldspar present, the peak at 27.5° 2θ is the most useful. But, when microcline is also present, the peaks at 29.7 and 30.7° 2θ are more useful because they are distinct from microcline peaks.

Two useful peaks in the plagioclase series not found in the alkali series, in addition to the ~28° 2θ peak, are at 22.05 and 24.90° 2θ. The peak at 23.50°

Table 7.9 Diffraction data for feldspars (2θ for CuKα radiation)

Microcline(Tr)		Ortho.(Mono)		Albite		Anorthite		Hi Sanidine	
d	2θ	d	2θ	d	2θ	d	2θ	d	2θ
4.22	21.06	4.22	21.05					4.24	20.95
				4.03	22.05	4.04	22.00		
3.80	23.40								
				3.68	24.90				
		3.47	25.66						
3.29		3.31	26.75*			3.33	26.75		
3.24	27.52	3.24	27.53			3.28	27.15		
								3.23	27.60
				3.19	27.95	3.19	27.95		
						3.18	28.05		
				3.15	28.25				
3.03	29.50	2.99	29.85					3.00	29.78
2.96	30.20			An33		An67			
				3.01	29.68	3.05	29.28		
2.89	30.95								
		2.77	32.35						

*The most intense peaks are underlined.

2θ is seldom useful because it is too close to peaks from the alkali series. Under the best conditions you may be able to approximate the composition of the plagioclase by noting the position of the peak between 29.20 and 29.90° 2θ. In a separate box in Table 7.9, the values for this peak are given for two specific values in the plagioclase series. Unfortunately, the variation in peak position does not vary regularly with composition. Under quite rare conditions you may be able to distinguish triclinic from monoclinic K-feldspar by comparing the microcline triplet at 29.50, 30.20, and 30.95° 2θ, to that of orthoclase at 29.85, 30.85, and 32.35° 2θ. At least two peaks are included in the table because they are so close to diagnostic peaks that you will probably not be able to resolve them. They are the anorthite peak at 28.05°, which should appear as a high-angle shoulder on the most intense peak for anorthite, and the microcline peak at 30.70°, which will probably be a bit more intense than the 30.95° 2θ peak.

The high-temperature, ordered, monoclinic K-feldspar sanidine is common in bentonites. Its pattern is similar to that of orthoclase. A useful way to infer the composition of the feldspars in the clay-size fraction is to study the silt fraction of a rock with a petrographic microscope or a microprobe.

There must be someone who loves feldspars, but to the clay mineralogist they are an unmitigated headache. They have many reflections that interfere with almost anything you want to do beyond simple qualitative analysis. Separation of very fine particle sizes (<0.1 μm or less) seems to be the only way to get rid of them.

Zeolites

Zeolites are a surprisingly common constituent in clay-bearing rocks and they are frequently present in the clay-sized fraction. Table 7.10 gives the X-ray powder data for four of the most common minerals of this group. The data in Table 7.10 should be used with these important qualifications. Some zeolites are very susceptible to change under relatively gentle lab treatments. For example, heulandite will change structurally if it dehydrates, and it can dehydrate at < 130°C. Clinoptilolite, classified by Breck (1974) as belonging to the same group as heulandite, is stable up to 700°C. A second qualification is that many zeolites are members of a solid-solution series. For example, analcime is the sodium-rich end of a series, the calcium-rich end of which is wairakite. In the structural position for a large cation, phillipsite can contain varying amounts of sodium, potassium, and calcium. With the variation in these, there are changes in the silicon-to-aluminum ratio in the tetrahedral sites. All this is to say that both peak positions and intensities should be expected to differ somewhat from the values in Table 7.10. Values for *d* and *I* in the table are rounded off from those in the original source, and only those peaks with an intensity of 10 or more are listed, except for the lowest-angle peaks for phillipsite and analcime. A strong clue for zeolites is the presence of

Table 7.10. Diffraction data for zeolites

Heulandite[a]			Clinoptilolite[b]			Phillipsite[c]			Analcime[d]		
d	I	2θ	d	I	2θ	d	I	2θ	d	I	2θ
8.96	100	9.9	8.95	13	9.9	8.11	8	10.9	9.14	2	9.7
7.94	12	11.1				7.18	63	12.3			
						7.16	66	12.4			
						6.42	17	13.8	5.60	60	15.8
5.26	10	16.8				5.38	19	16.5			
5.10	70	17.4				5.07	23	17.5	4.85	20	18.2
4.65	32	19.1	4.65	19	19.1	4.94	27	18.0			
3.98	65	22.3	3.98	61	22.3	4.13	36	21.5			
3.90	43	22.8	3.91	63	22.7	4.12	41	21.6			
3.84	11	23.2				4.06	18	21.9			
3.43	21	26.0							3.43	100	26.0
3.41	15	26.1				3.27	37	27.3			
3.18	19	28.1	3.17	16	28.2	3.21	100	27.8			
3.13	22	28.5	3.12	15	28.6	3.14	35	28.4			
3.07	12	29.1	3.00	18	29.8	3.13	34	28.5	2.93	50	30.5
2.99	29	29.9	2.97	47	31.0	2.93	15	30.5			
2.97	91	31.0	2.79	16	32.1	2.76	20	332.6	2.69	16	33.3
2.81	23	31.8				2.75	36	32.6			
2.73	20	32.8				2.70	16	32.1	2.51	14	35.8
						2.68	21	33.4	2.23	40	40.5

These data have changed substantially from the 1st edition. It isn't that the minerals have changed; resolution and detection have improved. [a]PDF 41-1357. [b] PDF 39-1383. [c]PDF 39-1375; see PDF 12-195 also. [d]PDF 19-1180; see PDF 41-1478 also.

sharp reflections in the clay mineral range, roughly the range below about 12° 2θ. Among the common minerals, only gypsum has a reflection there. For additional information on zeolites, see Mumpton (1977).

Carbonates

Carbonates, especially calcite and dolomite, are commonly associated with clay minerals. Peak positions will vary with the limited solid-solution series between calcite and dolomite and between ankerite and dolomite. They also will vary for the complete solid-solution series between siderite and magnesite and between siderite and rhodocrosite and for the limited substitution of Ca^{2+} for Fe^{2+} in siderite. Other carbonates that you could encounter in sedimentary rocks are smithsonite, rhodocrosite, and kutnahorite. The orthorhombic carbonates tend to have fewer ionic substitutions, therefore there is less variation in their peak positions. Strontianite and all the rhombohedral carbonates but magnesite form concretionary masses in shales. Vaterite, though it is unstable, is included here because it is produced by some immature forms of brachiopods and molluscs and may therefore be found in Recent sediments. X-ray data for eight carbonates are given in Tables 7.11A and 7.11B. Notice that, except for siderite, all of them have one very strong

Table 7.11A. Diffraction data for the rhombohedral carbonates

Calcite[a]			Ankerite[b]			Dolomite[c]			Siderite[d]		
d	I	2θ	d	I	2θ	d	I	2θ	d	I	2θ
3.86	12	23.0	2.91	100	30.72	2.89	100	30.98	3.59	25	24.78
3.04	100	29.43	2.20	5	41.04	2.67	4	33.56	2.80	100	32.02
2.50	14	36.00				2.41	7	37.39	2.35	20	38.37
2.29	18	39.43				2.19	19	41.18	2.13	20	42.35
2.10	16	43.18				2.02	10	44.99	1.96	20	46.42
			1.797	6	50.85	1.787	13	51.10			

[a]PDF 5-586. [b]PDF 41-586 (the 3 most intense peaks!!). [c]PDF 36-426. [d]PDF 29-696.

Table 7.11B. Diffraction data for the orthorhombic carbonates plus vaterite

Aragonite[a]			Strontianite[b]			Witherite[c]			Vaterite[d]		
d	I	2θ	d	I	2θ	d	I	2θ	d	I	2θ
3.40	100	26.24	4.37	14	20.34	4.56	9	19.5	4.23	25	21.0
3.27	50	27.25	3.54	100	25.19	3.72	100	23.9	3.57	60	24.9
2.73	9	32.80	3.45	70	25.82	3.67	53	24.3	3.29	100	27.1
2.70	60	33.18	3.01	22	29.64	3.22	15	27.75	2.73	90	32.8
2.48	40	36.21	2.84	20	31.52	2.66	11	33.75	2.11	20	42.86
2.41	14	37.36	2.60	12	34.55	2.63	24	34.12	2.06	60	43.96
1.98	55	45.82	2.55	23	35.14	2.59	23	34.63			

[a]PDF 41-1475; [b]PDF 5-418; [c]PDF 5-378; [d]PDF 33-268.

peak and not much else. This makes their identification uncertain unless they are abundant enough to produce detectable weak reflections. What if you have to know? Treat the sample with hot, 1N HCl, wash by centrifugation, and analyze again. If the peak is still there, it is not due to a carbonate. We recommend starting with Reeder (1983) for additional information on any carbonates you may encounter.

Apatite, Pyrite, and Jarosite

Table 7.12 gives diffraction data for apatite, pyrite, and jarosite. Considerable substitution in apatite causes variation in peak positions, so the values given are guidelines. The JCPDS publication *Selected Powder Diffraction Data for Minerals* (1974) gives 26 cards for 26 different apatites; the 1993 version list them only by varietal name. Shown in Table 7.12 are two carbonate apatites. Specimen A is a natural specimen and specimen B is synthetic. Pyrite, if present in detectable amounts, produces a relatively high background with older machines because the incident CuKα radiation is ideal for exciting fluorescent radiation from iron. However, most of the excess background can be eliminated by using a single-crystal monochromator (p.50). Jarosite,

Table 7.12. Diffraction data for two carbonate apatites, pyrite, and jarosite

Apatite A[a]			Apatite B[b]			Pyrite[c]			Jarosite[d]		
d	I	2θ	d	I	2θ	d	I	2θ	d	I	2θ
8.13	18	10.9	3.46	25	25.8	3.13	35	28.54	5.93	45	14.9
4.06	10	21.9	3.04	10	29.4	2.71	85	33.07	5.72	25	15.5
3.43	16	26.0	2.78	100	32.2	2.43	65	37.02	5.09	70	17.4
3.08	25	27.0	2.68	40	33.4	2.21	50	40.79	3.65	40	24.4
2.81	80	31.83	2.62	10	34.20	1.92	40	47.45	3.11	75	28.7
2.77	25	32.27	2.23	16	40.43	1.63	100	56.34	3.08	100	29.0
2.72	100	32.97							2.97	15	30.14
2.63	12	34.13							2.86	30	31.26

[a]PDF 21-145. [b]PDF 19-272. [c]PDF 6-710. [d]PDF 22-827.

$KFe_3(SO_4)_2(OH)_6$, may be an unfamiliar name to you, but it is a very common, yellow, weathering product of black shales and mudstones, particularly those in the arid western United States. Some have confused it with the uranium mineral carnotite, which it resembles. It forms by the oxidation of pyrite, which produces sulfuric acid that attacks illite and K-feldspar, liberates K, and forms the compound indicated by the formula.

Gypsum, Anhydrite, Celestite, and Barite

Data for these minerals are shown in Table 7.13. Gypsum is the likely phase in surface outcrops, but anhydrite is frequently found in subsurface samples such as well cuttings and cores. Gypsum is easily identified by its strong peak at $d = 7.56$ Å. Anhydrite has a similarly diagnostic reflection from a 3.50 Å spacing. If gypsum has been heated, even to 60°C (e.g., to dry it), bassanite, $CaSO_4 \cdot 1/2H_2O$, may form. Its most intense reflection is a 3.00 Å spacing (29.80° 2θ) but it has a diagnostic peak at $d = 6.01$ Å (14.75° 2θ). The peak positions for these minerals will not vary much because there isn't much solid solution in them.

Table 7.13. Diffraction data for sulfates

Anhydrite[a]			Gypsum[b]			Celestite[c]			Barite[d]		
d	I	2θ	d	I	2θ	d	I	2θ	d	I	2θ
3.88	5	22.9	7.61	45	11.7	4.23	11	21	4.40	16	20.2
3.50	100	25.46	4.28	90	20.8	3.77	35	23.6	4.34	30	20.47
2.85	29	31.40	3.07	30	29.1	3.43	30	25.95	3.90	50	22.81
2.47	7	36.33	2.87	100	31.2	3.30	98	27.06	3.77	12	23.58
2.33	20	38.68	2.68	50	33.5	3.18	59	28.09	3.58	30	24.89
2.21	20	40.87				2.97	100	30.07	3.45	100	25.86
						2.73	63	32.79	3.32	70	26.86
						2.67	49	33.51	3.10	95	28.77
									2.84	50	31.55

[a]PDF 37-1496. [b]PDF 21-816. [c]PDF 5-593. [d]PDF 24-1035.

Lepidocrocite, Goethite, Gibbsite, and Anatase

Lepidocrocite, goethite, and gibbsite are common in soils and some non-marine sedimentary rocks. Anatase is frequently encountered in shales, sandstones, and bentonites. Except for anatase, these minerals can produce broad reflections that resemble those of the clay minerals. Such peaks make it difficult to pick exact peak positions. These minerals lose water on heating to form new minerals. Therefore, heat treatment may provide useful confirmation of identifications.

Lepidocrocite on heating to 230 to 280°C forms maghemite γ-Fe_2O_3; and to hematite upon further heating to 400 to 500°C. Goethite forms hematite (α-Fe_2O_3), which gives broad peaks. Heating to 900°C yields well-crystallized hematite with sharp peaks. Gibbsite on heating to 150 to 200°C changes first to boehmite γ-$AlO(OH)$, and then to a compound that doesn't occur as a

Table 7.14. Diffraction data for lepidocrocite, goethite, gibbsite and anatase

Lepidocrocite[a]			Goethite[b]			Gibbsite[c]			Anatase[d]		
d	I	2θ	d	I	2θ	d	I	2θ	d	I	2θ
6.26	100	14.2	4.98	12	17.8	4.85	100	18.3	3.52	100	25.3
3.29	90	27.1	4.18	100	21.24	4.37	70	20.3	2.43	10	36.98
2.47	80	36.4	3.38	10	26.34	4.32	50	20.6	2.38	20	37.83
2.36	20	38.1	2.69	35	33.27	3.36	17	26.5	2.33	10	38.61
2.09	20	43.3	2.58	12	34.73	3.18	25	28.1	1.89	35	48.09
1.94	70	46.90	2.49	10	36.09	2.46	25	36.5			
			2.45	50	36.68	2.38	55	37.8			
			2.25	14	40.02	2.16	27	41.8			
						2.05	40	44.2	1.98	10	45.8

[a]PDF 8-98. [b]PDF 29-713. [c]PDF 33-18. [d]PDF 21-1272.

Table 7.15. Diffraction data for lepidocrocite, goethite, and gibbsite after heat treatment[a]

(Lepidocrocite) γ-Fe_2O_3[b]			(Goethite) α-Fe_2O_3[c]			(Gibbsite) χ-Al_2O_3[d]		
d	I	2θ	d	I	2θ	d	I	2θ
5.90	5	15.0	3.67	35	24.3	4.52	30	19.6
4.82	5	18.4	2.69	100	33.3	2.38	70	37.8
3.73	5	23.9	2.51	75	35.8	2.12	80	42.7
3.40	5	26.2	2.20	25	41.0	1.90	30	47.9
3.20	10	27.9	1.84	30	49.5			
2.95	30	30.3						
2.51	100	35.8						

[a]All data from Rooksby (1961). [b,c]Formed by heating for 1 h at 300°C. [d]Formed by heating for 1 h at 300°C; results could be indeterminate.

mineral so far as we know χ-Al_2O_3. Data for XRD tracings of the phases resulting from such treatments are given in Table 7.15. The results from heating will not produce spectacular XRD tracings; gibbsite may be indeterminate, but in any event the gibbsite pattern will be destroyed.

Almost certainly we have left out your favorite mineral, but those given should do as starters. Brown and Brindley (1980, pp. 348ff) have provided an extensive table that is helpful or even indispensable for preliminary identifications. Copy this table and glue it to the wall of your X-ray diffraction laboratory.

SUMMARY

Discrete clay minerals are best identified from diffraction tracings of oriented aggregates. They show a *rational* series of 00*l* peaks. Clay and nonclay minerals are frequently mixed. The peaks of the nonclay minerals are usually sharper and narrower than those of clay minerals. A collection of diffraction tracings of the common, discrete clay minerals is probably the most useful tool for identification.

REFERENCES

Bailey, S. W. (1980) Structures of layer silicates: in Brindley, G. W., and Brown, G., editors, *Crystal Structures of Clay Minerals and Their X-Ray Identification*, Monograph **5**, Mineralogical Society, London, pp. 1-123.

Bohor, B. F., and Triplehorn, D. M. (1993) Tonsteins: Altered Volcanic-Ash Layers in Coal-Bearing Sequences: *Special Paper* **285**, Geological Society of America, Boulder, Colorado, 44 pp.

Borg, I. Y., and Smith, D. K. (1969a) Calculated X-ray powder patterns for silicate minerals: *Geol. Soc. Amer. Memoir* **122**, 896 pp.

Borg, I. Y., and Smith, D. K. (1969b) Calculated powder patterns. Part II. Six potassium feldspars and barium feldspar: *Amer. Minerl.* **54**, 163-81.

Breck, D. W. (1974) *Zeolite Molecular Sieves*: Wiley, New York, 771 pp.

Brindley, G. W. (1980) Order-disorder in clay mineral structures: in Brindley, G. W., and Brown, G., editors, *Crystal Structures of Clay Minerals and Their X-Ray Identification*, Monograph **5**, Mineralogical Society, London, pp. 125-95.

Brown, G. (1980) Associated minerals: in Brindley, G. W., and Brown, G., editors, *Crystal Structures of Clay Minerals and Their X-Ray Identification*, Monograph **5**, Mineralogical Society, London, pp. 361-410.

Brown, G., and Brindley, G. W. (1980) X-ray diffraction procedures for clay mineral identification: in Brindley, G. W., and Brown, G., editors, *Crystal Structures of Clay Minerals and Their X-Ray Identification*, Monograph **5**, Mineralogical Society, London, pp. 305-59.

Byström-Brusewitz, A. M. (1975) Studies of the Li test to distinguish beidellite from montmorillonite: *Proceedings, Internat. Clay Conf.*, 1972, Mexico City, Applied Publishing, Wilmette, Ill., pp. 419-28.

Calvert, C. S. (1984) Simplified, complete CsCl-hydrazine-dimethylsulfoxide intercalation of kaolinite: *Clays and Clay Minerals* **32**, 125-30.

Chen, P-Y. (1977) Table of Key Lines in X-Ray Powder Diffraction Patterns of Minerals in Clays and Associated Rocks: *Occasional Paper* **21**, Geological Survey of Indiana, Bloomington, Indiana, 67 pp.

Churchman, G. J., Whitton, J. S., Claridge, G. G. C., and Theng, B. K. G. (1984) Intercalation method using formamide for differentiating halloysite from kaolinite: *Clays and Clay*

Minerals **32**, 241-48.

Drits, V. A., Plançon, B. A., Sakharov, B. A., Besson, G., Tsipursky, S. I., and Tchoubar, C.(1984) Diffraction effects calculated for structural models of K-saturated montmorillonite containing different types of defects: *Clay Minerals* **19**, 541-61.

Drits, V. A., and Tchoubar, C. (1990) X-Ray Diffraction by Disordered Lamellar Structures: Springer-Verlag, New York, 371 pp.

Frondel, C. (1962) *Silica Minerals*, Vol. III, *The System of Mineralogy*: Wiley, New York, 334 pp.

Greene-Kelly, R. (1952) Irreversible dehydration in montmorillonite: *Clay Mineral Bull.* **1**, 221-27.

Greene-Kelly, R. (1953) Irreversible dehydration in montmorillonite. Part II: *Clay Mineral Bull.* **2**, 52-6.

Guthrie, G. D. Jr., Bish, D. L., and Reynolds, R. C., Jr. (1995) Modeling the X-ray diffraction pattern of opal: *Amer. Minerl.* **80**, 869-72.

Jones, J. B., and Segnit, E. R. (1971) The nature of opal I. Nomenclature and constituent phases: *J. Geol. Soc. Australia* **18**, 57-68.

JCPDS (1993) *Mineral Powder Diffraction File Databook*: Joint Committee on Powder Diffraction Standards, Swarthmore, Pa., 781 pp.

MacEwan, D. M. C., and Wilson, M. J. (1980) Interlayer and intercalation complexes of clay minerals: in Brindley, G.W., and Brown, G., editors, *Crystal Structures of Clay Minerals and Their X-Ray Identification*, Monograph **5**, Mineralogical Society, London, pp. 197-248.

Mumpton, F. A., ed. (1977) *Mineralogy and Geology of Natural Zeolites*: Short Course Notes, Vol. **4**, Mineralogical Society of America, Washington, D.C., 233 pp.

Reeder, R. J., ed. (1983) *Carbonates: Mineralogy and Chemistry: Reviews in Mineralogy*, Vol. **11**, Mineralogical Society of America, Washington, D.C., 394 pp.

Rooksby, H. P. (1961) Oxides and hydroxides of aluminum and iron: in Brown, G., ed., *The X-Ray Identification and Crystal Structures of Clay Minerals*: Mineralogical Society, London, pp. 354-92.

Reynolds, R. C., Jr. (1985) *NEWMOD$^{©}$ a Computer Program for the Calculation of One-Dimensional Diffraction Patterns of Mixed-Layered Clays*: R. C. Reynolds, 8 Brook Rd., Hanover, NH, 03755.

Reynolds, R. C., Jr. (1986) The Lorentz-polarization factor and preferred orientation in oriented clay aggregates: *Clays and Clay Minerals* **34**, 359-67.

Środoń, J. (1980) Precise identification of illite/smectite interstratifications by X-ray powder diffraction: *Clays and Clay Minerals* **28**, 401-11.

Walker, G. F. (1958) Reactions of expanding lattice minerals with glycerol and ethylene glycol: *Clay Miner. Bull.* **3**, 302-13.

Chapter 8
Identification of Mixed-Layered Clay Minerals

X-ray diffraction patterns of mixed-layered (or interlayered or interstratified) clay minerals pose some of the most difficult problems of interpretation, particularly if, as is often the case, the mixed-layered species are present in physical mixtures that include the simple clay types. Multiple analyses are the rule. They use all the available strategies such as heat treatments and solvation with water and ethylene glycol (EG). To identify mixed-layered species successfully, you need to consider the entire diffraction pattern, because the breadth and symmetry of each diffraction peak may be diagnostic, together with the peak position and intensity. This chapter contains representative examples of diffraction patterns, but more extensive coverage of each of the mineral types is discussed in Reynolds (1980). The theory upon which calculated diffraction patterns are based was developed by MacEwan (1958), augmented by Reynolds (1967), applied by Reynolds and Hower (1970), and described in detail by Reynolds (1980). All calculated diffractograms shown here have been calculated with the computer program NEWMOD© (Reynolds, 1985, see the Appendix). They are all calculated based on CuKα radiation, unless noted otherwise. No attempts have been made to portray correct absolute intensities or correct relative intensities between patterns. The vertical scales have been selected either to scale the strongest peaks to 100% or to emphasize the details of the weaker reflections.

A complete description of a mixed-layered clay mineral requires identification of the types of layers involved (e.g., illite and smectite), the proportions of each, and the type of order or lack thereof in the stacking sequence of the two-layer types along the Z direction (the Reichweite; see Chapter 5, p. 173). We limit the discussion here to two-component systems because multicomponent interstratification seems to be exceedingly rare in sedimentary rocks. In addition, consideration is given only to illite, smectite, vermiculite, kaolinite, chlorite, and serpentine (Al-lizardite and bertherine). These mixed-layered mineral components are those that the shale petrologist is most likely to encounter.

The easiest mixed-layered clay minerals to identify contain two components in equal proportions with $R1$ ordering. Such minerals have been assigned mineral names and can be thought of as non-mixed-layered minerals that have unit cells whose 001 spacings are equal to the sums of 001 spacings of the two components involved. Large cells such as these are called *superstructures*, and it is convenient to designate their spacings as $d(00l)^*$. (This asterisk should not be confused with the symbol for reciprocal space, which is a widely used mathematical device in X-ray crystallography.) Diffraction patterns for well-oriented samples of these two-component systems consist of a set of basal reflections whose spacings follow the Bragg law, $d = d(001)^*/l$, where l is the integral order. If this relation is closely obeyed, the pattern of peaks is called *rational*. Corrensite is a common example, one type of which is made up of equal molecular proportions (50/50) of chlorite and smectite unit cells that are stacked in perfect alternation ($R1$). You might question whether corrensite should be considered a mixed-layered clay mineral, given its regular structure and rational pattern. After all, pure chlorite consists of a regular alteration (in 50/50 proportions) of silicate (talc-like) and hydroxide (brucite-like) "unit cells," yet we usually do not think of chlorite as an interstratified mineral that consists of talc and brucite. Mineral species such as corrensite, however, need to be placed somewhere, and traditionally they are included with the mixed-layered clay minerals.

Other mixed-layered clay minerals do not share such rigid structural constraints. In these, proportions of end-members are not equal, and ordering may be imperfect. Consequently, randomness occurs in the stacking sequences, leading to structures that can be described only statistically. The stacking sequences in such structures, therefore, are not periodic, and, as you might expect, they produce aperiodic $00l$ diffraction patterns. These patterns we call *irrational*. Irrationality is evidence of some randomness in the interstratification, and the approach to rationality is quantified by the percent standard deviation about the mean of a set of $d(001)$ values calculated from the various reflections. The number is reported as the coefficient of variation

Table 8.1. Coefficient of variation for $00l$ diffraction patterns of EG-corrensite and of randomly interstratified chlorite/EG-smectite containing 70% chlorite

	Corrensite			Chlorite/EG-smectite	
l	d	$l \cdot d(001^*)$	l	d	$l \cdot d(001)$
1	31.31	31.31	1	15.29	15.29
2	15.73	31.46	2	7.41	14.82
4	7.78	31.12	3	4.74	14.22
5	6.23	31.15	4	3.49	13.96
6	5.19	31.14	5	2.83	14.15
7	4.46	31.22	7	2.05	14.35
8	3.89	31.12	9	1.56	14.04
9	3.46	31.14			
10	3.11	31.10			
Mean		31.20			14.40
CV		0.38%			3.34%

(CV). By definition (Bailey, 1982), values < 0.75% define a mineral that qualifies for a specific name, such as corrensite. If CV > 0.75%, the structure is sufficiently aperiodic to limit the designation to mixed-layered nomenclature, such as chlorite/smectite. Table 8.1 shows CV values for corrensite and for a randomly interstratified chlorite/smectite.

The corrensite spacings follow the Bragg law as well as can be expected, given the usual inaccuracies in measuring d. On the other hand, the diffraction pattern of randomly interstratified chlorite/smectite is demonstrably irrational and can be confidently ascribed to a mixed-layered species with at least some randomness in the stacking sequence. The test, which is easy to apply, is definitive—if you are certain that all peaks considered belong to the same mineral species. Let us see why some types of mixed-layering produce irrational patterns.

MÉRING'S PRINCIPLES AND MIXED-LAYERED NOMENCLATURE

The model conceived by Méring (1949) provides the best insights into the character of mixed-layered clay mineral diffraction patterns. According to this concept, reflections occur between the nominal positions of the $00l$ peaks of both members of the mixture, and the positions of these intermediate reflections are fixed by the proportions of the end-members in the interstratification. The observed reflections are thus composites, or compromises, of the positions and, to a certain extent, the intensities of two end-member reflections. For the example of illite/EG-smectite, a reflection will be present between the illite 002 and EG-smectite 003 (15.8° 2θ) peak positions, and it is labeled the illite/EG-smectite 002/003. Remember, you will not see these end-member reflections in the diffraction pattern; they simply represent diffraction components that together produce a maximum. The maximum is broad if end-member reflections are far apart and sharp if they are close together. The diffraction pattern is usually unlike that of either of the end-members. The diffractogram for randomly interstratified clay minerals thus differs from that of a normal clay mineral in two important respects: (1) members of the set of so-called $00l$ peaks do not have the same breadths, and (2) the reflections have irrational spacings.

Our nomenclature for mixed-layered clay minerals is as follows. We name a mineral by listing the species with the smallest $d(001)$ (in the natural state) first and separate the names with a slash. For example, mixed-layered illite-smectite is called illite/smectite (IS or I/S, as you may find in the literature). The mineral name is preceded by the Reichweite, which describes the ordering type (see Box 5.3, p. 173), and a number in parentheses following the first name gives the decimal fraction of that type in the mixed-layered mineral. For example, the designation $R1$ illite(0.7)/smectite means a composition of 70 mol % illite with 30 mol % smectite. The layer types are

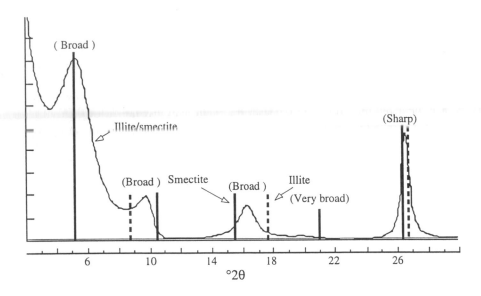

Fig. 8.1. *R*0 illite(0.5)/EG-smectite and the locations of the 00*l* reflections for illite and smectite.

stacked along *Z* in such a way that smectite layers never follow smectite layers.

Figure 8.1 demonstrates Méring's principles. The continuous profile illustrates a randomly interstratified (*R*0) illite/EG-smectite, 50% illite. Using the terminology given above, this becomes *R*0 illite(0.5)/EG-smectite. The solid vertical bars mark the positions of the smectite 00*l* reflections, and the broken bars designate the same for illite. Peaks whose positions depend on the proportions of illite and smectite occur near 10° 2θ for the composite illite 001 and EG-smectite 002 reflections and 16° for the illite 002 and EG-smectite 003 reflections. What reflections do you think contribute to the peak at 26.6° 2θ? Notice that the "peak" near 20° 2θ is very weak and broad. It is formed by the composite of the illite 002 and the smectite 004 reflections (the 002/004), which are widely separated. The reflection near 16° 2θ is stronger, yet also quite broad because it is a composite of the illite peak near 17.8° and the smectite peak near 15.8° 2θ. On the other hand, the peak near 26.6° 2θ owes its sharpness to the close juxtaposition of the illite 003 and the smectite 005 reflections. Note that a strong but broad reflection occurs at 5.2° 2θ at the smectite 001 position. The smectite 001 is not displaced because no illite peak is nearby. Instead, it becomes progressively broader and weaker as the proportion of illite increases.

Ordering causes the appearance of new reflections containing components of the 00*l* series of the superstructure. Figure 8.2 compares diffraction patterns for EG-K-rectorite (dashed line) and *R*1 illite(0.7)/smectite. The patterns are similar except for some peak displacements, line breadths, and the near disappearance of the superstructure 001* reflection for the 70% illite pattern. The position of 001* is marked only by a slight change in slope of the "background." Think of the 70% composition as a random interstratification of *rectorite* with enough illite layers to bring the composition from 50% illite to 70% illite. The 002* is weakened and broadened for two reasons. First, the rectorite component is less abundant than it was in the 50/50 composition, and, second, interference between the 002* and the illite 001 reflections has broadened the composite diffraction reflection, thus reducing its height. This ordered example contains some randomness in the stacking sequence along the Z axis because some smectite layers are separated by more than one illite layer. *Randomness must be present if the mineral contains component proportions that differ from 50%. But there need not be ordering at the 50/50 composition.*

If you think about the peak positions for the 70% composition, they will make sense. Figure 8.2 has vertical lines that depict the positions of the illite 001 and 002 reflections. The second-order superstructure reflection (002*) at about 6.6° 2θ has been displaced slightly toward the illite 001 position, and the 003* has moved in the opposite direction, also because of interference with the illite 001 peak. The K-rectorite 005* reflection (the 004* and 006*

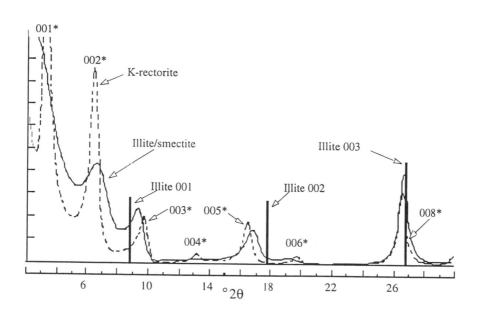

Fig. 8.2. *R*1 illite(0.7)/EG-smectite and EG-K-rectorite[*R*1 illite(0.5)/EG-smectite] and the positions of the peaks for pure illite.

are barely visible) near 16.2° 2θ has been similarly affected by proximity to the illite 002 and lies near 16.8° (the 002/005* reflection) for the 70% composition. The illite 003 is almost coincident with the K-rectorite 008* reflection, so the peak near 26.6° is essentially unaffected. With increasing illite content, the superstructure peaks that are far from the illite reflections (the 001*, 002*, 004*, and 006*) become progressively weakened, and the other reflections migrate toward the nominal positions of the pure illite pattern. You will be able to see how these principles operate in the following discussion of the diffraction patterns of mixed-layered clay minerals.

The Q Rule, a Broadening Descriptor

The different line breadths that are evident in Fig. 8.1 are quantifiable, and indeed, we can invent a rule for detecting small amounts of interstratification—amounts so small that they can escape detection by a consideration of peak positions. The rule is a quantification of Méring's principles. It works this way. Suppose that small (a few percent) amounts of mica are suspected in a chlorite. Simply take the ratio of the mica $d(001)$ divided by the chlorite $d(001)$, or in this case, $10/14.2 = 0.704$, and multiply this ratio by the l values for the chlorite reflections (Table 8.2). Then for each such value, calculate the deviation from the nearest integer and call the result Q. The values for Q predict the different line breadths. In this example, the broadest of the "chlorite" reflections will be the 002 and 005, and the 003, 007, and 00,10 will be sharp. A Q value of 0.000 designates a reflection that has no mixed-layered broadening, that is, the line breadth is controlled only by crystallite thickness (and instrumental effects). A value of 0.5 identifies a peak that has the maximum breadth possible for that mixed-layered mineral.

Figure 8.3 shows a plot of these Q values versus the calculated line breadths for a randomly interstratified $R0$ biotite(0.1)/chlorite. The line breadths have been multiplied by cosθ to eliminate angle-dependent particle-size broadening, e.g., see the Scherrer equation, Eq. (3.8), p. 87. For the calculated pattern, a mean defect-free distance of 10 unit cells was used in conjunction with a maximum crystallite thickness of 50 unit cells (see Eq. A.8 and related discussion, p. 370). The calculated data show some scatter and the curve is not quite linear, but nevertheless you can see that Q is well correlated with line breadth. The broken line on Fig. 8.3 shows the approximate relation for $R0$ biotite(0.5)/chlorite. Suppose that the chlorite contained no biotite at all. Then a plot of Q versus line breadth would yield points distributed along a line parallel to the vertical axis of Fig. 8.3 because all the peaks would have identical breadths (Q ceases to have any meaning for such a case), and these are controlled by particle size broadening, the angular dependence of which has been removed by multiplying each of the line breadths by cosθ. We conclude that the *slope* of the curve of Fig. 8.3 can be used to estimate the proportion of biotite in a mixed-layered biotite/chlorite. Such a slope-composition relation is highly nonlinear with greatest sensitivity at small

Table 8.2. Calculation of the broadening descriptor, Q

Order	Order x 0.714	Q
1	0.704	0.296
2	1.408	0.408
3	2.113	0.113
4	2.817	0.183
5	3.521	0.479
6	4.225	0.225
7	4.930	0.070
8	5.634	0.366
9	6.338	0.338
10	7.042	0.042

contents of one of the interstratified species. A few percent or less of a minor component can be identified and estimated by this method, and such a small amount might escape detection if you paid attention only to peak positions. Note—in the extreme example given here (reflections out to the chlorite 00,10), you would have to correct the experimental data for instrumental effects, most importantly, for $CuK\alpha_1$-$CuK\alpha_2$ broadening.

The relation between Q and line breadth needs calibration by means of known standards (unlikely) or from calculated line breadths, which can be obtained, as in the example of Fig. 8.3, from NEWMOD© (Reynolds, 1985). The Q principle is also useful for the *identification* of mixed-layered clay minerals. Suppose you suspect that your sample contains a paragonite [$d(001)$ = 9.7 Å] interstratified with very small amounts of 2-water-layer smectite [$d(001)$=15 Å]. You could verify this identification by a measurement of the line breadths.

The absolute line breadths of all the reflections must be correctly modeled if line broadening is used for a quantitative analysis of layer proportions. If the discussion above dealt with a mineral of much smaller or larger crystallite thickness, the slope of the line on Fig. 8.3 would be different, and for very thin crystallites, the scatter of points about the line would be increased.

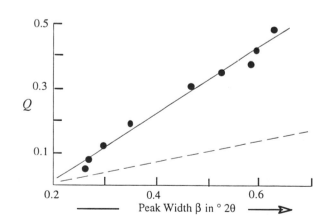

Fig. 8.3. The correlation between Q and line-breadth for R0 biotite(0.1)/chlorite. The broken line locates the relation for R0 biotite(0.5)/chlorite.

Think of the breadth of a large-Q peak as a composite of mixed-layered broadening and crystallite thickness broadening. The mixed-layered component is fixed for a given layer proportion, but the crystallite size component becomes large for thin crystallites and overwhelms the mixed-layered effect. There are always errors in the measurement of peak breadth, and at some level of small crystallite size, the additional effect of mixed layering on peak breadth disappears in the experimental error. Alternatively, the Q rule works best for very thick crystallites because most of the breadth of large-Q reflections is due to mixed-layered broadening.

The rationale behind the meaning of Q is as follows. If Q is very small, it means that a reflection from one interstratified component is almost superimposed on a reflection from the other component, and as Méring's rules tell us, that results in a sharp reflection. If Q is 0.5, that means that the reflection whose order is indicated by the "Order" column in Table 8.2 is equidistant from two flanking orders of the other component, and that causes maximum line broadening.

Let's see what can be done with line broadening in a more exhaustive application. Figure 8.4A shows three experimental diffraction patterns of chlorite that has been (1) heated at 250°C and analyzed in an enclosed chamber through which was streamed tank nitrogen to maintain dehydration (Chapter 6); (2) solvated by EG; and (3) analyzed in the air-dried condition. The data were obtained with two Soller slits and a 0.05° detector slit to minimize instrumental broadening. A thick slide of a well-oriented specimen was used in conjunction with long count times for the step scan procedure, producing a pattern with a high peak-to-noise ratio.

Some shift is noticeable among the 002 reflections of the three patterns, but one would be hard put to assign a quantitative value to any smectite present. (The EG-treated preparation shows a superstructure reflection that indicates the physical admixture of a small amount of corrensite, but the strengths of corrensite's other reflections are insufficient to affect the chlorite peaks very much.) Peak breadths for all the chlorite reflections from the three preparations were measured as the width, in ° 2θ, of the peaks at half-height. These measurements were made from plots of the diffraction patterns at greatly expanded horizontal scales. Trial and error calculations were performed by NEWMOD© for different values of the distribution of N values (the number of unit cells in coherent scattering array) and percent randomly interstratified smectite, and the same values for each of these were applied to all three final model structures.

Figure 8.4B shows comparisons between experimental and calculated (model) line breadths for five reflections from each of the three preparations. The agreement between experiment and model is excellent, and a conclusion based on 15 reflections gives confidence that the procedures used are valid and that the amount of smectite interstratified with this chlorite must be very close to 7%. Notice that for some preparations, the differences in breadths are

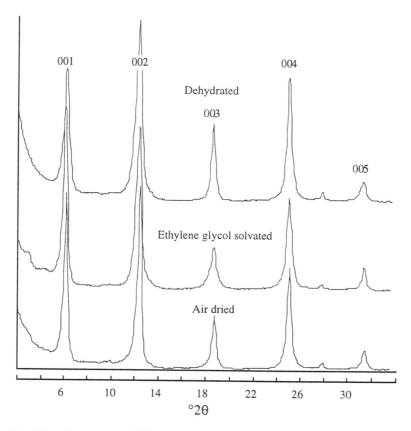

Fig. 8.4A. Experimental diffraction patterns for different preparations of a chloritic mineral.

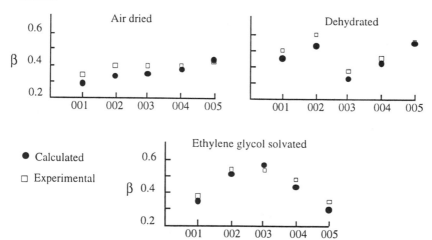

Fig 8.4B. Comparisons between experimental and calculated line breadths (β) for different diffraction orders of the preparations shown by Fig. 8.4A.

as large as a factor of two, and that differences between all peak breadths are much larger than the error of measurement which is estimated to be 0.02° 2θ. These observations suggest that we could go to much smaller proportions of smectite and still quantify the result—say, down to 2% if the instrumental contributions to the peak breadths were removed by deconvolution procedures.

The line-broadening method of analysis cannot be used for compositions that exceed about 10% to 15% of one component unless the interstratification is random. In any event, the peak position method is the one to use for compositions nearer the middle of the range. A *caveat* is in order here. The mineral line breadths can be swamped by instrumental broadening if the diffractometer is poorly focused (one Soller slit, large detector slit, large beam slit, or misalignment). A refined quantitative analysis of mixed-layering based on line breadths may require that instrumental broadening effects be removed mathematically from the diffraction profiles by the process of deconvolution (see Klug and Alexander, 1974), and that is beyond the scope of this text, though it is not particularly complicated or time consuming even with PC-type computers. For many applications, however, it is sufficient to use two Soller slits, a 1° beam slit, and a 0.05° (or smaller) detector slit and to limit the analysis to the lower diffraction angles (<40° 2θ), where $K\alpha_1/K\alpha_2$ broadening is minimal.

The successful application of Méring's principle in this rigorous manner suggests that it has a fundamental reality that has not been generally appreciated. It is impressive what the great ones were able to do without computers.

MIXED-LAYERED CLAY MINERALS

Illite/Smectite

Illite/smectite is the most abundant, diverse, and widespread of the mixed-layered clay minerals in sedimentary rocks and soils. You will need to become proficient in identifying it in all its structural ramifications if you intend to do much work on clay minerals from soils or rocks. The best method of analysis is to study and compare diffraction patterns produced from both air-dried and EG-solvated preparations.

Most of the attention here is given to the interpretation of diffraction patterns obtained from EG-solvated specimens because these patterns are the most diagnostic. For patterns from the air-dried condition, we assume that a smectite interlayer contains two planes of water molecules coordinated about exchangeable Ca ions, producing a smectite $d(001)$ of 15 Å. Unfortunately, the basal spacing is a fairly sensitive function of ambient humidity and the type of exchange cation. Thus, unless you saturate your samples with Ca and maintain strict control of humidity (a difficult thing to do), you should not use peak positions from air-dried preparations to obtain composition estimates. Air-dried sample data are important to use as baselines from which to observe the changes brought about by EG solvation.

Suppose that diffraction patterns are recorded from a mixed-layered clay mineral in both the air-dried and EG-solvated conditions. When you see that EG solvation has caused significant changes in the diffraction pattern, a smectite component is surely present, because EG solvation has little effect on vermiculite, and other likely layer types are not affected at all by this treatment. So you have a "something"/smectite, and that something may be illite, chlorite, kaolinite, talc, serpentine, vermiculite, etc. The interstratification is random and probably rich in smectite if EG solvation produces a peak near 5.2° 2θ. Concentrate now on the EG-solvated case. Examine the region near 16 to 17.7° 2θ, and if you note a reflection there (the 002/003), the diagnosis is likely illite/smectite. You can estimate the proportion of illite from the data in Table 8.3, which have been tabulated from calculated diffractograms for each specific composition. If you want to work with only one reflection, the peak near 17° 2θ is the best choice. The peak near 9° is more sensitive to shifts caused by small crystallite size.

You can confirm your qualitative identification of the mixed-layered species by heating the sample to 375°C for 1 h. The diffraction pattern should resemble a pure illite (10 Å structure) with a relatively weak 003 reflection.

The quantity Δ2θ (Table 8.3) is useful because, for some mixed-layered minerals, it is the most "robust" descriptor of composition, as the mathematicians say. Środoń (1980) has demonstrated the power of differential 2θ measurements for accurately estimating the composition of illite/smectite. His procedures involve several reflections and require very good diffraction patterns free from peak interferences.

The advantages of using Δ2θ are several. It is a differential measurement; therefore, results based on it are relatively insensitive to goniometer zero-alignment problems and specimen displacement errors. The latter are very likely despite your best efforts. For illite/smectite and some other mixed-layered clay mineral analyses, the sensitivity of the composition estimate is enhanced because the two peaks involved in the measurement move in opposite directions as the composition changes. Finally, analytical errors caused by sample-to-sample variations in the thickness of the EG interlayer in smectites (Środoń, 1980) are at least minimized by the use of Δ2θ because changes in EG thickness cause both reflections to be displaced in the same direction. The differential 2θ value is thus relatively unaffected by variations in thicknesses of the interlayer solvation complex.

Let us now see how all this is applied to the diffraction pattern shown in Fig. 8.5. EG solvation has produced a strong but broad reflection near 5° 2θ (*d* = 17 Å). A good guess at this point is that you have a randomly interstratified illite/smectite. The other peak positions, near 9° and 16° 2θ, support this supposition. Quantify the result using their positions (Table 8.3).

You could make a better estimate of composition, using the $\Delta 2\theta$ value in Table 8.3, assuming that interferences from other clay minerals do not make this measurement impractical.

Figure 8.6 shows diffraction patterns for $R1$ illite(0.7)/smectite. Again, concentrating on the EG-solvated case, the high-angle portion of the pattern is similar to the random case in Fig. 8.5, except that the peaks are sharper for the ordered structure. Most important, a strong reflection is present at low diffraction angles (6.5° 2θ, 13.3 Å) that contains a dominating component of the second-order superstructure reflection (002*). This reflection should shout ORDERING! to you. *The interpretive importance of the low-angle region cannot be overemphasized.* A reflection at 5° 2θ indicates random interstratification, and one near 6.5° 2θ indicates $R1$ ordering. The latter reflection becomes broad and weak and does not move very much as the composition becomes more illitic.

A noticeable peak at angles between 7 and 8° 2θ (about 8° in Fig. 8.7) indicates long-range ordering, that is, Reichweite values greater than unity. Figure 8.7 shows a calculated pattern for $R3$ illite(0.9)/smectite. In the EG-solvated state, the broad shoulder at 8° 2θ, or $d = 11.1$ Å, contains components of the illite 001 reflection and of the fourth order of a 47 Å superstructure peak whose unit cell consists of three 10 Å illite layers and one 17 Å EG-smectite layer. The reflection is called the 001/004*. This peak is diagnostic for long-range ordering, that is, $R > 1$. It denotes the presence, in the mixed-layered crystallites, of the frequent unit cell sequence ISII mixed with additional illite layers.

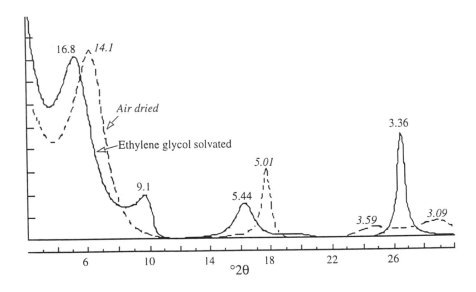

Fig. 8.5. EG-solvated and air-dried $R0$ illite(0.5)/smectite.

We draw three conclusions that are crucial to the correct identification and description of mixed-layered illite/smectite.

1. The diffraction patterns of air-dried illite/smectite are altered significantly by solvation with EG, and such behavior leads to the provisional identification of illite/smectite. Heating to 375°C for 1 h provides confirmation of the identification; the result is a diffraction pattern similar to that of illite.

2. The Reichweite (ordering type) is determined by the position of the reflection between 5 and 8.5° 2θ for EG-solvated preparations.

3. Percent illite can be determined by the position of the reflection near 16 to 17° 2θ, but a better way is to base the estimate on a value for Δ2θ (Table 8.3).

Characterization of I/S is routine and definitive if you are armed with high-quality diffraction patterns from air-dried, EG-solvated, and dehydrated preparations of a pure I/S sample. Interpretations of poor patterns from mineral mixtures are another matter. Mixtures of discrete (detrital?) illite with I/S in shales are the norm, and if the I/S is illite-rich, the criteria described above are difficult to apply. Figure 8.8 shows air-dried and EG-solvated simulations of R1 illite(0.65)/smectite mixed with 40 wt % illite. The top and bottom traces are typical calculated patterns. The two middle traces were prepared by superimposing the calculated traces on a noisy background produced by a pseudorandom number generator. These are meant to show the effects of the pattern degradation that results from poor (average?) laboratory procedures.

Table 8.3. The positions (CuKα) of useful reflections for estimating percent illite in illite/EG-smectite

% Illite	Reichweite	001/002 d(Å)	° 2θ	002/003 d(Å)	° 2θ	° Δ2θ
10	0	8.58	10.31	5.61	15.80	5.49
20	0	8.67	10.20	5.58	15.88	5.68
30	0	8.77	10.09	5.53	16.03	5.94
40	0	8.89	9.95	5.50	16.11	6.16
50	0	9.05	9.77	5.44	16.29	6.52
60	1	9.22	9.59	5.34	16.60	7.01
70	1	9.40	9.41	5.28	16.79	7.38
80	1	9.64	9.17	5.20	17.05	7.88
90	3	9.82	9.01	5.10	17.39	8.38

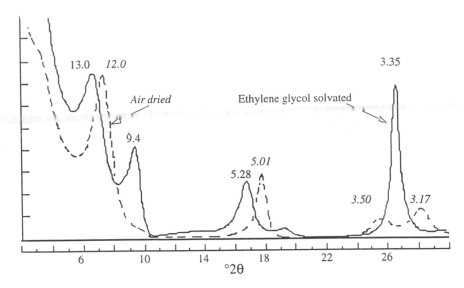

Fig. 8.6. EG-solvated and air-dried $R1$ illite(0.7)/smectite.

First of all, how do we know that I/S is present? Well, the hypothetical sample has responded to EG solvation, so it certainly contains an expandable component. Notice the very sharp and intense reflection near 17.7° for the air-dried structure. The character of this peak is due to the almost perfect superposition of the two-water-layer ($d = 15$ Å) smectite 003 component on the second-order illite contribution at 5 Å ($Q = 0$). Thus both members of the

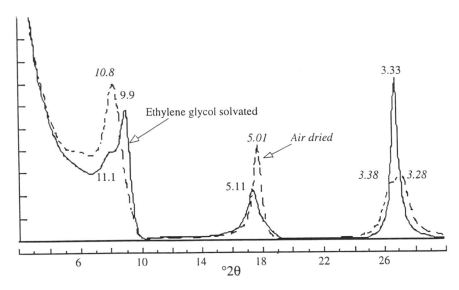

Fig. 8.7. EG-solvated and air-dried $R3$ illite(0.9)/smectite.

interstratification have a peak there, so there is (1) no mixed-layered line broadening or distortion, (2) no peak shift, and (3) the amplitudes are additive. The discrete illite also has a peak at that position, which adds to the peak height but little to the breadth of the aggregate reflection. The aggregate peak becomes broad after EG solvation because the smectite 003 component is no longer coincident with the illite 002 position, resulting in mixed-layered broadening and a shift in peak position away from the 002 peak of the discrete illite. So, we are pretty sure that the sample contains I/S. Confirmation of the diagnosis is provided by the high-angle shoulder on the

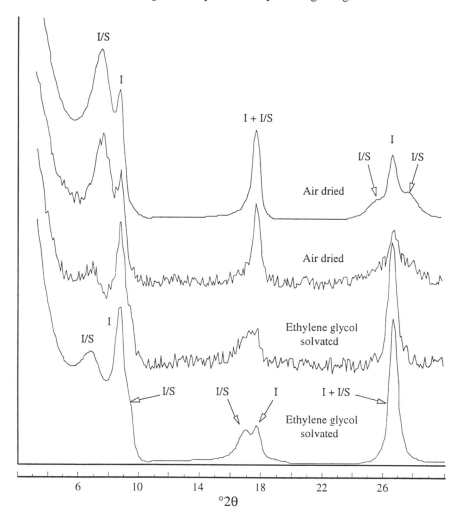

Fig. 8.8. Calculated patterns of EG-solvated and air-dried *R*1 illite(0.65)/smectite mixed with 40 wt % illite. Shown are calculated patterns and the same patterns modulated by random noise to simulate poor data and the problems of analysis faced because of poor XRD tracings.

discrete illite 001 reflection that appears upon EG solvation. This unresolved mixed-layered reflection is the 001/002. You should become sensitive to the shape of the discrete illite 001 because if no I/S is present, the peak should be concave on the high-angle side; I/S of considerable expandability is present if EG solvation produces that bulge on the high-angle side, and an increase in the asymmetry of the illite 001 peak in the low-angle direction indicates the presence of I/S of low (10% or less) expandability.

The broad reflection near 6.7° that has shifted to that position with EG treatment can only be a strong component of the 002*—at least that is the only conclusion consistent with the other observations. Its position between about 6.6° and 6.8° indicates R1 ordering, and its relatively low intensity, compared to the other unresolved I/S peaks, suggests that the I/S is illite-rich—perhaps between 20% and 35% expandable.

We can't do much more with this hypothetical sample if all we have are the two noisy diffraction patterns. We can't use the position of the 002/003 or the Δ2θ method because the required reflections are not resolved. We conclude that the I/S is R1 ordered and contains between 20% and 35% expandable layers.

Better diffraction patterns usually help a great deal in characterizing I/S in admixture with discrete illite, but not always. If discrete illite and/or I/S consists of very thin crystallites, then the peaks are broadened so much that problems of resolution remain regardless of the noise levels in the diffraction patterns. But at least if you work for high-quality patterns, you will be confident that you have all the information that can be obtained by XRD from any given sample.

Chlorite/Smectite and Chlorite/Vermiculite

Chlorite/smectite is the second most abundant mixed-layered clay mineral in sedimentary rocks and soils. Reported occurrences indicate that it lacks the diversity that we find in illite/smectite because structures with $R > 1$ and minerals that contain much more than about 50% smectite have not been identified with certainty, and even random structures seem to be limited to only slightly expandable compositions. In short, it seems that the majority of structures that you will encounter will be either $R1$ 50% chlorite compositions or $R0$ minerals with maybe 10% smectite layers or less. There is room for much disagreement here, but we will treat what we believe are the common types of chlorite/smectite: low-charge corrensite (chlorite/smectite) and chlorite/smectite, $R0$, > 50% chlorite. Unlike I/S, however, ordering type is difficult to ascertain if compositions are less than about 30% smectite, so ordered structures may be more common than we think.

Figure 8.9 shows diffraction patterns for low-charge corrensite (trioctahedral chlorite/smectite) in the air-dried and EG-solvated conditions. The 50/50 composition and $R1$ ordering produce a superstructure $d(001)^*$ of $14.2 + 16.9 = 31.1$ Å for the EG-solvated case. Other peaks are positioned according to the Bragg law, and you can see the regular spacing of the various orders of reflection. The 003* is missing because it is almost an extinction; i.e., the scattering amplitudes from the unit cell cancel out. The air-dried form also produces a rational pattern, and in this case the superstructure 001* has a spacing of 14.2 + 15 = 29.2 Å. In our experience, the single most diagnostic criterion for corrensite lies in the appearance, upon EG solvation, of the 004* at about 11.3° 2θ or 7.8 Å. This identifier is useful for typical samples that have interferences on many of the corrensite lines. (You might think, what about the superstructure reflection at very low diffraction angles? Isn't that definitive? The trouble is that there are several instrumental effects that can produce a spurious reflection there—the so-called "Texas Clay Peak," among others—see D. R. Pevear for this terminology. If this very low angle peak is very intense, the diagnosis of corrensite is correct. If it is weak, beware of instrumental artifacts. Caution: calculated tracings of corrensite that deviate from 50:50 by as little as 2-3% sharply diminish the intensity of this peak, a possible reason that chlorite/smectite close to 50:50, but not corrensite, is seldom recognized.) Confirmation of a corrensite identification can be made by dehydrating the sample, causing the collapse of the smectite component to approximately 9.7 Å, and leading to a new superstructure spacing of 9.7 + 14.2 = 23.9 Å (Fig. 8.10). As before, the spacings of the reflections produce a rational diffraction pattern, but now, the 001* is absent.

Chlorite/vermiculite is a species closely related to chlorite/smectite. The difference between them detectable by X-ray diffraction methods is the extent of expansion when treated with EG or glycerol. In the air-dried condition, low-charge corrensite (chlorite/smectite) has a nominal $d(001)*$ of 29.2 Å that expands to 31.1 Å upon EG solvation. The $d(001)*$ value for high-charge corrensite (chlorite/vermiculite) is 28.6 Å and is unaffected by EG solvation. Differentiation between the two minerals is based on quite small differences in peak position, which can be unreliable because the $d(001)*$ values given here are nominal ones and there is a bit of variation from sample to sample. Heat treatments do not resolve the difficulties because the smectite and vermiculite components in the two collapse to approximately 10 Å, particularly if the sample has been K-saturated. The best diagnosis is based on Mg saturation and *glycerol* solvation (Walker, 1957). The smectite component of chlorite/smectite expands with this treatment to 17.7 Å, but vermiculite in chlorite/vermiculite remains at about 14.4 Å. Even these criteria can be ambiguous because some samples of chlorite/smectite absorb EG or glycerol very slowly, and days or weeks may be required for full expansion. A normal 12-h EG vapor treatment may, therefore, fail to disclose expansion and lead to the misidentification of chlorite/smectite as chlorite/vermiculite. To make matters worse, there may be interstratified expandable layers with charges that are intermediate between those of smectite and vermiculite, so a clean cut operational definition may be impossible.

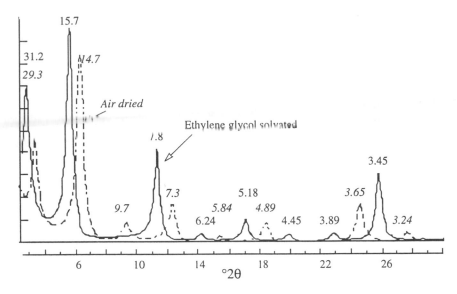

Fig. 8.9. Trioctahedral low-charge corrensite. All octahedral sheets contain 1 Fe per 3 sites.

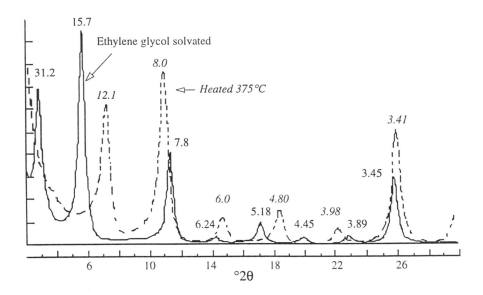

Fig. 8.10. Trioctahedral low-charge corrensite. All octahedral sheets contain 1 Fe per 3 sites.

The best advice that we can give is as follows: The mineral is chlorite/smectite if the diffraction pattern indicates expansion with EG; the result is indeterminate if EG solvation does not cause apparent expansion; chlorite/smectite is indicated if the sample shows expansion with Mg saturation and glycerol solvation; and the best diagnosis is chlorite/vermiculite if the diffraction pattern is essentially unaffected by this treatment.

Reported instances of chlorite/swelling-chlorite (Lippmann, 1956; Vivaldi and MacEwan, 1957) serve to cloud the picture further. Swelling-chlorite is such a poorly described phase that we have not attempted to calculate its diffraction characteristics. Just hope that any &corrensite™ you encounter expands with EG solvation to about 31 Å and collapses (in the K-saturated form) to $d(001)* = 24$ Å upon heat treatment. Figure 8.11 shows calculated patterns for $R1$ chlorite(0.5)/vermiculite in the air-dried condition and after heating. You should compare it to the corrensite of Figs. 8.9 and 8.10.

$R0$ chlorite(0.8)/smectite is shown in Fig. 8.12. The patterns superficially resemble chlorite except that the peaks have different breadths, the spacings are irrational, and solvation of the air-dried sample with EG has caused changes in the diffractogram. The relative intensities of the basal series depend on the content and location of the Fe substitution within the two possible chlorite octahedral sheets. The effects of Fe on intensity have been described for pure chlorite in Chapter 7. Confirmation of a chlorite/smectite diagnosis is provided by the diffraction pattern for the heated condition (Fig. 8.13). Notice that heating has caused sharpening of the reflection near 18.5

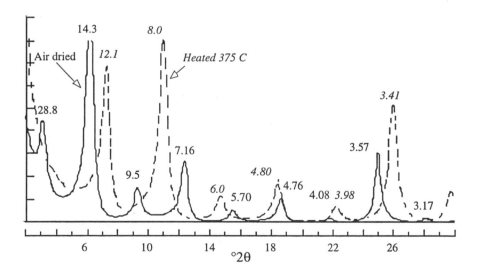

Fig. 8.11. Trioctahedral high-charge corrensite. All octahedral sheets contain 1 Fe per 3 sites.

2θ and migration of the 001/001 peak. The heat test is important for such minerals because of the possible misidentification of vermiculite/smectite as chlorite/smectite. Had the mineral been vermiculite/smectite, heat treatment would have collapsed the structure to about 10 Å and the diffraction pattern would have resembled that of either glauconite or illite.

The proportions of chlorite and smectite in mixed-layered chlorite/smectite can be estimated by the use of the data of Table 8.4. Again, we recommend the Δ2θ parameter for the best estimate of composition. Small (< 10%) amounts of smectite in smectite/chlorite are difficult to measure by peak migration characteristics, so the line-broadening analysis with application of the Q rule discussed earlier is the way to go for these compositions.

A common and vexing problem is identifying mixed-layered chlorite/smectite, of any type, in samples that contain discrete smectite or chlorite or both, and these seem to be most (all?) samples. This difficulty doubtless is responsible for the controversy concerning the occurrence of mixed-layered chlorite/smectite whose composition is not near the 50/50 value that defines corrensite (see Hillier, 1995 and Robinson and Bevins, 1994, for opposite sides of this discussion). The minerals corrensite, mixed-layered chlorite/smectite, smectite, and chlorite have unit cells that in the air dried and EG-solvated states have thicknesses of approximately 14, 15, and 17 Å. These are similar enough to cause problems of multiple mineral peak

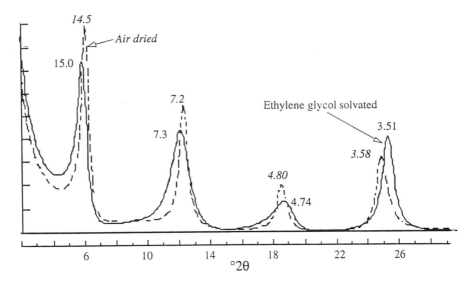

Fig. 8.12. *R*0 chlorite(0.8)/smectite, air dried and glycol solvated. All octahedral sheets contain 1 Fe per 3 sites.

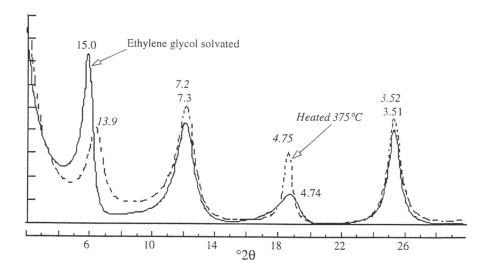

Fig. 8.13. *R*0 chlorite(0.8)/smectite, glycol solvated and heated to 375°C. All octahedral sheets contain 1 Fe per 3 sites.

resolution from such a mixture. But if the sample is dehydrated (heated for 1 h at 250°C in the Na-saturated condition) and analyzed in a stream of dry nitrogen to keep it dehydrated, expandable layers in the smectite and mixed-layered chlorite minerals are collapsed to about 9.6 Å, and that produces very different looking diffractograms. Little can be learned from EG and air-dried preparations of these mineral mixtures.

Figure 8.14 compares experimental diffractograms from Ca-saturated EG-solvated and Na-saturated dehydrated preparations of the clay-sized fraction from a sample of the Point Sal ophiolite. The power of the dehydration

Table 8.4. The positions (CuKα) of useful reflections for estimating percent chlorite in *R*0 chlorite/EG-smectite

% Chlorite	002/002		004/005		
	d(Å)	°2θ	*d*(Å)	°2θ	Δ° 2θ
10	8.39	10.54	3.39	26.29	15.75
20	8.29	10.68	3.40	26.21	15.54
30	8.15	10.86	3.42	26.05	15.19
40	7.98	11.09	3.43	25.98	14.89
50	7.80	11.34	3.45	25.82	14.48
60	7.59	11.66	3.47	25.67	14.01
80	7.40	11.96	3.50	25.45	13.49
90	7.18	12.33	3.53	25.23	12.90

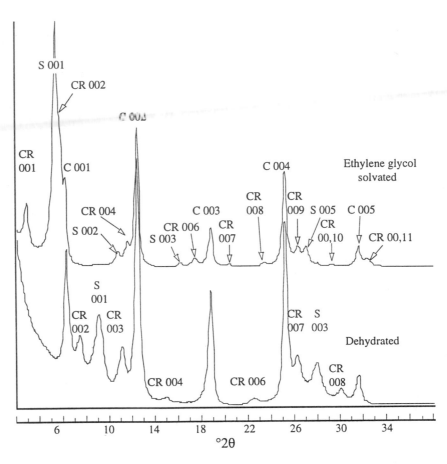

Fig. 8.14. Experimental diffraction patterns of a natural mixture of smectite, chlorite, and corrensite. Peak positions marked as orders of 00*l* reflections.

technique is demonstrated by the complete separation of all the peaks from the different minerals. The strong reflections in the low-angle region particularly are now all in the clear. Let's use everything we have learned at this point (and a little more) to squeeze all the information possible from the diffraction pattern. All five chlorite reflections are now visible, and they allow the following interpretations: (1) the chlorite shows no line broadening pattern caused by interstratification; (2) the nearly equal intensities of the 001 and 003 reflections indicate that the substitution of Fe is symmetrical between the silicate layer and the hydroxide interlayer; and (3) the relative intensities of the five basal chlorite reflections are consistent with a composition of 1.6 Fe atoms per 6 octahedral sites, based on comparisons with calculated diffraction patterns.

Still referring to Fig. 8.14, the dehydrated smectite has its 003 reflection at $d = 3.20$ Å, and that value is consistent with a composition that contains no interstratified chlorite. The value of $d = 9.7$ Å for the 001 smectite peak might suggest chlorite interstratification, but that shift from the predicted value of 9.6 Å is also present in the calculated pattern. The shift is caused by the sensitivity of this reflection to particle size effects (Reynolds, 1968; Trunz, 1976). The smectite pattern of the dehydrated sample has fairly strong 001 and 003 peaks, but no detectable 002 reflection, suggesting that it is trioctahedral (or nontronite, a possibility that we will ignore).

Table 8.5 characterizes the corrensite based on the dehydrated preparation (Fig. 8.14). Calculated and observed intensities are in good agreement for a composition of 0.8 Fe atoms per 3 octahedral sites in the chlorite component and 0.8 Fe atoms per 3 octahedral sites in the smectite component. It is interesting that this is the same composition estimated for the chlorite in the sample. The coefficient of variability is 0.29% based on the 002* to the 008* reflections from the dehydrated preparation, indicating almost perfect $R1$ ordering with a 50/50 proportion of chlorite to smectite.

We know enough about the minerals to compute a quantitative analysis. The integrated areas were measured for the chlorite 002, the smectite 001, and the corrensite 003* reflections, all from the dehydrated preparation, and using the methods outlined in Chapter 9, we get chlorite 83%, smectite 10%, and corrensite 8% (by weight). Mineral intensity factors were calculated by NEWMOD© for chlorite and corrensite using the estimated Fe contents. The intensity ratio of the 002 and 003 reflections from the EG preparation (Fig. 8.14) was matched by a model saponite containing 0.35 Fe per three octahedral sites, and that composition was used in the calculation of the mineral intensity factor for the dehydrated smectite. (See Eq. 9.11 and related discussion for details of this procedure.) Unfortunately, the quantitative analysis presented above is suspect because the sample preparation method used (pipetted glass slide) tends to emphasize the finer particle sizes. It would have been better had we used the filter transfer method described in Chapter 6. In addition, considerable uncertainty lies in the mineral intensity factor for the smectite because it is based on the relative intensities of only two reflections.

Table 8.5. Characterization of corrensite, 0.8 Fe per 3 octahedral sites. Data for the dehydrated preparation (see Fig. 8.14)

Reflection	Integrated intensity (calc.)	Integrated intensity (obs.)
001	2	Not detected
002	125	113
003	173	183
004	26	26
006	19	34
007	115	91
008	39	53
009	3	Not detected

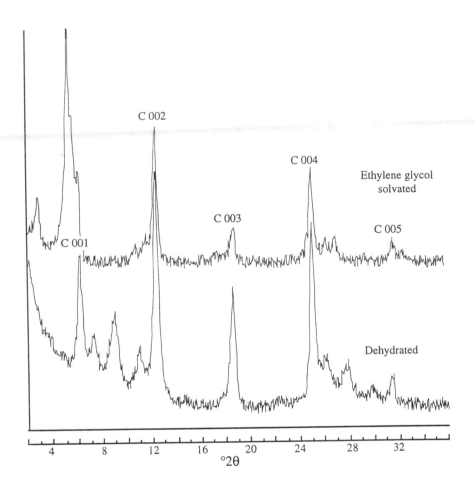

Fig. 8.15. The experimental diffractograms of Fig. 8.14 to which computer-generated noise has been added to simulate poor data.

Figure 8.15 illustrates the experimental diffraction patterns of Fig. 8.14 that have been superimposed on a noisy background to simulate poorer data. Many of the weaker reflections are lost, but you can see that the low-angle region of the dehydrated preparation retains well-resolved diagnostic peaks for the three minerals. A qualitative analysis based on the upper (EG-solvated) trace would inspire less confidence.

Kaolinite/Smectite
The older literature contains few reports of kaolinite/smectite, and all the well-documented occurrences are randomly interstratified (Schultz et al., 1971; Wiewiora, 1971; Sakharov and Drits, 1973; Hughes et al., 1987), except for one example from sandstone that seems to be *R*1 kaolinite(0.5)/smectite (Thomas, 1989). However, a recent publication by

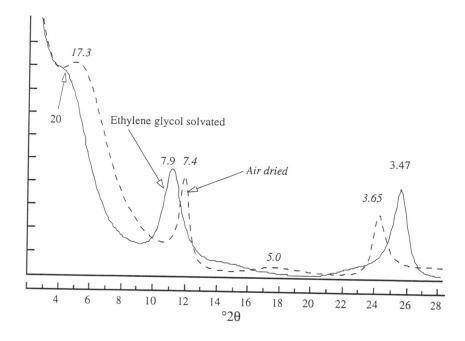

Fig. 8.16. *R*0 kaolinite(0.7)/smectite, glycol-solvated and air-dried; peak positions in Ångströms.

Hughes et al. (1993) reports the widespread occurrence of kaolinite/smectite in soils and paleosols ranging in age from Pennsylvanian to Holocene, and they find evidence of *R*1 ordering in some of the samples. They suggest that kaolinite-rich kaolinite/smectite may have been overlooked by earlier workers who misidentified some occurrences as either "poorly crystalline" kaolinite or dehydrated halloysite.

Randomly interstratified kaolinite/smectite presents few problems unless it is present as a minor constituent in a clay mixture. Solvation with EG causes easily detected changes in the diffraction pattern with respect to the air-dried preparation (Fig. 8.16). Heat treatments are necessary for definitive interpretations. They change the pattern quite drastically, producing changes that are unintuitive. For the example of *R*0 kaolinite(0.7)/smectite (Figs. 8.16, 8.17), the strong reflections at *d* = 7.9 Å (EG solvated) and *d* = 7.4 Å (air dried) *increase* to 8.1 Å upon heating; so what do we have here—a clay mineral that expands upon heat treatment? No, the reason for the peak displacements is that heating has created 9.7 Å layers (9° 2θ) from 17 and 15 Å layers for the smectite, respectively, and the dehydrated smectite 001 peak has a smaller diffraction angle (larger *d*) than the 002 reflections from the EG or air-dried preparations. Therefore, the mixed-layered reflection is pulled (i.e., the Méring pull) to a lower angle (larger *d*) upon collapse of the structure. In case of small amounts of expandable layers, < 10 to 15%, look for a slight smearing of the 17 Å peak in the sense described by Fig. 5.8 and associated text (p. 161).

Figures 8.18 and 8.19 show calculated diffraction patterns for smectite-rich *R*0 kaolinite/smectite. Again, the increase in *d* from 8.3 Å in the EG treated state to 9.4 Å is evident upon heating to 375°C.

The peak positions for EG-solvated kaolinite/smectite produce conventional Méring-type migrations with changing composition, and, as with the other species described above, we recommend the use of the Δ2θ parameter for determining composition. Table 8.6 contains values for the positions of peaks that are useful for determining the composition of randomly interstratified kaolinite/smectite.

The calculated diffraction patterns of Fig. 8.20 give an inkling of why kaolinite/smectite may be under-reported. Suppose that you had run only an air-dried preparation, and further suppose that equilibration had not occurred between the sample and laboratory humidity, giving a random mixture of one- and two-water-layer smectite interlayers. The trace on Fig. 8.20 labeled "air dried*" is a pattern for kaolinite/smectite with 20% of such mixed expandable layers. Above it, is a pattern for pure kaolinite based on very thin crystallites. Except for the low-angle background, which could be obscured by pure smectite in a real sample, the two patterns have little to distinguish them, and

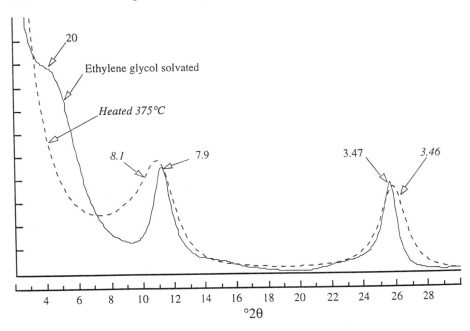

Fig. 8.17. *R*0 kaolinite(0.7)/smectite, glycol-solvated and heated to 375°C. Peak positions in Ångströms.

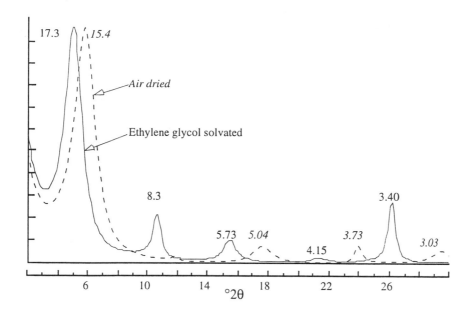

Fig. 8.18. *R*0 kaolinite(0.3)/smectite, glycol-solvated and air-dried. Peak positions in Ångströms.

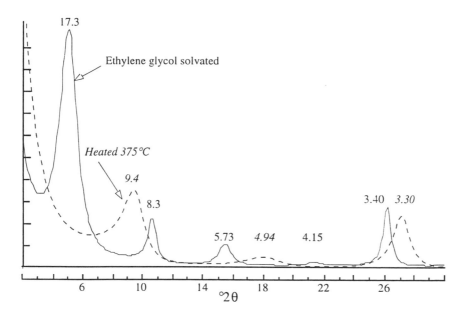

Fig. 8.19. *R*0 kaolinite(0.3)/smectite, glycol-solvated and heated to 375°C. Peak positions in Ångströms.

Table 8.6. The positions (CuKα) of useful reflections for estimating percent kaolinite in *R*0 kaolinite/EG-smectite

% Kaolinite	001/002 d(Å)	001/002 ° 2θ	002/005 d(Å)	002/005 ° 2θ	Δ° 2θ
10	8.46	10.46	3.39	26.29	15.83
20	8.38	10.56	3.39	26.29	15.73
30	8.31	10.65	3.40	26.21	15.56
40	8.23	10.75	3.41	26.13	15.38
50	8.10	10.92	3.43	25.98	15.06
60	7.98	11.10	3.44	25.90	14.80
70	7.90	11.19	3.47	25.67	14.48
80	7.65	11.57	3.50	25.45	13.88
90	7.43	11.91	3.54	25.16	13.25

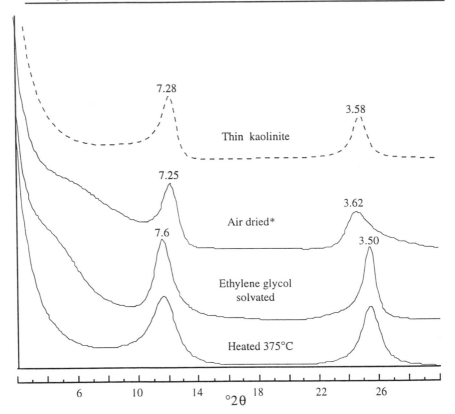

Fig. 8.20.   Calculated diffraction patterns for different preparations of *R*0 kaolinite(0.8)/smectite compared to pure but finely crystalline kaolinite. The air-dried* simulation depicts a three-component system of randomly interstratified kaolinite (0.8), two-water-layer smectite (0.1) and one-water-layer smectite (0.1). Peak positions in Ångströms.

a diagnosis of kaolinite or dehydrated halloysite (because of the broad reflections) is a reasonable one. Alternatively, you may have data only for heated and EG-solvated states (lower two traces on Fig. 8.20). The mineral contains almost no expandable component because dehydration and EG solvation have only slight to no differential effects—correct? Good thinking, but incorrect in this case. By chance, low-expandability kaolinite/smectite gives almost identical diffraction results for these two structural states. However, assuming that you are unaware of this coincidence, you should be suspicious because the two peaks on the lower traces of Fig. 8.20 are not rationally related. Two times 3.50 Å equals 7.0 Å, and the first peak has a spacing of 7.6 Å. This is an impossible situation, so something must have gone wrong in running the sample. Maybe the goniometer is off. When there is enough time, we will rerun the sample.

The dilemma disappears if all the diffraction data of Fig. 8.20 are considered. There are large changes in d between the peaks of the air-dried and the other preparations. These confirm the diagnosis of kaolinite/smectite. Once that is established, go to Table 8.6 and work out the percent of kaolinite in the mixed-layered mineral. We have not shown diffraction patterns for pure one- and two-water-layer structures, but they too show large differences in *d* compared to the EG-solvated and dehydrated cases. This example demonstrates the need for diffraction patterns of all three preparations if kaolinite/smectite is to be confidently identified.

Figures 8.21 and 8.22 show patterns for regularly interstratified (*R* 1) kaolinite(0.5)/smectite. This mineral would qualify for a mineral name like corrensite or rectorite if a specimen could be found that is sufficiently pure to allow complete characterization. Its diffraction characteristics should look like those of the calculated patterns. Be on the lookout for it, for there is much validity in the idea that you only find what you look for, or as someone has said, I will see it when I believe it.

Serpentine/Chlorite

Recall our discussion of this puzzling mineral in Chapter 5 (pp. 185ff) and our decision to call it serpentine/chlorite. However, the calculated X-ray diffraction patterns of serpentine/chlorite and kaolinite/chlorite are indistinguishable at the low concentrations of 7 Å layers so far reported for this mineral. The structures of 7 Å/chlorite mixed-layered clay minerals have unique characteristics that place them in their own diffraction class. The basal spacing of chlorite is very nearly twice that of the 7 Å component, so that reflections from the 7 Å component fall almost exactly on all even-order chlorite peaks. This structure provides the paradoxical situation of a randomly interstratified mineral that produces a rational 00*l* diffraction pattern whose peak positions are independent of composition.

Because the 1:1 layer is ~7 Å and chlorite is ~14 Å, when they are interstratified, X-ray diffraction tracings of the odd-numbered 00*l* peaks are

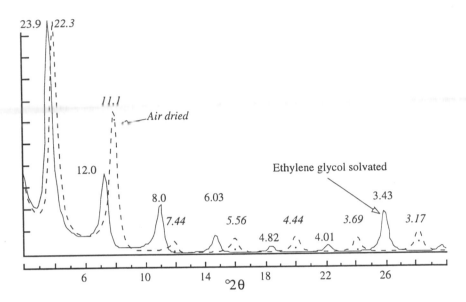

Fig. 8.21. *R*1 kaolinite(0.5)/smectite, glycol-solvated and air-dried. Peak positions in Ångströms.

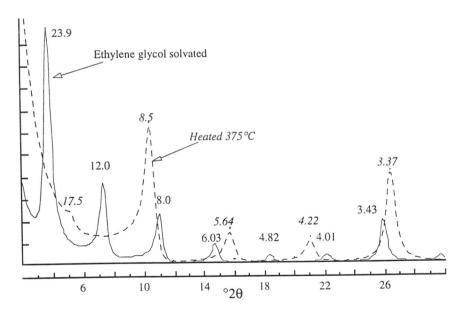

Fig. 8.22. *R*1 kaolinite(0.5)/smectite, glycol-solvated and heated to 375°C. Peak positions in Ångströms.

broadened. Can you see why? Recall that the 001 peak of chlorite represents diffraction from a 14 Å spacing and that the 002 peak represents the 14 Å spacing behaving as if it were a 7 Å spacing, i.e., the n of $n\lambda = 2d\sin\theta$ is one for the 001 peak and two for the 002 peak. Can you see from this that all even-numbered reflections will behave like 7 Å/n spacings, and that the 7 Å 1:1 layers diffract constructively with these even-ordered reflections from the 14 Å layers? (It may help to return to Chapter 3, where Fig. 3.13 is discussed on pp. 85ff.) In the case of what would otherwise be the chlorite $d(001)/1$, the interstratified 7 Å 1:1 layers break up the stack of 14 Å layers by acting as defects, but the 7 Å spacing of the 1:1 layers matches the repeat distance along the Z^* for the reflection from the $d(001)/2$ spacing. So, effectively, N is relatively small for the odd-numbered peaks, and they, therefore, have broad peaks. And N is large for the even-numbered ones and will have relatively sharp peaks (Reynolds et al., 1992). Do we need to remind you of the Scherrer equation (Eq. 3.8)?

We also can use Méring's principles in order to explain the character of the diffraction pattern. Remember, peaks are broad if end-member reflections are widely separated. Conversely, sharp peaks indicate the close juxtaposition of reflections from both components. For serpentine/chlorite, all odd-order chlorite peaks are well separated from any serpentine reflection, and chlorite even-order reflections are virtually superimposed on the serpentine peak positions (Fig. 8.23).

Reynolds et al. (1992) discussed in detail the X-ray diffraction characteristics of one specimen of serpentine/chlorite from the Tuscaloosa sandstone of the Gulf Coast region, and developed an empirical formula for estimating the percent serpentine present (Eqs. 8.1 and 8.2). Its application requires that the diffractograms be obtained using two Soller slits (2°), a 1° beam slit, and a 0.05° detector slit. Finer slits can be used but are unnecessary. Reynolds et al. recommend using the 004 and 005 chlorite reflections.

$$\beta_r = \left(\beta_{005}^{1.25} - \beta_{004}^{1.25}\right)^{\frac{1}{1.25}}$$

(8.1)

and

$$\% \text{ Serpentine} = -0.51 + 24.27\beta_r$$

(8.2)

The quantity β_r in Eqs. (8.1 and 8.2) denotes the breadth of the 005 reflection, corrected for the baseline breadth of the 004, which contains no mixed-layered broadening. You might look at Eq. (8.2) and think, this can't be correct. What if there is no mixed-layering? Then β_r is zero, and if I substitute zero into Eq. (8.2), the result is -0.51% serpentine. Your reasoning is correct. But remember that these equations are empirical, or in other words,

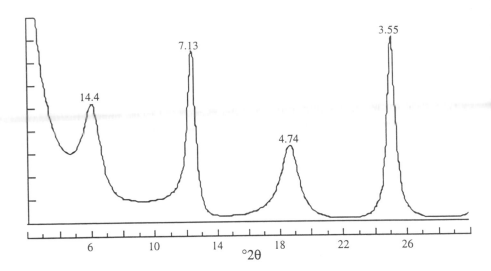

Fig. 8.23. Trioctahedral $R0$ serpentine(0.2)/chlorite, Mg end-members. Peak positions in Ångströms.

they are not based on first principles. Clearly, they cannot apply to very low proportions of serpentine. In fact, they apply very well only to the range of 1-20% serpentine ($R0$), which is the likely range for natural occurrences of this mixed-layered mineral. Your measurement of line breadth is not going to be accurate enough anyway to allow you to deal with compositions that have less than 1% serpentine.

In summary, serpentine/chlorite or kaolinite/chlorite have peaks with rational positions that are fixed despite the proportions of end-members, but increasing percentages of interstratified serpentine or kaolinite cause increased broadening of odd-order chlorite reflections with respect to even orders (Fig. 8.23). The minerals are unaffected by liquid solvation and/or moderate heat treatments.

Mica/Vermiculite

Mixed-layered trioctahedral mica/vermiculite is uncommon in sedimentary rocks, but it is widespread in hydrothermal and soil clay mineral suites. The regularly interstratified phase, containing 50% mica, was one of the first ordered interstratified clay minerals to be recognized. It was described by Gruner (1934) who called it hydrobiotite. Its calculated diffraction pattern is shown in Fig. 8.24. The regular Bragg series of spacings results from a $d(001)^*$ of 24.4 Å in the air-dried state. Collapse of the vermiculite component by heat treatment produces a rational pattern that looks like that of a simple mica with $d(001) = 10$ Å. In fact, the dehydrated vermiculite spacing may be a bit more than 10 Å because heat-treated vermiculites rarely collapse

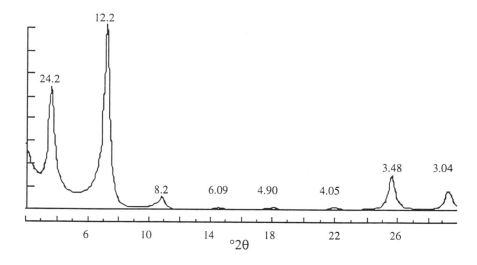

Fig. 8.24. Hydrobiotite (trioctahedral) air-dried [*R*1 biotite(0.5)/vermiculite]. Silicate octahedral sheets contain 1 Fe per 3 sites. Peak positions in Ångströms.

that much; a value of 10.3 Å for the apparent 001 reflection squares better with experience, so d(001) may be closer to 10.15 Å. The collapsed phase produces a very weak reflection near $d = 5$ Å because of heavy out-of-phase (with respect to the rest of the silicate skeleton) octahedral scattering caused by the trioctahedral character and augmented by the likelihood of significant amounts of Fe in those sites. Its diffraction pattern resembles that of glauconite. Proof of a vermiculite component requires that you saturate with Mg, completely solvate with glycerol, and obtain a diffraction pattern that is essentially unchanged from that of the air-dried condition.

Figure 8.25 shows the diffraction pattern for trioctahedral *R*1 mica(0.65)/vermiculite. You can see the similarities to the hydrobiotite pattern in this diffractogram, but, as you would expect, peaks have shifted as a result of the increased mica content and the randomness that is present because the composition is not 50/50. The superstructure reflection (001*) is reduced to a low shoulder on the low-angle background. Estimates of composition for these minerals are usually based on the position of the 001/ 002* reflection because, for most samples, that is the only one you are likely to see. Table 8.7 shows the position of this reflection as a function of composition together with peak positions for the *R*0 mica/vermiculite 001/002 and 003/004 reflections.

Figures 8.26 and 8.27 give examples of randomly and *R*1 ordered, interstratified trioctahedral mica/vermiculite. The patterns for *R*0 are simpler than those for *R*1 because there are no superstructure components in the reflections. The low-angle reflection (001/001) is usually the most practical to use for estimates of composition (Table 8.7).

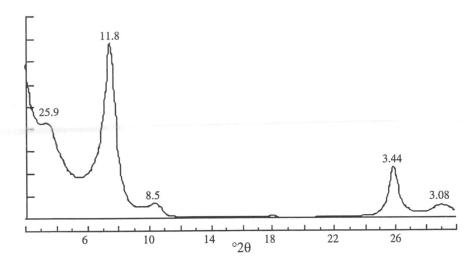

Fig. 8.25. Trioctahedral *R*1 mica(0.65)/vermiculite. Silicate octahedral sheets contain 1 Fe per 3 sites. Peak positions in Ångströms.

Identifying *R*0 versus *R*1 ordering is difficult if the mica content is greater than about 80%. Figure 8.27 shows a pattern for 80% mica, *R*1. Its identity is indicated by the subtle high-angle shoulder on the 001/001 reflection. This shoulder is the 001/003*, and it is more obvious at compositions that contain less than 70% mica (see Fig. 8.25). (Here is a good example of how subtle features of peak shape can be important.)

We will not discuss the diffraction characteristics of dioctahedral chlorite/smectite and dioctahedral illite/vermiculite because the basal diffraction patterns for these resemble those of the trioctahedral varieties.

Table 8.7. The positions (CuKα) of useful reflections for estimating percent vermiculite in trioctahedral mica/vermiculite (air dried)

	RO				R1	
	001/002		003/004			
001/002*						
% Vermiculite	d(Å)	° 2θ	d(Å)	° 2θ	d(Å)	° 2θ
10	10.27	8.61	3.35	26.61	10.38	
8.52						
20	10.90	8.11	3.38	26.37	11.13	7.94
40	12.53	7.05	3.45	25.82	11.97	7.39
50	13.02	6.79	3.48	25.60	12.15	
7.28						
60	13.43	6.58	3.51	25.37		
70	13.75	6.43	3.53	25.23		
80	13.97	6.33	3.55	25.08		
90	14.17	6.24	3.57	24.94		

*Indicates that the 002 is a basal spacing for a superstructure.

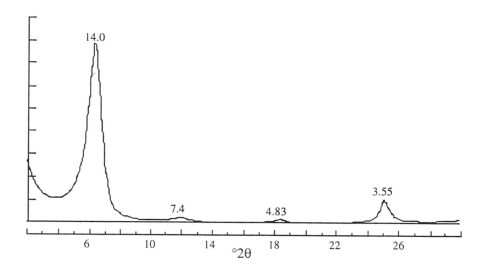

Fig. 8.26. Trioctahedral *R* 1 mica(0.2)/vermiculite. Silicate octahedral sheets contain 1 Fe per 3 sites. Peak positions in Ångströms.

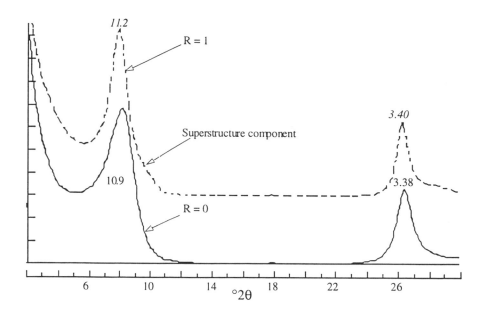

Fig. 8.27. *R* 1 and *R* 0 trioctahedral mica(0.8)/vermiculite. Silicate octahedral sheets contain 1 Fe per 3 sites. Peak positions in Ångströms.

Peak intensities differ, but peak position is not much affected by the composition of the octahedral layers. The identification of dioctahedral types is best based on the position of the 060 reflection (Chapter 7).

SUMMARY

Identifying clay minerals from their diffraction tracings is something of a *Gestalt* process, i.e., identifying the whole—being able to say that it is an illite/smectite because it has an illite/smectite pattern. The mixed-layered clay minerals with any appreciable degree of randomness to their interstratification show an irrational series of peaks. Relative broadness of peaks on a single pattern varies for these clay minerals, unlike peaks for discrete clay minerals or nonclay minerals. And the peaks of mixed-layered clay minerals can be much broader than those of discrete clay minerals. Mixed-layered line broadening is a function of the separation of the peaks of the two components that are incorporated into one peak (Fig. 8.1). The most useful tool for identifying mixed-layered clay minerals is the modeling of their 00l diffraction tracings for comparison to experimental diffractograms.

REFERENCES

Bailey, S. W. (1982) Nomenclature for regular interstratifications: *Amer. Minerl.* **67**, 394-98.

Brindley, G. W., and Gillery, F. H. (1953) A mixed-layer kaolin-chlorite structure: in Swineford, A., and Plummer, N., editors, *Proceedings*, 2nd National Conf. on Clays and Clay Minerals, Columbia, Missouri, Pub. 327, Natl. Academy of Sciences–Natl. Research Council, 349-53.

Gruner, J. W. (1934) The structure of vermiculites and their collapse by dehydration: *Amer. Minerl.* **19**, 557-78.

Hillier, S. (1995) Mafic phyllosilicates in low grade metabasites: Characterization using deconvolution analysis discussion: *Clay Minerals* **30**, 67-73.

Hughes, R. E., DeMaris, P. J., White, W. A., and Cowin, D. K. (1987) Origin of clay minerals in Pennsylvanian strata of the Illinois Basin: in Schultz, L. G., van Olphen, H., and Mumpton, F. A., editors, *Proceedings*, Internatl. Clay Conf., Denver, 1985, The Clay Minerals Society, Bloomington, Indiana., 97-104.

Hughes, R. E., Moore, D. M., and Reynolds, R. C., Jr., (1993) The nature, detection, occurrence, and origin of kaolinite/smectite: *in* Murray, H., Bundy, W., and Harvey, C., editors, *Kaolin: Genesis and Utilization*, Special Pub. No. 1, The Clay Minerals Society, 291-323.

Klug, H. A., and Alexander, L. E. (1974) *X-Ray Diffraction Procedures*: J. Wiley, New York, 966 pp.

Lippmann, F. (1956) Clay minerals from the Roet member of the Triassic near Goettingen, Germany: *J. Sed. Pet.* **26**, 125-39.

MacEwan, D. M. C. (1958) Fourier transform methods for studying X-ray scattering from lamellar systems. II. The calculation of X-ray diffraction effects for various types of interstratification: *Kolloidzeitschrift* **156**, 61-7.

Méring, J. (1949) L'Inté reference des Rayons X dans les systems à stratification dé sordonnée: *Acta Crystallogr.* **2**, 371-77.

Reynolds, R. C., Jr. (1967) Interstratified clay systems: calculation of the total one-dimensional diffraction function: *Amer. Minerl.* **52**, 661-73.

Reynolds, R. C., Jr. (1968) The effect of particle size on apparent lattice spacings: *Acta Crystallogr.* **A24, pt. 2**, 319-20.

Reynolds, R. C., Jr. (1980) Interstratified clay minerals: in Brindley, G. W., and Brown, G., editors, *Crystal Structures of Clay Minerals and Their X-Ray Identification*: Monograph No. 5, Mineralogical Society, London, 249-303.

Reynolds, R. C., Jr., and Hower, J. (1970) The nature of interlayering in mixed-layer illite-montmorillonite: *Clays and Clay Minerals* **18**, 25-36.

Reynolds, R. C., Jr. (1985) *NEWMOD, A Computer Program for the Calculation of the Basal Diffraction Intensities of Mixed-Layered Clay Minerals*: R. C. Reynolds, 8 Brook Rd. Hanover NH.

Reynolds, R. C., Jr., Distefano, M. P., and Lahann, R. W. (1992) Randomly interstratified serpentine/chlorite: Its detection and quantification by powder X-ray diffraction methods: *Clays and Clay Minerals* **40**, 262-67.

Robinson, D., and Bevins, R. E. (1994) Mafic phyllosilicates in low grade metabasites: Characterization using deconvolution analysis: *Clay Minerals* **29**, 223-37.

Sakharov, B. A., and Drits, V. A. (1973) Mixed-layer kaolinite-montmorillonite. A comparison of observed and calculated diffraction patterns: *Clays and Clay Minerals* **21**, 15-17.

Schultz, L. G., Shepard, A. O., Blackmon, P. D., and Starkey, H. C. (1971) Mixed-layer kaolinite-montmorillonite from the Yucatan Peninsula, Mexico: *Clays and Clay Minerals* **19**, 137-50.

Środoń, J. (1980) Precise identification of illite/smectite interstratifications by X-ray powder diffraction: *Clays and Clay Minerals* **28**, 401-11.

Thomas, A. R. (1989) A new mixed layer mineral—Regular 1:1 mixed layer kaolinite/smectite: *Programs with Abstracts*: 26th Annual Meeting The Clay Minerals Society, Sacramento, California, 69.

Trunz, V. (1976) The influence of crystallite size on the apparent basal spacings of kaolinite: *Clays and Clay Minerals* **24**, 84-7.

Vivaldi, J. L. M., and MacEwan, D. M. C. (1957) Triassic chlorites from the Jura and Catalan Coastal Range: *Clay Min. Bull.* **3**, 177-83.

Walker, G. F. (1957) On the differentiation of vermiculites and smectites in clays: *Clay Min. Bull.* **3**, 154-63.

Wiewiora, A. (1971) A mixed-layer kaolinite-smectite from Lower Silesia, Poland: *Clays and Clay Minerals* **19**, 415-16.

Chapter 9
Quantitative Analysis

The intensity of a diffraction peak from a particular mineral seems to be simply related to the abundance of that mineral in a mixture. Unfortunately, that simplicity is deceptive. You will see, as you read on, that quantitative analysis by X-ray diffraction (XRD) is a complicated business requiring attention to many details. You will encounter samples that produce inaccurate results even if you work hard to deal with all the possible variables.

A most important problem, over which little control can be exercised, is choosing a standard mineral whose diffraction characteristics are identical to those of that same mineral in the unknowns. For example, all specimens of illite are not identical. In fact, a class of compounds can have a good deal of chemical variation and still be called by a single mineral name. You saw in Chapter 3 that the intrinsic diffraction intensity of a given peak from a specific mineral depends on its chemical composition (e.g., the Fe content of chlorite). The analyses of Mg-rich chlorites will therefore be incorrect if an Fe-rich chlorite is used as the basis of standardization.

No invariant methodology is possible for the quantitative analysis of clay minerals by XRD methods. Instead, you must select the best techniques consistent with the characteristics presented by the sample. Optimal procedures are often inappropriate because of peak interferences from other minerals present in the sample. These two factors, standard suitability and peak interference, constitute the last two unsolved problems in quantitative analysis. With care, all other difficulties can be minimized or eliminated. But if you are naive, you can fail to pay attention to even trivial sources of error and produce highly precise analyses so inaccurate that they are meaningless.

This discussion of quantitative analysis is summarized from Reynolds (1989). An excellent review of principles and useful information is given by Brindley (1980), and there are many other papers that deal with the subject (e.g., Schultz, 1960; Chung, 1974; Cody and Thompson, 1976). There is room for a good deal of individuality in designing quantitative procedures, and we do not claim that our suggestions are definitive or exhaustive. They will work as well as any others, however, and their discussion in some detail will provide you with the background necessary for developing your own methods, which can be tailored to your specific applications.

How good can quantitative analysis of clay minerals based on XRD be? Well, if precision is the criterion, the answer is very good indeed. A set of replicate analyses should provide a standard deviation of perhaps ±5% (or less) of the amounts present if the constituent minerals are all present in reasonably large quantities, say 20% or more. Accuracy is another matter. Quantitative analysis should be considered good if errors amount to ±10% of the amounts present for major constituents, and ±20% for minerals whose concentrations are less than 20%. These numbers are only "off-the-cuff" approximations, but you get the idea that quantitative analysis of minerals by XRD does not approach the accuracy that we expect from quantitative chemical analysis. We use the term *precision* here to signify repeatability, that is, the ability to get the same answer (correct or incorrect) from each of a set of multiple measurements. *Accuracy* is the difference between your results and the *correct* answer. Griffiths (1967) has discussed these concepts in detail.

REQUIRED SAMPLE CHARACTERISTICS

A sample of mineral powder must possess certain characteristics if it is to produce the most accurate results. The following outline summarizes the key points we will document. See if you can understand the reasons for each of them as you read on.

1. The sample must be longer than the spread of the incident beam at the lowest diffraction angles used. Therefore, two quantities need attention: (1) the sample length, and (2) the angular divergence of the beam or size of the divergence slit.
2. The sample must be infinitely thick at the highest diffraction angles used.
3. The sample must be mounted in the diffractometer so that, for all diffraction angles, the angle between the sample surface and the incident beam is equal to the angle between the sample surface and the diffracted beam.
4. There must be no particle-size gradient between the top and bottom of the sample.

These factors can be difficult to control, but their effects on the accuracy and precision of analyses are minimized if you adhere to two important principles.

1. Avoid the use of low-angle (<10° 2θ) reflections for quantitative analysis.
2. Select analytical peaks that are as close together (in ° 2θ) as possible.

Sample Length

Diffraction intensity is proportional to the volume of the sample irradiated. The goniometer geometry fixes the theta-independent width of the sample area irradiated. The irradiated length of the sample is proportional to $1/\sin\theta$, so the irradiated area is equal to a constant divided by $\sin\theta$. The integrated thickness (depth) exposed to the incident beam is proportional to $\sin\theta$. Volume is equal to area times thickness, so the product of these two cancels the angle-dependent character of each if, and only if, the sample is infinitely thick and longer than the spread of the beam spot at the lowest diffraction angles recorded.

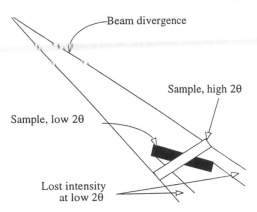

Fig. 9.1. Incident intensity loss at low diffraction angles.

The sketch in Fig. 9.1 shows a sample at high and low diffraction angles. The sample does not intercept the entire spread of the incident beam at the low angle shown, and some diffracted intensity will be lost. If R_0 is the goniometer radius in centimeters and α is the angular aperture of the divergence slit in degrees, then the length L of the area irradiated is

$$L = R_0 \tan\alpha/\sin\theta \qquad (9.1)$$

The sample must be at least this long or intensity will be lost, and that loss will increase with decreasing 2θ. Consider a specific case. If $R_0 = 20$ cm, and a $1°$ slit is used, the sample must be 8 cm long for accurate intensities at $5°$ 2θ. A 4-cm sample will show a 50% intensity loss at this diffraction angle. With a $0.3°$ slit, the required sample length is diminished to a reasonable 2.4 cm.

Small (2.5 cm diameter) circular sample holders are now widely used, particularly with automatic sample changers. The beam loss at low angles is severe because of the small sample area, and the mathematical treatment of beam-loss effects is complicated by the circular area.

If you wish to appreciate the intensity loss for a given sample area and at a specific diffraction angle, calculate values from Eq. (9.1) and divide them by the sample areas. Table 9.1 shows several examples of such calculations.

So what can we do about the problem of short samples? You might suggest corrections based on Eq. (9.1) and illustrated by Table 9.1. The problem with this approach is that it assumes a strictly uniform intensity of illumination throughout the incident-beam cross section, and such a beam

Table 9.1. Sample shape and size and the loss of diffraction intensity as a function of 2θ

°2θ	Intensity, rectangular sample (4 cm long)	Intensity, circular sample (2.5 cm diam.)
2	0.20	0.10
4	0.40	0.20
6	0.60	0.29
8	0.80	0.39
10	1.00[a]	0.49
15	1.00	0.73
20	1.00	0.88
25	1.00	0.95
30	1.00	0.95

Goniometer radius = 20 cm; incident beam slit = 1°. [a]Full intensity.

"footprint" is unlikely. You could try to maintain all samples at some fixed length and standardize procedures with standard mineral slides of the same length, keeping the divergence slit fixed, of course. This is fine for your instrument, but you will not be able to use values published by others for mineral intensity parameters unless the published values have been measured with identical geometrical conditions. The best solution is to use fine slits at low diffraction angles, where the intensity loss due to such slits is unimportant, and larger slits, consistent with the relations of Eq. (9.1), for the higher angles. Record overlapping angular ranges with the two-slit systems and then normalize one pattern to the same intensity base as the other. This procedure will produce a single diffractogram that contains no intensity distortions due to sample length. Still another alternative is to avoid the use of low-angle reflections for quantitative analysis.

Sample Thickness

A different error affects high-angle peaks that can be serious and difficult to control. The error arises from samples that are too thin, and, as you will discover with a little laboratory experience, it is very difficult to make samples thick enough if you use procedures that are optimum in other respects.

Reproducible measurements of diffraction intensity from thin preparations require that all samples have exactly the same thicknesses and identical mass absorption coefficients. The mass absorption coefficient of a mixture depends on its composition, which is the unknown quantity here, so these requirements are manifestly impossible. The mass absorption coefficient is controlled by the *chemical composition* of the material, so you might suggest obtaining separate chemical analyses for your unknowns and using these data as the basis for intensity corrections. Fair enough, but you still are left with the difficult or impossible problem of producing uniform thicknesses for all the

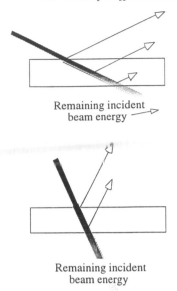

Fig. 9.2. Increased loss of X-ray energy at high diffraction angles.

preparations. The sample can be made very thin, so thin that absorption in it can be neglected over the 2θ range studied. This procedure is sound, but it may produce diffraction intensities that are too feeble to be useful for the identification and quantification of some kinds of clay minerals, particularly the mixed-layered species. The best approach is to make the sample so thick that adding to its thickness causes no increase in diffraction intensity. To put it another way, make the sample so thick that the incident beam energy is "used up" by diffraction and absorption processes within the sample. This condition is termed *infinite thickness*, and the standard equations that describe diffraction from a mineral mixture assume powder aggregates that are this thick. Let us examine some of the theory that underlies these comments.

Samples that are too thin produce weakened peaks at high diffraction angles and correct intensities below some low value of 2θ. Figure 9.2 shows why. You can see that the path length, and hence opportunity for absorption and diffraction, increases as 2θ diminishes. Consequently, a sample may be infinitely thick at, say, 5° 2θ and much too thin for maximum diffraction effects at a high angle, such as 50° 2θ. This is a situation patently undesirable, and three approaches can be used to obviate it. First, prepare the sample in such a way that it is infinitely thick at the highest diffraction angles involved in the analysis. Second, use analytical peaks that are near neighbors. This procedure works because the effect of sample thickness changes only slightly over small 2θ increments. Third, you can measure a combined function of the thickness and absorption coefficient and correct the intensities to the values that they would have if the sample was infinitely thick. We need to consider the basic theory that describes the absorption of X-rays to see how this correction might be accomplished.

In Chapter 2 we demonstrated that the absorption of X-rays follows the Lambert law, $I/I_0 = \exp(-\mu t)$, where μ is the linear absorption coefficient and t is the thickness (in cm). The linear absorption coefficient is not very useful because it depends on density. A better quantity is the mass absorption coefficient μ^*, which describes the absorption per gram or, more precisely, the absorption per gram per square centimeter. We substitute these into the Lambert law and obtain $I/I_0 = \exp(-\mu^* g)$, in which μ^* is the fraction of the radiation absorbed per gram centimeter squared and g is the mineral density in units of grams per square centimeter.

Figure 9.3 displays the absorption characteristics of a powder aggregate.

Consider the small imaginary crystallite nearest to the sample surface. The diffracted radiation from this crystallite has been diminished twice by absorption: once by absorption over the path of the incident beam, and once by the path length out of the sample. We will denote the depth of this increment not by distance but by g, the mass of material (expressed as grams per square centimeter) that lies above it. The sum of the path lengths in and out of the sample is proportional to $2g/\sin\theta$. So the intensity contributed by this increment is described by the equation $I/I_0 = \exp(-2\mu^*g/\sin\theta)$. Here, we intend to derive the total integrated diffraction energy that leaves the sample and is recorded as intensity. Pretty clearly, we need to integrate all increments x of diffraction such as those shown by Fig. 9.3. The limit of the integration must extend from $x = 0$ (the sample surface) to $x = g$ (the total mass of the sample) where Δx is the incremental mass.

Remember, intensity of diffraction is proportional to the volume of the sample irradiated. We showed before (Fig. 9.1) that the area of the sample that is bathed in the incident beam is proportional to $1/\sin\theta$. Equation (9.3) gives the diffraction due to the integrated depth of irradiation, and you can see that this quantity is *proportional* to $\sin\theta$ (the term $1/2\mu^*$ is constant with respect to θ). Therefore, the volume, which is the product of area times depth, is independent of 2θ. This is the great advantage of the diffraction geometry of the flat-specimen diffractometer, namely, it produces no 2θ-dependent absorption effect. This advantage is lost, however, if the sample is too short or if a theta-compensating slit is used (i.e., a slit that automatically changes as θ changes).

$$\frac{I}{I_0} = \int_0^g \exp\left(\frac{-2\mu^*x}{\sin\theta}\right)dx$$

(9.2)

The solution of the integral is

$$\frac{I}{I_0} = \frac{\sin\theta}{2\mu^*}\left[1 - \exp\left(\frac{-2\mu^*g}{\sin\theta}\right)\right]$$

(9.3)

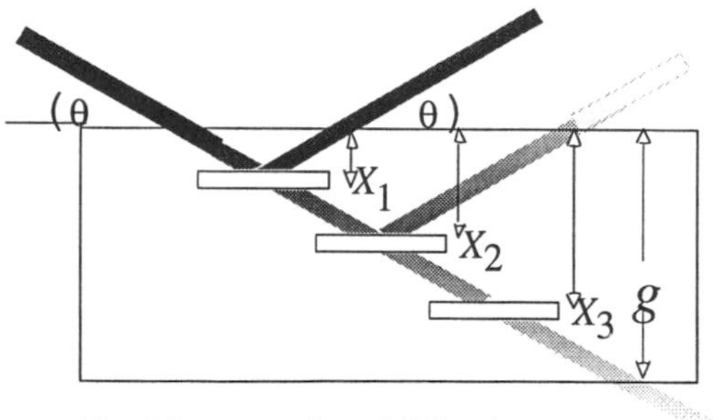

Fig. 9.3. Integration of diffraction energy over the sample mass g.

The formulation given here shows that the measured diffraction intensities are inversely proportional to μ^* because μ^* is not canceled by the product of the quantities that make up the irradiated volume. Here, μ^* is the mean *sample* mass absorption coefficient and is identical to the mineral mass absorption coefficient only if the sample is monomineralic.

Equation (9.3) has been solved for different sample thicknesses (g) and diffraction angles (2θ). The results, shown in Table 9.2, demonstrate the effects of thin samples on high-angle peak intensity. A clay mineral sample on a standard 47-mm membrane filter (see Chapter 6) covers about 11 cm^2, and 5 mg/cm^2 of sample amounts to 55 mg. This is by no means an exceptionally thin sample. Note that at 25° 2θ, intensity loss is 12% in comparison with the lower angles. If the amount of material is decreased to 33 mg, the errors become serious if analytical peaks are used that cover a large range of 2θ. The value of μ^* (45) used for these calculations is a good average for illites. If a sample consists of mostly montmorillonite or kaolinite, the thickness problem is more severe because these minerals have μ^* values of about 30 to 35.

Figure 9.4 shows the experimental arrangement that can be used to correct for the intensity loss caused by thin samples. We use the Drever (1973) method to prepare samples (see Chapter 6). The sample substrate is a 40 x 50 mm thin-section cover glass and is thin enough (0.2 mm) to allow measurable transmission of the incident X-ray beam. Mount a polycrystalline quartz aggregate in the diffractometer, and set the diffraction angle at the position of the strongest quartz peak (26.65° 2θ) in order to produce a strong monochromatic beam for the absorption measurements. First the substrate and then the sample plus substrate are treated as filters. A measurement of their filter factors will give us what we need. Measure the transmission through the cover glass (I_0) first, then add the clay mineral layer and measure the transmission of the quartz-diffracted beam again (Fig. 9.4), thus obtaining a value for I. The product μ^*g is obtained by substituting these values into a rearranged form of the Lambert law:

$$\mu^*g = -\ln(I/I_0) \tag{9.4}$$

Now that we know μ^*g, we can proceed to run the diffraction pattern on our unknown, measure the required peak intensities, and correct each of them, at their appropriate 2θ positions, by Eq. (9.3). The following example uses fictitious but reasonable numbers. The intensity of the strong quartz peak transmitted through the clean glass (Fig. 9.4) is 10,000 cps. The clay layer is added to the substrate and the same quartz reflection gives 6,000 cps. By the Lambert law (Eq. [9.4]),

$$\ln(6,000/10,000) = -\mu^*g = 0.51$$

Table 9.2.   Fraction of diffraction intensity for different sample thicknesses and diffraction angles (results apply to $\mu^* = 45$)

Thickness (mg/cm²)	I/I_0			
	$2\theta = 10°$	$2\theta = 15°$	$2\theta = 25°$	$2\theta = 35°$
1	0.64	0.50	0.34	0.26
3	0.95	0.87	0.71	0.59
5	0.99	0.97	0.88	0.78
10	1.00	1.00	0.98	0.95
15	1.00	1.00	1.00	0.99

The correction factor for the intensity of a reflection at 50° 2θ is obtained by Eq. (9.3) (0.423 is the sine of 50°/2 or 25°):

$$I/I_0 = 1 - \exp(-2 \cdot 0.51/0.423) = 0.91$$

The intensity of the reflection at 50° 2θ is 91% of the infinite thickness value, so we make a correction by dividing the measured diffraction intensity at 50° 2θ by 0.91.

Did you notice that the term $\sin\theta/2\mu^*$ in Eq. (9.3) has been neglected in this correction? The procedure is valid because the total sample geometry incorporates area as well as thickness effects, and, as we have seen, sample area is proportional to $1/\sin\theta$. Consequently, $\sin\theta$ cancels out because the volume irradiated is obtained by multiplying the functions that describe these two effects. Quantitative analyses are based on peak intensity ratios and not on absolute intensities, so the mass absorption coefficient likewise cancels.

These procedures provide the only possibility, to our knowledge, of realistic quantitative analysis of very small samples—samples so small that they will certainly be much thinner than the requirements of infinite thickness. These methods must also be used when the physical properties of a sample cause thicknesses to be less than about 10 mg of sample per square centimeter. Pure smectites are notable examples of such materials.

You may not wish to be, or cannot afford to be, bothered by the extra measurements just described. All right, there is a way out. Use peaks that are very close in 2θ for quantitative analysis, for the sample thickness effect will be minimal for

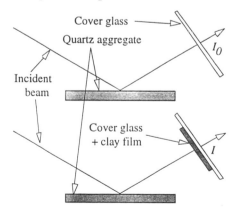

Fig. 9.4. Experimental arrangement for measuring $\mu^* g$.

Fig. 9.5. Definition of the sample position error ω.

them. How minimal is minimal, and how close in 2θ must the peaks be? Use Eq. (9.3) to find out, assuming that you have some idea of sample thickness, Another option is to weigh the glass substrate before and after application of the clay sample film and calculate the sample density. If the sample density exceeds 15 mg/cm² (Table 9.2), you don't have to be bothered by any of this, for your sample is infinitely thick and thickness at that point is irrelevant to good quantitative results.

Sample Position
The sample must be correctly mounted in the sample holder assembly of the diffractometer. Displacement of the sample surface away from the goniometer axis is the largest source of error in peak position or *d*. But such errors have little effect on diffraction peak intensity, and they are neglected here. The 2:1 alignment is another matter. By 2:1 alignment we mean positioning the sample so that the angle between the incident beam and the sample surface is identical to the angle between the sample surface and the diffracted beam. Both of these angles are called θ, and each must be exactly half of 2θ, which is the angle between the undeviated incident beam and its diffracted counterpart (see Fig. 3.7, p. 69).

The 2:1 alignment of the goniometer is carried out with precision slits and plates and is so accurate in a properly aligned instrument that it contributes no errors to peak intensity measurements. Clay mineral samples, however, are not flat and smooth. There may be a ridge or high spot of material on one end of the sample, or mineral debris on the reference pins or knife edges of the sample holder assembly. Displacements caused by these irregularities result in sample misalignment. Figure 9.5 shows the geometry of the 2:1 alignment error that we call ω. If ω is a constant positive value, in the sense shown by Fig. 9.5, intensity is diminished with decreasing 2θ. If ω is negative, intensity is enhanced with decreasing 2θ. Equation (9.5) is the mathematical expression that governs this effect:

$$\frac{I}{I_0} = \frac{1}{2}\left(1 + \frac{\sin(\theta - \omega)}{\sin(\theta + \omega)}\right)$$

(9.5)

Table 9.3 shows errors caused by different amounts of misalignment at different diffraction angles. Peak intensity ratios can contain serious errors if widely separated reflections are used and if errors in ω amount to 0.5° or more. A misalignment value of ω = 0.5° causes the two ends of a 4-cm sample to be displaced from each other by only 0.35 mm.

Table 9.3. The effect of sample 2:1 misalignment, ω, on intensity for different 2θ values

$^\circ 2\theta$	I/I_0		
	$\omega = 1^\circ$	$\omega = 0.5^\circ$	$\omega = 0.1^\circ$
3	0.60	0.75	0.94
5	0.71	0.83	0.96
10	0.83	0.91	0.98
18	0.90	0.95	0.99
26	0.93	0.96	0.99

Sample alignment errors can be avoided only by the use of samples with flat surfaces, a precise sample mounting procedure, and scrupulous attention to the cleanliness of the sample mounting assembly. If these conditions cannot be maintained, several methods can minimize or eliminate intensity errors. One method is to spin the sample during the diffraction scan to average any tilt that the sample may have with respect to the sample holder assembly. A modification of this method, which is almost as good, can be used if the sample is mounted in a separable carrier that is then locked into the goniometer sample holder. Simply run the sample, turn the carrier end for end, and run it again. The average of the two sets of data will eliminate alignment errors. If you cannot, or will not, go to this extra trouble, your only recourse is to use only peaks that are closely spaced and at high diffraction angles (i.e., >10° 2θ).

Homogeneity of the Sample
The preparation of samples that consist of oriented clay mineral crystallites usually involves centrifuging onto a substrate, making a smear, drying a suspension on a glass slide or disk, or accumulating the suspension on a filter using a vacuum or pressure apparatus (see Table 6.3, p. 215). These procedures take time, and during the period of preparation the coarser particles segregate at the bottom of the sample because they settle faster. The suspension from which the sample is prepared contains a range of particle sizes, and there is no reason to think that the different clay minerals in the suspension are uniformly represented in all particle-size increments. The sample, then, will often be enriched at the top in the finest constituents and on the bottom in the coarsest. (The Millipore® transfer filter method is an exception because the first increment of sediment is representative of the whole sample.) The incident X-ray beam is absorbed logarithmically in the sample, so the surface contributes a disproportionately large component to the measured diffraction. In short, this condition causes the overestimation of minerals in the fine-grained fraction and the underestimation of the coarsest constituents. The low-angle peaks are the most severely affected because the

path length within the sample is long at low diffraction angles (Fig. 9.2), causing the surface to dominate the diffraction effects more than it does at higher diffraction angles. But all peaks are affected to such an extent that we simply cannot tolerate a specimen that is not mineralogically homogeneous. In Chapter 6, we discussed the preparation of oriented aggregates for qualitative and quantitative analysis, so at this point we assume that you have prepared your clay samples by either the smear or the Millipore® transfer methods, because these two techniques cause the least segregation of particle sizes.

All right, we have a "perfect sample." It is thick enough, long enough, compositionally uniform throughout its thickness, and mounted perfectly in the diffractometer. What sources of error are still present? Why can't the analysis be perfect?

EQUATIONS FOR QUANTITATIVE ANALYSIS

Basic Quantitative Diffraction Equation

Our answer begins with a general equation that relates diffracted X-ray intensity to instrumental, geometric, and crystallographic parameters and is given by

$$I = K\,Wf\left(\frac{1}{V^2}\right)\left(\frac{1}{\rho}\right)|F|^2\,(1 + \cos^2 2\theta)\left(\frac{1}{\sin 2\theta}\right)\Psi\left(\frac{1}{\mu*}\right)$$

(9.6)

where I is the integrated intensity of a diffraction peak and Wf is the weight fraction of a given mineral in a sample. Wf is the quantity we are after. $|F|^2$ is the square of the modulus of the amplitude of scattering (the structure factor; see Chapter 3, p. 72) in the direction 2θ, V is the volume of the unit cell, ρ is the density of the mineral, $\mu*$ is the mean mass absorption coefficient of the *mineral mixture*, $1 + \cos^2 2\theta$ is the polarization factor, $1/\sin 2\theta$ is the single-crystal Lorentz factor, and ψ is the powder ring distribution factor. The value of ψ is controlled by the axial divergence (the Soller slit array), the diffraction angle 2θ, and the degree of preferred orientation of the crystallographic planes that produce the observed diffraction intensity. The mean orientation of these planes is parallel to the sample surface. The constant K is the product of a number of physical units such as the charge and mass of the electron, the velocity of light, and the wavelength of the incident X-rays. K also is affected by the efficiency of the X-ray and detector tubes, the X-ray tube operating conditions, the transmission characteristics of the β filter or monochromator, and the slit systems used—divergence, receiving, scatter, and Soller.

If all these quantities are known (most unlikely), quantitative analysis could be accomplished with errors governed only by the accuracy of peak intensity measurements, which can be reduced by proper experimental procedures to errors as small as ±2%. Fortunately, most of the parameters in Eq. (9.6) can indeed be dealt with satisfactorily, and the difficulties with quantitative analysis arise from just a few of them. Let us examine each of the parameters in some detail and isolate the troublesome ones.

Quantitative analysis requires that you use integrated intensity. Most nonclay minerals consist of large (i.e., hundreds of unit cells), coherent scattering domains. Diffraction peak breadths for these substances are a result only of the optical resolution of the diffraction apparatus. For such materials, the simple height of a diffraction peak is an adequate measure of its integrated intensity because all peak widths are identical. Clay minerals, on the other hand, give intrinsically broad lines because of small scattering domains, structural disorder, mixed layering, or some combination of these factors. The line breadths are thus characteristics of the crystallites (see the Scherrer equation, Chapter 3), and the effects of their variability must be eliminated by the use of peak areas. Area can be measured with a planimeter or integrated directly by using step-scanning procedures controlled by data acquisition and data processing computers, or peak deconvolution programs, if you have one. The area can be approximated very well by multiplying the peak height times the peak width at half-height. Nonlinear backgrounds and partial peak interferences may make such measurements difficult, but we never said that quantitative analysis was going to be easy.

There is a widespread belief that poor crystallinity causes weak integrated diffraction intensities. Yes, this is true in the rare (or even nonexistent) occurrences of crystals that contain regions of X-ray-amorphous or highly disordered matter on the unit-cell scale. Metals beaten by a hammer can be disordered on this scale, but it remains to be demonstrated whether or not clay minerals ever produce such diffraction effects. Poor crystallinity, due to defects between unit cells or small crystallite sizes, causes changes in peak height but not in integrated intensity (see James, 1965; Klug and Alexander, 1974). Rotational or *X-Y* translational periodicities along the *Z* axis in micas and chlorites that lead to the different polytypes have no effect on integrated 00*l* intensities or, for that matter, on their breadths. Clearly, integrated peak intensity is the experimental measurement upon which the quantitative analysis of the clay minerals must be based.

The quantities ρ and V for a given mineral are insensitive to the compositional variations common in most mineral species. Chlorite and biotite, however, have a sufficiently large range in composition to cause possible significant variations in ρ. The density of chlorite varies from 2.6 to 3.3 g/cm^3 or 12% about the midpoint of the range, so even an educated guess of the Fe content goes a long way toward reducing the effect of ρ in producing errors in the determination of chlorite.

Variations in the *mean sample* mass absorption coefficient μ^* are unimportant because μ^* cancels out of the intensity expression and need not be known if internal standard methods are used, or if intensity ratios of peaks from different minerals are used to produce results normalized to 100%. But absorption processes can cause errors in quantitative analysis for some highly absorptive minerals, particularly if they are present as large particles.

As it is used in Eq. (9.6), μ^* accurately describes absorption conditions only if all the grains in a sample are so small that absorption in a single grain is no more than about 1% of the energy incident on that grain. If the grains or crystallites absorb more than this, microabsorption causes peak intensities to become partly dependent on particle size and on the *individual* absorption coefficients of each of the constituents (Brindley, 1945). Bulk rock powders whose grain sizes are approximately 10 μm surely produce poor results for minerals with high absorption coefficients (for CuKα) like pyrite, siderite, and iron oxides (Brindley, 1961). The highly absorbing minerals will be underestimated, and for pyrite (compared to, say, quartz) the error is approximately a factor of two (for CuKα radiation). Most clay minerals, irradiated with CuKα X-rays, are unaffected by microabsorption because their crystallites are so thin. But Fe-rich chlorite and glauconite are heavy absorbers, and their concentrations may be underestimated unless particle thicknesses are equal to or thinner than 0.1 to 0.2 μm (Brindley, 1961).

Sample-to-sample variations in the structure factor F can be a major source of error in quantitative analysis. F is controlled by the chemical composition of the mineral and the atomic positions in the unit cell—the crystal structure (see Chapter 3). Variations in atomic position are small for a given mineral species, and have little effect on F at the moderate to low diffraction angles where useful analytical reflections are found. Variations in chemical composition, however, can cause large changes in F. The magnitude of F is the sum of the atomic scattering amplitudes of the atoms within the unit cell, the phase of each ray modified with respect to all others. The atomic scattering factor, at a given diffraction angle, is proportional to the number of electrons in an atom. Consequently, little change in F is caused by Mg-Al or Si-Al substitutions because all the fully ionized forms of these atoms have the same number of electrons. On the other hand, Fe substitution produces major changes in F because Fe^{2+} has 24 electrons, whereas Al^{3+} or Mg^{2+}, the usual ions replaced, have only 10 electrons. Similarly, because of the large atomic number of K, and therefore a high atomic scattering factor, the number of K^+ ions in illite and glauconite has a major control on F.

We conclude that two common ions exert important controls on the values of F for the clay minerals–Fe in octahedral coordination, and K in the interlayer. Review in your mind the 2:1 clay mineral structures described in Fig. 3.20 and Table 3.1, p. 100, and you will recall that both of these ions occupy sites that lie on centers of symmetry on projection to Z. Consequently, the full complement of the scattering from such sites is added to or subtracted from F, that is, their cosine terms in the structure factor calculation are equal to +1 or -1. The sensitivity of F to the composition of octahedral and interlayer sites thus arises not only from the possible variations in electron density of these sites but from their particular position in the layer structure. There is not a great deal that you can do about variations in F, but you can be forewarned that quantitative analyses of Fe-rich clay minerals are not going to be very satisfactory unless you have matching standards that have the same compositions as the unknowns. Glauconite is a particularly bad actor because of variability in both Fe and interlayer K concentrations.

The Lorentz-polarization factor, as in Eq. 9.6, is often written to include the polarization factor $1 + \cos^2 2\theta$, the single-crystal crystal factor $1/\sin 2\theta$, and the powder ring distribution factor ψ. The first two of these quantities have values that are independent of the instrument and the sample. They cause no difficulties in quantitative analysis. But the powder ring distribution factor depends on the sample and the instrument, and its uncontrolled variation can be a source of large analytical errors.

The factor ψ is related to the fraction of crystals in the sample that are oriented so that they diffract into the receiving slit at a specific angle 2θ, a Bragg angle. It is constant for a single crystal and proportional to $1/\sin\theta$ for random powders, but these are special cases that are almost never applicable to clay mineral samples. The distribution factor ψ can be computed for oriented clay mineral aggregates if the degree of preferred orientation is measured (Reynolds, 1986). Unfortunately, such procedures are impractical for routine analysis and indeed are difficult or impossible unless the clay minerals in the aggregate are very well oriented and you have access to a two-circle diffractometer. We are going to have to minimize errors due to variations in ψ by other means, which we will describe. But first let us see how ψ affects diffraction peak intensity.

Let σ^* be the standard deviation of the tilt angles of the crystallites about the mean crystallite orientation in an aggregate. The mean orientation is such that the 001 planes are parallel to the sample surface. For a single crystal $\sigma^* = 0$, and for a random powder $\sigma^* = \infty$. Typical methods for the preparation of oriented aggregates produce σ^* values from $4°$ to $30°$. You probably know that "good orientation" produces an intense $00l$ diffraction pattern and that poor orientation is the cause of poor patterns that have a good deal of random noise in the background and on the peak profiles. Peak intensity is proportional to $(1/\sigma^*)^2$, so the estimated limiting values of σ^* alluded to above will produce a 56-fold difference in basal or $00l$ intensities. This potential variability causes no difficulties in quantitative analysis if an orienting internal standard is used or if the analysis is based on ratios of peak intensities from different minerals, always assuming that all the minerals in a given sample orient similarly. But the variability of σ^* absolutely precludes the external standard method for the quantitative analysis of the clay minerals. You can't with any confidence measure, say, the illite 001 from a pure illite and use this value as a basis of comparison to the illite 001 from an unknown, because you will not know the degrees of preferred orientation of each of the two illites in your experiment.

All methods of quantitative XRD analysis assume that the orientations or lack thereof are identical for all analytical phases in the sample. Thus *it is futile to attempt quantification of quartz, feldspar, calcite, etc., along with the platy clay minerals in an oriented clay aggregate, for these nonclay minerals certainly do not have the same orientation characteristics as the clay minerals.* For the same reason, fibrous clay minerals, such as halloysite and polygorskite, cannot be analyzed in an oriented aggregate with other clay minerals. The only way to analyze these phases is to use random powder preparations, for then the orientation characteristics of the different minerals are identical because they are random, and ψ for each mineral is simply $1/\sin\theta$.

Preferred orientation has another effect on peak intensities that can present serious problems to the analyst because it makes the intensity depend on diffraction angle. If you had a diffraction pattern of a hypothetical substance that produced values of F^2 that were identical for many peaks, you would observe that the measured intensities decreased in a nonlinear fashion as the diffraction angle increased. Furthermore, for a similar, hypothetical, oriented clay mineral aggregate, the rate of falloff of intensity (of the 00l peaks) with increasing 2θ increases as the preferred orientation becomes poorer, that is, as σ^* becomes larger. Consequently, for samples of 50/50 mixtures of chlorite and illite with different preferred orientations, the intensity ratio of (e.g.) the chlorite 001 and illite 005 reflections will depend on the preferred orientation (σ^*) within each sample. If, as is usually the case, a calibration standard has better (or worse) orientation than an unknown, an analysis of the unknown will be inaccurate if it is based on a peak intensity ratio from the standard.

The effect of orientation on angle-dependent intensity is also modified by the Soller slit arrangement of the diffractometer. Even if you could maintain uniform values for σ^* from sample to sample, you could not apply quantitative calibration intensities to data taken from a diffractometer that incorporates a different Soller slit array. Many of the newer instruments use only one Soller slit, whereas the older ones have two. Consequently, you should be careful in applying calibration intensity data from the older literature to results obtained in your laboratory if your diffractometer has only one Soller slit.

Only two methods can be used to eliminate errors caused by sample-to-sample differences in preferred orientation. The first is to use random powders that are random enough to eliminate this source of error. You will pay the price of sensitivity and accuracy, however, because the basal clay reflections will be very weak. In addition, necessary techniques such as ethylene glycol solvation are difficult to accomplish. The other alternative, and the one we recommend, is to *use oriented aggregates, but restrict the peak selection to those peaks that are very close together with respect to 2θ* because intensity ratios from such peaks are essentially independent of Soller slit collimation and preferred orientation. In this regard, the chlorite 001 to illite 005 peak ratio used as an example is among the worst of possible choices. Much better alternatives are the chlorite 001 to illite 001, the chlorite 003 to illite 002 and the chlorite 004 to illite 003 peak intensity ratios. Of these, the last two are superior because other factors make the extreme low-angle region a poor choice. If you use the illite 003 reflection, you must allow for quartz interference if quartz is present in your sample.

Derivation of a Working Form of the Equation for Analysis

The most commonly used method for quantitative analysis involves peak intensity ratios. This method gives the relative proportions of the clay minerals normalized to 100%. For this procedure, we will need a simplified version of Eq. (9.6), and, inasmuch as peak ratios are used, some of the terms of Eq. (9.6) will cancel. We collect terms that are fixed for a specific reflection of a mineral of known composition and set their product equal to a constant [U in Eq.(9.7)]. Let us assume that the orientation is random so that we can avoid the complexities of the powder ring distribution factor. This simplification is valid even for oriented aggregates if the analytical peaks are close together with respect to 2θ, for, as we have seen, under these circumstances, the preferred orientation has little effect on peak intensity *ratios*, so long as σ^* is the same for all the minerals.

$$U_1 = \left(\frac{1}{V_1}\right)^2 \left(\frac{1}{\rho_1}\right)|F_1|^2 (1 + \cos^2 2\theta_1)\left(\frac{1}{\sin 2\theta_1}\right)$$

(9.7)

The subscripts refer to mineral type 1, which is mixed with mineral type 2. Substitute U_1 in Eq. (9.6) to get

$$I_1 = KWf_1\, U_1\,(1/\mu^*)$$

(9.8)

where μ^* refers to the mixture. A similar equation is derived for mineral 2, and these two equations are divided to get

$$I_1/I_2 = (Wf_1/Wf_2)(U_1/U_2)$$

(9.9)

in which K and μ^* have canceled, or

$$Wf_1/Wf_2 = (I_1/I_2)(U_2/U_1)$$

(9.10)

where the ratio of the intensities is equal to the ratio of the weight fractions times the ratio U_2/U_1. Recall that U_1 and U_2 are constants. Therefore their ratio is constant and Eq. (9.10) is the well-known result that the ratio of the weight fractions of minerals 1 and 2 is proportional to the intensity ratio of diffraction peaks from each of them. These two minerals can be mixed in equal proportions by weight, in which case the ratio of weight fractions = 1. If mineral 2 is assigned the role of a reference mineral, or normalization base for which $U = 1$, Eq. (9.10) can be rewritten as

$$I_1/I_2 = MIF_1/1 \qquad (9.11)$$

The quantity MIF (mineral intensity factor) is a calibration constant for the diffraction peak of mineral 1 that allows the quantitative estimation of the proportion of mineral 1 present in a mixture with mineral 2.

Equal-weight mixtures can be made of mineral 2 with different minerals, and each of these mixtures can be analyzed by X-ray diffraction to provide a series of MIF constants for each of the different minerals. All these MIF constants are based on unit intensity for the mineral selected as the base of normalization. (This is analogous to setting the specific gravity of water to 1 as a standard for all other specific gravities.) This procedure is laborious, and it is impractical for rare mineral species absent from an investigator's sample standard collection. We believe that the simpler and more accurate method is to calculate appropriate U values and from these to compute the various MIF constants. Alternatively, one can calculate mineral reference intensities that are theoretical values in counts per second (cps) for peak intensities of specific minerals under specified instrumental operating conditions. We must

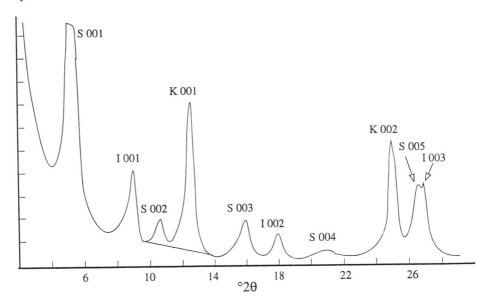

Fig. 9.6 Clay mineral mixture of illite, kaolinite, and EG-smectite. Background drawn under S 002 and K 001.

Table 9.4. Representative calculation of the abundance of each phase in Fig. 9.6

Mineral	Reflection	MIF	Intensity (cps)	Intensity/MIF	wt %[a]
Illite	002	0.51	1920	3,765	32
Smectite	003	0.81	3174	3,919	33
Kaolinite	002	2.19	9000	4,110	35
Illite	001	1.00			0
				Sum = 11,794	
Illite	002	0.51	1920	3,765	33
Smectite	003	0.81	3174	3,919	35
Kaolinite	001	3.08	11174	3,628	32
Illite	001	1.00			0
				Sum = 11,312	

[a]Weight % = 100 · (intensity/MIF)/sum.

emphasize again that for this procedure to be valid, σ^* must be the same for the minerals grains in a given sample. If we cannot assume this, we cannot do quantitative analysis. (This same system has been used for many of the common minerals in the JCPDS database using the strongest peak of corundum as the reference intensity. Why wouldn't this work as a reference for clay minerals?)

Application of the equations developed above is illustrated by the following hypothetical but realistic quantitative analysis. Figure 9.6 shows a calculated diffraction pattern of an oriented clay mineral aggregate that has been treated with ethylene glycol. Using criteria discussed in Chapter 7, we have decided that the sample contains kaolinite, illite, and dioctahedral smectite. We must choose the peaks to use. None should be from the low-angle region, and all should be close together. Unfortunately, the strong illite 003 reflection is almost perfectly superimposed on the smectite 005, so neither of these is useful. Good choices are the smectite 003, illite 002, and kaolinite 001 or 002 peaks. MIF values for these must have been measured or calculated, and for the present, assume that they are accurately represented by the data in Table 9.4. The MIF values have been normalized so that the intensity of the illite 003 is taken as unity. The illite 003 reflection cannot be used for the analysis we describe here because of interference problems, but the 002 peak is in the clear and it nominally has about half the intensity of the 003, so we will use the illite 002 reflection and assign it the value of 0.51. Table 9.4 contains the integrated peak areas or intensities (cps) for the relevant peaks in Fig. 9.6 and demonstrates how the quantitative analysis is computed. Peak areas are the product of the peak height times peak width at half-height (see Figs. 9.7 and 9.8).

Two examples are given in Table 9.4: one that uses the kaolinite 001 reflection, and one that uses the kaolinite 002 reflection. The results are not identical, probably because of uncertainties in the background under these two kaolinite peaks (Fig. 9.6). The procedure forces the analysis to total 100%, hiding general errors or a low total due to the presence of noncrystalline matter. A better procedure is to use an orienting internal standard, such as pyrophyllite, which is added in known amounts to each sample (Mossman et al., 1967). We discuss this technique next.

The Method of the Orienting Internal Standard

The orienting internal standard method is based on ratios of intensities of peaks from minerals of the original sample to the intensity of a single reflection from a mineral that has been mixed into each unknown in known amounts. The mineral added as an internal standard must have the same crystallite morphology as the unknown minerals, in this case a platy habit; its particle-size range should be similar to the unknown crystals to minimize problems of particle-size segregation during sample preparation; and its diffraction pattern should contain no reflections that interfere with useful reflections from the unknown minerals. Pyrophyllite is an excellent choice for clay mineral analysis.

Mixing dry clay powders with pyrophyllite is not feasible. Dried clay minerals often aggregate so strongly that redispersion is difficult and thorough mixing of clay-sized grains is time consuming. Instead, add a calibrated aqueous suspension of pyrophyllite by pipette to a calibrated suspension of clay to produce a mixture that is suitable for sample preparation by the Millipore® filter transfer method (Drever, 1973). Calibrate suspensions by extracting a known volume by pipette and adding it to a weighed evaporating dish. Dry the liquid in an oven set for hot but not boiling temperatures. Cool the dish in a desiccator and weigh it. The difference between the original dish weight and the weight of the dish plus clay provides a figure for the milligrams of clay per milliliter of suspension. It is an easy matter to add any given amount of pyrophyllite to each unknown after such calibrations have been made for the unknowns and pyrophyllite suspensions.

A pyrophyllite peak must be selected for an analysis standard, and the 003 reflection is a good choice (at ~29° 2θ it is out of the way of the strong quartz peak and the illite 003 peak). Mix equal proportions of pyrophyllite with a pure clay mineral such as illite and measure the peak intensity ratio of the pyrophyllite 003 to the illite 003. The illite 003 reflection has been arbitrarily set equal to unity, so this intensity ratio will provide an MIF for the pyrophyllite 003 reflection.

In this method, the transformation of the analytical measurements of intensity into a quantitative analysis is accomplished in exactly the same fashion as that described by Table 9.4, except that the data are not normalized to 100%. Assume that 20% by weight pyrophyllite has been added to each sample. A normalization constant Y is calculated as follows:

$$Y = (20) (\text{MIF}_p/I_p)$$

where MIF_p and I_p refer, respectively, to the pyrophyllite mineral intensity factor and the intensity of the standard pyrophyllite reflection for this hypothetical analysis. The quotients I/MIF for all the other minerals are multiplied by Y to provide values for their weight percent abundance in the sample. If the entire procedure is error-free, and if there is no amorphous material present, the sum of the unknown minerals will add to 80%. For a real analysis, sums between 70% and 90% should be considered "good." The analysis can be recast to a pyrophyllite-free format by multiplying all the mineral percentages by 100/80.

There are two advantages to the orienting internal standard method. Of most importance is the analysis total. This provides a figure of merit for the analysis. If it is near 100%, then large errors are probably absent and minerals have neither been missed nor misidentified. Canceling errors could be present, and if you suspect such a situation, you will have to devise other tests that are beyond the scope of this discussion. Another advantage of this method lies in its suitability for analyses of only one or two minerals in a mixture with others. Polycomponent mixtures may present such severe interference problems that some minerals cannot be determined. For example, smectite would be easy to measure in a mixture of smectite, chlorite, and kaolinite, but chlorite and kaolinite mixtures may present insurmountable difficulties. If you are lucky, perhaps all you really need to know for your application is the amount of smectite in each sample from a suite. If this is the case, the orienting internal standard method is the best procedure.

There are also some disadvantages to this method: It is difficult to grind pyrophyllite to $< 2\mu m$ (reducing its megascopic crystals to the clay-sized range requires extensive wet grinding and ultrasonic treatment); once pyrophyllite has been mixed into a sample, it can not be removed; and the minerals composing a natural sample are probably better and more uniformly oriented than laboratory mixtures.

MINERAL REFERENCE INTENSITIES

General Comments

The widespread availability of computers makes it an easy matter to calculate MIF for Eq. (9.11). The appendix describes a commercially available computer program (NEWMOD©, Reynolds, 1980; Reynolds, 1985) that can be used for such calculations, and it is suitable for simple clay minerals as well as for mixed-layered species. Mineral intensity factors can be computed much faster than they can be measured—indeed, empirical calibration is needlessly time-consuming. In addition, empirical calibration requires that the analyst have on hand a very large number of pure standard minerals, some of which may be difficult or impossible to obtain. If mixed-layered clay minerals are to be analyzed, the requirements of such a sample library become formidable because of the scarcity of such minerals in pure form. Calculated MIF factors do not eliminate most of the sources of error in quantitative analysis, but they add no new ones and they provide the opportunity to optimize a particular MIF for a particular analysis suite. Consider a given sample that produces an illite pattern for which the intensity ratio 002/003 is very small, suggesting a high Fe content. The analyst can turn to a computer program that calculates the intensities of the basal reflections of illite and adjust the Fe content until the observed 002/003 ratio is attained. At that point a "customized" MIF has been calculated whose use will at least be an improvement over the application of a standard nominal value.

Reynolds (1989) has shown that excellent agreement is attainable between measured and calculated MIF values for kaolinite, illite, smectite, and mixed-layered illite/smectite. It is likely that similarly acceptable results would have been obtained for other clay minerals had they been available for study. We believe that the field of clay mineralogy and its associated technology has progressed sufficiently to render empirical calibration obsolete.

Calculated Mineral Reference Intensities

Tables 9.5 and 9.6 show calculated diffraction intensity data (from NEWMOD©) for a variety of minerals. The values are absolute intensities that would be achieved experimentally if you had good control of the variables in Eq. (9.6) and estimated the constant in that equation by means of the intensity from a suitable standard, such as randomly oriented quartz. We call these quantities *mineral reference intensities*, and they can be converted to MIF values simply by dividing each, for example, by the calculated integrated intensity for the illite 003 reflection (3940 cps). But the absolute base for a set of such values makes no difference in their application. You use the numbers in Tables 9.5 and 9.6 in exactly the same way as the demonstration case treated by Table 9.4 and Fig. 9.6. If you measure such numbers empirically in the laboratory, the MIF or peak ratio form is obtained; but if reference intensities are calculated, the results are in cps and can be used in that form for quantitative standardization.

Reflections were selected from the range of intermediate diffraction angles to avoid the many problems encountered in the low-angle region. You may have interferences on these peaks and be forced to use others. Reference intensities for other reflections from a given mineral can be easily calculated from these by using experimentally observed values of relative intensity for the 00*l* series. For example, if a standard illite gives I(005)/I(003) = 0.5, then MIF = 0.5 for the illite 005 reflection, and the mineral reference intensity for the 005 reflection is 0.5 x 3940 = 1970 cps. The data of Tables 9.5 and 9.6 cannot be used in their present form for quantitative analysis by diffractometers that incorporate a θ-compensating slit. A perfectly adjusted

Table 9.5. Calculated integrated intensities for 00*l* reflections of various clay minerals

Mineral	Reflection	Intensity[a](cps)	Compositional parameters per $(Si,Al)_4O_{10}(OH)_2$
Montmorillonite	003	3110	0.1 Fe; ethylene glycol
Montmorillonite	005	4070	0.1 Fe; ethylene glycol
Saponite	003	1920	0.0 Fe; ethylene glycol
Saponite	005	4950	0.0 Fe; ethylene glycol
Saponite	003	620	1.0 Fe; ethylene glycol
Saponite	005	6070	1.0 Fe; ethylene glycol
Illite	002	1900	0.75 K; 0.1 Fe
Illite	003	3940	0.75 K; 0.1 Fe
Chlorite	003	4430	1.0 Fe 2:1 layer and
			1.0 Fe interlayer
Vermiculite	005	4140	0.0 Fe; Mg-wtr interlayer
Vermiculite	005	5000	1.0 Fe; Mg-wtr interlayer
Kaolinite	002	8610	Composition can't be changed
Kaolinite	003	830	Composition can't be changed

[a]The constant in Eq. (9.6) has been adjusted to provide 3940 cps for the illite 003 peak assuming conditions of $\mu^* = 45$, $\sigma^* = 12°$, and CuKα radiation.

θ-compensating slit causes an intensity variation that is proportional to sin θ. Consequently, valid reference intensities should be obtained by multiplying each of the intensities of Tables 9.5 and 9.6 by an appropriate value of sin θ. Mixed-layered clay minerals are often difficult to deal with because of peak interferences from other clay minerals. For example, illite/smectite is often mixed with discrete illite, and if the composition of the illite/smectite is near the illite end-member, resolution of the requisite reflections may be impossible by ordinary XRD methods. Similarly, corrensite usually contains discrete chlorite that causes the same kinds of interference problems.

Several methods can be used to determine the percent illite in illite/smectite, and we describe two of them that are based on the ethylene glycol-solvated condition (Table 8.3). The intensity of the illite/smectite 003/005 mixed-layered peak is almost independent of composition (see Table 9.6) and ordering type. This peak is the best to use if no interference is present due to discrete illite, which, unfortunately, is very common. Quartz also has a significant interference on this peak, but that can be eliminated by measuring the integrated area of the quartz 101 at 20.85° 2θ, multiplying by 4.3, and subtracting that quantity from the integrated area of the clay mineral reflection near 26.65°. If both illite (or smectite) and illite/smectite are present, the illite/smectite 002/003 reflection can be used, although its

Table 9.6. Calculated integrated peak intensities for illite/smectite (ethylene glycol)[a]

Percent illite	Reichweite	002/003[b] (cps)	003/005 (cps)
10	0	1620	4050
20	0	1540	4010
30	0	1380	3990
40	0	1180	4010
50	0	1130	4010
60	1	1050	3950
70	1	1000	3920
80	1	890	3860
90	3	880	3940

[a]The constant in Eq. (9.6) has been adjusted to provide 3940 cps for the illite 003 peak assuming conditions of $\mu^* = 45$, $\sigma^* = 12$, and $CuK\alpha$ radiation. 0.1 Fe per $(Si,Al)_4O_{10}(OH)_2$ applies to all compositions. [b]Applies to the low-angle half of the peak; see Fig. 9.9.

intensity is very sensitive to Fe content. (Note that the value of 4.3 for the ratio of the quartz intensities correlates better with experimental data than the intensities given by JCPDS card 33-1161. Based on experience in your lab, you may use a slightly different value.)

The high-angle side of the illite/smectite 002/003 reflection contains a component from a superstructure for ordered interstratifications and is markedly asymmetrical for random types. But there is a way to remove the sensitivity of this peak area to ordering, and that is to use the integrated area

Table 9.7. Recommended reflections for the quantitative analysis of various mixtures[a]

Mixture	Reflections	Comments
Illite, kaolinite	I(003), K(002)	Remove quartz interference[b]
Smectite, kaolinite	S(005), K(002)	Remove quartz interference[b]
Illite, chlorite	I(003), C(003)	Remove quartz interference[b]
Smectite, chlorite	S(005), C(003)	Remove quartz interference[b]
I/S[c], kaolinite	I/S (003/005), K(002)	Remove quartz interference[b]
I/S, chlorite	I/S (003/005), C(003)	Remove quartz interference[b]
Chlorite, kaolinite	C(003), K(003)	K(003) is weak
Smectite, illite	S(003), I(002)	
I/S, illite	I/S (002/003), I(002)	See Fig. 9.9
I/S, smectite	I/S (002/003), S(003)	See Fig. 9.9

[a]All smectite components assumed to be solvated with ethylene glycol, and chlorite must have symmetrical Fe substitution, as evidenced by approximately equal intensities for the 001 and 003 reflections. [b]The removal of quartz interference on the reflection near $26.65° \, 2\theta$ is accomplished by subtracting 4.3 times the integrated intensity of the quartz 101 reflection at $20.85° \, 2\theta$ (CuKα). [c]I/S = illite/smectite.

of the low-angle half of the peak (see Fig. 9.9). This procedure also minimizes interference from the low-angle tail of the discrete illite 002 reflection, although this interference makes this method unworkable if the illite/smectite is rich in the illite end-member. If discrete illite and discrete smectite are not present, the best procedure is to use the illite/smectite 003/005 values shown in Table 9.6.

Practical Examples of the Application of Reference Intensities

Table 9.7 shows useful reflections for the analysis of various mixtures of clay minerals and comments on their application. The chosen peaks are compromises between optimal choices and interference problems. They will be applicable to many samples of clay minerals, although difficulties can be encountered caused by excessive line breadth, which exacerbates interference problems, and by increased Fe substitution, which changes the reference intensities.

MEASUREMENT OF PEAK INTENSITY

Computer methods for curve fitting and peak decomposition show great promise, but they are so easy to use and produce such apparently robust peak resolutions that they can give a sense of confidence where such is not warranted. Someone has said that "you ought to have to have a license to own a pair of vice-grips." Some tools in the hands of a novice are capable of great damage, though they can be powerful when wielded with skill. Peak decomposition algorithms in their most general application generate various numbers of peaks that have specific shape functions, positions, breadths, and intensities, and sum these curves point-by-point on a 2θ-intensity scale. The parameters are refined until a minimum least-squares value is attained between the calculated and experimental profiles. The problems arise when the several reflections necessary to fit a given profile are accepted uncritically as real peaks, even though the crystal structures of the suspected phases cannot be reconciled with the results. For example, various kinds of disorder, including mixed-layering, produce markedly asymmetrical reflections for very understandable reasons; but left to its own devices, a decomposition algorithm may find that other weaker (fictitious) peaks are necessary to explain an asymmetrical tail on an experimental profile.

So how should these potentially powerful tools be used? The more constraints you place on the peak decomposition routine, the more believable will be the results. In other words, the fewer the parameters refined, the better. Confine the problem to sensible limitations and force a solution based on a specific number of peaks, some of which may be at diffraction angles that are known and can be specified and thus need not be refined by the calculations. If you cannot obtain a satisfactory least-squares agreement, then your qualitative model may be incorrect. Try another. Finally, it may be better to accept poorer least-squares agreement based on a justifiable model than better statistics based on a wholly empirical one.

Having suggested the pitfalls, let's consider several commonly encountered problems that are nicely solvable by peak decomposition methods. One is the interference between the chlorite 004 and kaolinite 002 reflections. You know that there are only two peaks involved, and you have a good idea of the position of the kaolinite reflection. Specify that. Furthermore, if you can see the chlorite 001, measure its breadth, and give the computer that value for the chlorite 004 peak because it will be a very good estimate.

Another more involved procedure requires breaking out the illite 002 peak from the mixed-layered R1 I/S peak near it (see Fig. 9.9). Let the computer deal with perhaps four reflections, but specify that one must be at 17.75° 2θ (the discrete illite 002 peak). In addition to the illite 002 peak, the result should find a strong reflection near 17° 2θ (Fig. 9.9) and other weaker ones that make up the high-angle tail of the profile. Ignore these weaker peaks as artifacts of the process. Use the position of the peak near 17° 2θ to get percent expandability and use the area of its low-angle half (Table 9.6), in conjunction with the illite 002 area, to obtain the ratio of illite to I/S in the sample.

The algorithm can easily solve the interference problems presented by Figs. 9.7 and 9.8 because there are only two peaks involved and their positions can be satisfactorily obtained by direct measurement. For all such analyses, it helps a great deal to have the excellent peak-to-background resolution produced by long count times during the step scanning procedure.

Before leaving the subject of computer processing of diffraction profiles, we will try to clear up some confusion between the processes of profile decomposition and deconvolution because the two are very different. Decomposition consists of finding a suitable number of reflections and their characteristics whose intensities, when added on a point-by-point 2θ scale, produce an intensity-2θ profile that matches the experimental one. Deconvolution involves extracting every part of a particular intensity-2θ profile, no matter how complicated, from every part of an experimental pattern. This is traditionally done in Fourier space, but Ergun has developed an iterative procedure (Klug and Alexander, 1974) that is powerful and very easy to do with a computer program containing perhaps a dozen or so lines of text. Additional explanation follows.

The mathematical theory behind deconvolution is complicated and beyond the scope of this text, so we will have to settle for a verbal description of what it is and what it does. Let's consider a two-dimensional case. Suppose you had a poor camera and took a picture of a perfectly orthogonal grid of straight, thin lines. But the picture is very different from the grid. The lines are no longer straight and orthogonal. Perhaps they form a mesh of intersecting arcs, and the arc portions near the center of the image are much broader than they are at the extremities of the picture. You have *convolved* a complicated two-dimensional distorted function that is characteristic of your poor lens onto the image of a sharp orthogonal grid. Every part of the image of the sharp grid passed through every part of your cheap lens to produce the final convolution. Then you photograph a friend and obtain a strange portrait indeed. Can it be cleaned up? Yes, by the process of *deconvolution*. If the original grid is very sharp and strictly orthogonal, the poor image of that grid is solely a signature of the lens. That distorted signature can be removed from the friend's portrait by deconvolution (not simple subtraction) to produce an image that, ideally, is what the picture would have looked like had the lens been distortion-free. The word "ideally" is used here because the mathematical process of deconvolution produces intensity ripples that combine to produce a nonuniform intensity background and can blur the edges of some portions of the sharp image.

No diffraction goniometer is perfectly focused. A profile of a diffraction peak that is infinitely sharp (from an infinitely thick perfect crystal) will show $CuK\alpha_1$-$CuK\alpha_2$ asymmetrical broadening, and asymmetrical and symmetrical broadening caused by a number of instrumental factors (see Klug and Alexander, 1974). Your instrument has *convolved* these onto the diffracted intensity. But that experimental reflection (which is your instrument's signature) can be deconvolved from any portion of another experimental diffraction pattern to produce the pattern from a perfect diffractometer. This

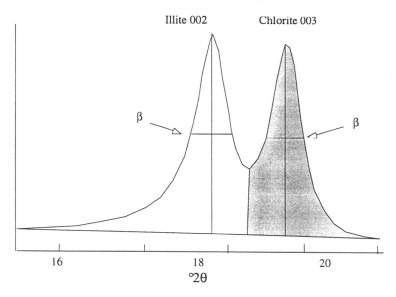

Fig. 9.7. Peak areas for 67% illite and 33% chlorite. Vertical lines represent peak height above background. Horizontal lines, labeled β, depict the peak width at half-height.

procedure does not sharpen diffraction patterns of clay minerals because their lines are intrinsically broad and asymmetrical, and these distortions are large compared to instrumental effects. For example, the poor resolution of the peaks of Figs. 9.7, 9.8, and 9.9 would be essentially unchanged by deconvolution of the instrumental signature of a modern, well-aligned diffractometer.

Peak decomposition is an important technique for our purposes; deconvolution is not. But the latter is crucial for extracting the pure shapes of high-order reflections like the chlorite 00,14. We go on now to a discussion of the more homely but so far most widely used methods for measuring integrated intensity from poorly resolved reflections.

The simplest case involves measuring the integrated areas of peaks that have no interferences and are superimposed on flat linear backgrounds. Intensities of such peaks can be measured by (1) subtracting a linear background and performing a numerical summation of the background-corrected intensities by computer methods, (2) multiplying the peak height above background times the peak width (in ° 2θ) at half the peak height, or (3) measuring the area of the peak above background with a planimeter. A more complicated situation arises when a peak is superimposed on a nonlinear background. Then you must use "artistic license" and estimate a realistic background so that it can be subtracted from the overall diffraction profile. The worst, and unfortunately very common, problem is the occurrence of two peaks that are incompletely resolved. Then you will have to estimate not only the background but the shapes of the peaks in the overlap region. If, in addition, one or both of the peaks are known to be asymmetrical, the measured peak areas will only be guesses, and the analysis needs to be reported with reservations. We will illustrate these conditions and suggest some ways to deal with them.

Figure 9.7 shows a portion of the diffraction pattern for a calculated mixture of illite and chlorite. The areas of the two peaks were measured before the patterns were added together, so in this case the relative areas are known. If two peaks are of approximately equal height and breadth and are symmetrical, the simplest way to measure their areas is to integrate from the tail of a reflection to the midpoint of the saddle between them. This is shown

Table 9.8. Peak areas measured by integration and by height times width. Reflections are the illite 002 and the chlorite 003, as shown in Fig. 9.7

Mineral	Correct areas (cps)	Measured areas (cps)	Ht x width (β) arbitrary units
Illite 002	1260	1485	1225
Chlorite 003	1390	1195	1185
Illite/chlorite ratio	0.91	1.24	1.03

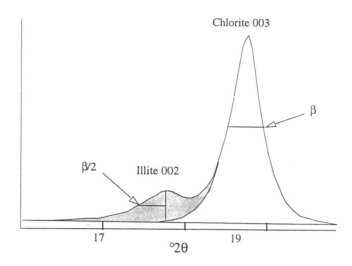

Fig. 9.8. A mixture of 75% chlorite and 25% illite. Vertical lines represent peak height above background. Horizontal lines, labeled β, which is equivalent to FWHM and depict the peak width at half-height.

by the two areas, one of which is shaded. The results from this integration are shown in Table 9.8. The vertical lines through the peaks represent their heights, and the horizontal lines through them, labeled β, mark the width at half-height. Relative areas were obtained by multiplying the widths by the heights. Table 9.8 gives a comparison between the correct relative peak areas and those obtained by the two different methods described.

The results are poor for integrated area probably because, for this case, both of the peaks are asymmetrical toward the low-angle direction. Note that the peak-height-times-β results are better. This may seem to be a "rough-and-ready" method, but, as we will see later, it often produces the best results of any of the techniques save those that require sophisticated computer techniques of curve-fitting.

Figure 9.8 depicts a more difficult analysis. The calculated diffraction pattern is based on 75% chlorite and 25% illite. The illite 002 reflection is weak and distorted by the tail of the intense chlorite peak. The peak areas were isolated by projecting the low-angle tail of the chlorite reflection under the illite peak to produce the background for the illite reflection. Integrated peak areas were measured by planimeter, divided by the appropriate mineral reference intensities (Table 9.5), and normalized (Table 9.4), yielding chlorite 70% and illite 30%. The composition of the simulated mixture also was determined by multiplying peak heights by β. Clearly, the illite 002 peak is too distorted to allow a direct measurement of β, so a value was obtained for β/2 (see Fig. 9.8) and doubled to give β. The composition based on this method is 73% chlorite and 27% illite, which is closer to the correct value than the result based on integrated areas. Again we see the superiority of the peak-width method, even for a case such as this that involves a very poorly resolved and weak reflection.

The diffraction profile in Fig. 9.9 represents a mixture of 75% *R*1 illite (0.8)/EG-smectite and 25% illite. A good deal of guesswork was required to estimate the background under the illite reflection because the illite/smectite peak is asymmetrical. The relative peak areas were measured by planimeter, but only the low-angle half of the illite smectite reflection was used because of uncertainties in the shape of the high-angle portion. This procedure is invariably necessary because the ubiquitous presence of illite causes interference on this, the most useful of the illite/smectite peaks. Note that mineral reference intensities in Table 9.6 are based on the integrated intensities of the low-angle portion of the illite/EG-smectite 002/003. The illite peak area and the illite/smectite "half-area" give 85% I/S and 15% illite, which is in only fair agreement with the nominal values. Much better results (78% I/S and 22% illite) were obtained by using the product of $\beta/2$ and the height for the illite/smectite 002/003 and that of β and the height for the discrete illite reflection.

Problems in analyzing mixtures of illite/smectite and illite diminish as the illite/smectite becomes more expandable because the 002/003 reflection is displaced farther from the illite 002. But as you can see from Fig. 9.9, illite/smectite minerals containing more than approximately 80% illite cannot be accurately determined in the presence of moderate to large amounts of discrete illite.

We have seen that superior results are obtained for measurements of peak area by the use of the peak-height-half-width method. Use this technique whenever you can. However, it is not always applicable. For example, it

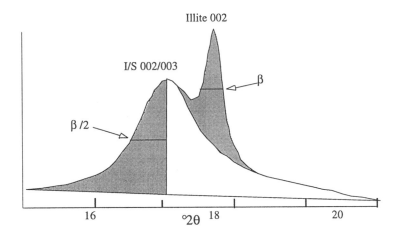

Fig. 9.9. A mixture of 25% illite and 75% R1 illite (0.8)/EG-smectite. The low-angle half of the illite/smectite peak is measured and doubled to avoid the effects of asymmetry of the peak itself and interference from the discrete illite peak.

cannot be used for a peak that lies on or near the saddle between two large reflections because the central peak will be distorted on both sides and thus provide no opportunity to use the half-width ($\beta/2$) as we did for our examples.

But why is the height-width method less sensitive to errors in selecting background than the more straightforward measurements of integrated area? To answer this question, we need to understand that a very large portion of the area of a peak lies in its "tails," which are just the regions where the most uncertainty will occur in fitting a profile to its underlying background. Errors in height or β give proportional errors in their product, but errors in selecting background produce exponentially enlarged errors in integrated area.

COMMENTS AND SUMMARY

We have treated quantitative analysis in more depth than the other subjects in this book for two reasons. Quantitative analysis is likely to be poorly understood or misunderstood, compared to the other XRD subjects we have discussed. In addition, quantitative analysis contains many pitfalls whose deleterious effects on the data are not obvious from an examination of the final results. Precision is easy, but as R. L. Snyder (a colleague at Alfred University, Alfred, NY) has remarked, "accuracy comes from God." The excellent repeatability that you can easily attain may lull you into thinking that the answers are correct, when, in fact, internal consistency may be a manifestation only of your ability to obtain a wrong answer every time and to do it with great precision.

No one can control perfectly the many variables in quantitative XRD mineral analysis. But the obvious pitfalls can usually be avoided, and even when they cannot, you should be able to recognize the samples for which the difficulties are insurmountable. At that point, the only valid solution is to throw the results out and report that the analysis simply can't be done by XRD methods alone.

The application of the analyses is important to keep in mind: How good do the data have to be? Is it really important to distinguish 10% kaolinite from 12%? Perhaps a practical application, or a theoretical one, requires only that you know that kaolinite is undetectable or a major component, or that you can demonstrate precision but not accuracy, an approach called quantitative representation QR (Hughes et al., 1994). Any analytical problem can probably be surmounted if enough resources are committed. One could, for example, employ combinations of XRD with other analytical methods. But at some point the slope of the cost-benefit curve must be considered, and working very hard to obtain results better than you need is inefficient. On the other hand, if very accurate quantitative mineralogical data are required, say, accurate to a few percent of the amounts present, then you will probably be unable to obtain them by X-ray methods alone, and you will need to search elsewhere.

Assuming, however, that the kind of accuracy attainable by XRD methods is applicable to your technical needs, we reiterate that:

1. The sample must be long enough for the divergence slit used and for the lowest diffraction angle recorded.

2. The sample must be thick enough at the highest diffraction angle recorded, or a correction must be applied to high-angle diffraction intensities.

3. The sample must be located as accurately as possible in the 2:1 diffraction geometry of the diffractometer.

4. The sample preparation procedure must allow only minimal particle-size segregation. We recommend the Millipore® transfer method.

5. The best diffraction peaks to use are as close together as possible. This condition minimizes errors caused by items 1 through 4.

6. Avoid the low-diffraction-angle region (below ~12° 2θ) because most of the factors discussed have their worst effects there.

7. External standards should not be employed because their orientation characteristics will certainly differ from your unknowns.

8. Quantitative analyses of platy (clay) and nonplaty (nonclay) minerals in the same sample cannot be achieved by any of the XRD methods that have been discussed because these two classes of mineral morphologies cause them to have very different preferred orientations. If random orientation for all phases can be achieved, then such analyses are feasible.

9. You must use integrated intensities for quantitative clay mineral analysis. At the present level of technology, the old-fashioned method that relates the area to the peak height times the width at half-height seems to be freer of errors than some of the more elegant methods.

10. You can obtain calibration factors for relative mineral intensities by measuring peak intensities from artificial mixtures. But why, when calculated calibration factors are more widely applicable, easier to obtain, and just as good or perhaps better?

REFERENCES

Brindley, G. W. (1945) Effect of grain or particle size on X-ray reflections from mixed powder: *Phil. Mag.* **36**, 347-69.

Brindley, G. W. (1961) Quantitative analysis of clay mixtures: in Brown, G., ed., *The X-Ray Identification and Crystal Structures of Clay Minerals*: Mineralogical Society, London, 489-516.

Brindley, G. W. (1980) Quantitative X-ray mineral analysis of clays: in Brindley G. W., and Brown, G., editors, *Crystal Structures of Clay Minerals and Their X-Ray Identification*: Monograph **5**, Mineralogical Society, London, 411-38.

Chung, F. H. (1974) Quantitative interpretation of X-ray diffraction patterns of mixtures. I. Matrix flushing method for quantitative multicomponent analysis: *J. Appl. Crystallogr.* **7**, 519-25.

Cody, R. D., and Thompson, G. L. (1976) Quantitative X-ray powder diffraction analysis of clays using an orienting internal standard and pressed disks of bulk clay samples: *Clays and Clay Minerals* **24**, 224-31.

Drever, J. I. (1973) The preparation of oriented clay mineral specimens for X-ray diffraction analysis by a filter-membrane peel technique: *Amer. Minerl.* **58**, 553-54.

Griffiths, J. C. (1967) *Scientific Method in Analysis of Sediments*: McGraw-Hill, New York, 508 pp.

Hughes, R. E., Moore, D. M., and Glass, H. D. (1994) Qualitative and quantitative analysis of clay minerals in soils: in Amonette, J. E., and Zelazny, L. W., editors, *Quantitative Methods in Soil Mineralogy*: SSSA Miscellaneous Publications, Soil Science Society of America, Madison, Wis., 330-59.

James, R. W. (1965) *The Optical Principles of the Diffraction of X-Rays*: Cornell University Press, Ithaca, New York, 664 pp.

Klug, H. P. and Alexander, L. E. (1974) *X-Ray Diffraction Procedures*: Wiley, New York, 996 pp.

Mossman, M. H., Freas, D. H., and Bailey, S. W. (1967) Orienting internal standard method for clay mineral X-ray analysis: in *Clays and Clay Minerals*, Proc. 15th National. Conf., Pittsburgh, Pa., Pergamon Press, New York, 441-53.

Reynolds, R. C., Jr. (1980) Interstratified clay minerals: in Brindley, G. W., and Brown, G., editors, *Crystal Structures of Clay minerals and Their X-Ray Identification*: Monograph 5, Mineralogical Society, London, 249-303.

Reynolds, R. C., Jr. (1985) *NEWMOD©, a Computer Program for the Calculation of One-Dimensional Diffraction Patterns of Mixed-Layered Clays*: R. C. Reynolds, Jr., 8 Brook Rd., Hanover, NH.

Reynolds, R. C., Jr. (1986) The Lorentz factor and preferred orientation in oriented clay aggregates: *Clays and Clay Minerals* **34**, 359-67.

Reynolds, R. C., Jr. (1989) Principles and techniques of quantitative analysis of clay minerals by X-ray powder diffraction: in Pevear, D. R., Mumpton, F. A., editors, *Quantitative Mineral Analysis of Clays*: Workshop Lectures, Vol. **1**, Clay Minerals Society, Boulder, CO, 4-36.

Schultz, L. G. (1960) Quantitative determination of some aluminous minerals in rocks: *Clays and Clay Minerals* **7**, 216-24.

If you have read this far, you know that the overwhelming emphasis in this book is on the interpretation of basal or 00*l* diffraction patterns from oriented clay mineral aggregates. This approach is justified because these patterns are the most diagnostic for both the simple and the mixed-layered clay minerals. Basal reflections from well-oriented samples are strong, producing excellent diffractograms with little effort on the part of the experimentalist. The peaks are strong because many crystallites are oriented so that they contribute to a given 00*l* reflection. Three-dimensional studies, on the other hand, require randomly oriented aggregates, and all the peaks are weak because only a small fraction of the crystallites have the required orientation for any given reflection.

Interferences from nonclay minerals are more severe in diffractograms from randomly oriented aggregates. Consider a mixture of quartz and a clay mineral. The quartz is randomly oriented in both oriented and randomly oriented preparations, so peak intensities are the same for both. But a clay mineral gives very strong reflections from the oriented aggregate, producing a high peak intensity ratio of clay mineral to quartz. In random orientation, peak intensities are "correct" for both, weakening the clay mineral relative to quartz reflections. Another problem is that the similarities of atomic structures in the *X-Y* plane of many of the clay minerals cause severe peak interferences among the clay minerals in a mixture.

Three-dimensional studies are difficult and not routine at present. You would not log a core with them. And you will have to give up on many samples; they simply present too many interference problems to be amenable to study. What kinds of rocks can be dealt with? The easiest are the bentonites and K-bentonites because the clay-sized fractions are almost monomineralic. Clay minerals from sandstones can often be extracted by gentle disaggregation procedures to yield suitable materials. The clay-sized fractions from carbonate rocks are often relatively free from detrital components, and some contain simple clay mineralogies. Unfortunately, the most important sedimentary rocks, shales, usually have such complex mixtures of detrital and diagenetic minerals in their fine-grain sizes that many of them are so difficult to deal with by any XRD methods (but see Grathoff and Moore, 1996) that high-resolution transmission electron microscopy has been the method for their study.

The previous paragraphs make it clear why so little is known about the geological significance of disorder in illite and illite/smectite (I/S). It is true that we recognize a *1Md* variety of illite, but the term *1Md* is often used as a qualitative label for what is in fact a member of a continuous series. However, the various types and amounts of disorder can be identified and quantified by means of calculated three-dimensional diffraction patterns. This is analogous to the way that NEWMOD© has been used to unravel the one-dimensional character of mixed-layered clay minerals. In this chapter, you will see the results produced by a computer program called WILDFIRE© (Reynolds, 1993, 1994), which is the three-dimensional analog of NEWMOD© (Reynolds, 1985). For those of you who wish to pursue the subject of disorder in lamellar crystals and its mathematical development, the text by Drits and Tchoubar (1990) is state of the art on this subject.

We recognize four types of disorder in smectite, illite, and I/S. The first is mixed-layering of illite with smectite, and that has been discussed in earlier chapters. The other three are (1) turbostratic stacking in smectite and illite/smectite; (2) rotational disorder in illite, which leads in the limit to the *1Md* species; and (3) interstratification of two different kinds of illite layers. The last category means that much of the illite/smectite that we have been identifying for years is really mixed-layered illite$_1$/illite$_2$/smectite. Now that you have finally reached the point at which you understand illite/smectite, it seems unfair to tell you that the designation illite/smectite is really an oversimplification, but we think it is.

SMALL CRYSTALS IN RECIPROCAL SPACE

The diffraction model that follows for disordered structures in reciprocal space superficially looks like the traditional Ewald construction, but it is not (recall discussion of reciprocal space in Chapter 3, p. 80). It has been modified to eliminate the Ewald sphere and its relation to the diffraction angle, 2θ. In addition, the planar indices *h*, *k*, and *l* are treated as continuous variables; *i.e.*, they are not limited to integral values. This is the construction formulated by Méring and Brindley (1951). It deals with a randomly oriented multicrystallite *powder* in reciprocal space in such a way that intensity can be summed over all values of *h*, *k*, and *l* that produce the same diffraction angle.

First, let's consider diffraction in three dimensions. A simple formulation that can be transformed to deal with disorder but will not be here (see Reynolds, 1993, for that development), is as follows:

$$I(hkl) = \left| F(hkl) \right|^2 \frac{\sin^2(\pi N_1 h)}{\sin^2(\pi h)} \frac{\sin^2(\pi N_2 k)}{\sin^2(\pi k)} \frac{\sin^2(\pi N_3 l)}{\sin^2(\pi l)}$$

$$(10.1)$$

We neglect the Lorentz-polarization factor. The three N terms refer to the dimensions of a crystal expressed by the numbers of unit cells in the X, Y, and Z directions. The sine-squared quotients make up the three-dimensional interference function. The reciprocal intercepts of the diffracting plane, h, k, and l, can have any value. Thus we can deal with diffraction from a plane with intercepts [e.g., 1.05, 2.13, 4.88, which, of course, is not the (125) plane—but close]. Such nonintegral intercepts make no sense for thick crystals because they describe a plane that would produce no measurable diffraction (recall the discussion associated with Fig. 3.13, p. 85). But as the crystals become smaller, the peaks become broader, and that means that diffraction occurs when the (125) is not quite lined up, or, in other words, when the diffracting plane is not strictly the (125) but one of a continuous series of planes whose orientations are clustered about the (125), each of which makes its intensity contribution with an incremental change in the diffraction angle. The farther the diffraction angle is from the precise value for the (125), the weaker the intensity becomes due to the effect of the interference function. If the crystal size is so small that crystallites consist of only one unit cell, there is no Bragg diffraction and scattering will be recorded from any orientation of the crystal. In this case h, k, and l serve to locate a plane whose only requirement for producing recordable scattering is that it has the same angle of incidence, with respect to the primary beam, as its angle with respect to the "diffracted" beam.

The structure factor F is given by Eq. (10.2), which is the three-dimensional form of F for a centrosymmetric structure (recall discussion of the one-dimensional structure factor from Chapter 3, p. 91). The sum is taken over all atoms (j), where f is the temperature-corrected scattering factor for each and the quantities x, y, and z are the three-dimensional atomic coordinates expressed in Å.

$$F(hkl) = \sum_j f_j \cos \left\{ 2\pi \left(h\, x_j / a + k\, y_j / b + l\, z_j / c \right) \right\}$$

(10.2)

Let us return to Eq. (10.1). For very large crystals, each of the sine-squared quotients has the constant value of N^2, if h, k, and l are integers, and if all the peaks have the same breadths (excluding instrumental effects). A set of $F\,(hkl)$ values is all that is needed to characterize the diffraction characteristics of the crystal, and that is what is measured for a single crystal structure determination. But if the crystal is small, the sine-squared quotients have finite values when h, k, and l are nonintegral; intensity falls off to zero when $h = h_0 \pm 1/N_1$; $k = k_0 \pm 1/N_2$; and $l = l_0 \pm 1/N_3$ (zero subscripts denote the integral values of h, k, and l). Two questions for you: (1) What are the plus-or-minus limits for h, k, and l if the crystals are infinitely large ($N = \infty$)? and (2) Why should diffraction peak breadths be inversely proportional to crystallite dimensions as given by the Scherrer equation on p. 87?

One-dimensional intensity calculations are simple because for any value of l there is only one value of 2θ. Consequently, we calculate intensity over a range of l and convert these l values to 2θ equivalents by means of the Bragg law $l\lambda = 2d\sin\theta$, or $\sin(\theta) = l\lambda/2d$, or $2\theta = 2\sin^{-1}(l\lambda/2d)$. But how to deal with 2θ in Eq. (10.1)? There are an infinite number of h, k, and l nonintegral values that produce the same 2θ (or the same d). A summation is required over all values of h, k, and l that produce a given d or 2θ. Because computers cannot handle infinite numbers of calculations some cleverness is required in the algorithm. We cannot go into this here, but see Reynolds (1993) if you want the details of how this can be done.

An orthorhombic (no nonorthogonal axes) structure in two dimensions serves as a simplified example of such a summation. Figure 10.1 shows the two orthogonal reciprocal space dimensions Z^* and X^*. Visualize the axis Y^* as normal to the plane of the figure, and set the two-dimensional X^* - Z^* plane at $k = 0$. Remember that the increments labeled h and l are $1/d(100)$ and $1/d(001)$, respectively. This arrangement deals with reflections of the type $h0l$. Reciprocal lattice spots are shown as ellipses, not as the points that a large crystal would produce. The shape of the spots is the reciprocal of the crystal shape (what else would you expect in reciprocal space?), and for this example, the crystal is thicker in the X direction than it is in the Z direction. The spots show diagrammatically that the intensities are stronger near their centers. The structure factor, F, is not included in this construction, so the spots represent only the three-dimensional interference function part of Eq. (10.1). Including F would have made the spots differ from each other in absolute intensity (and some might be missing due to extinctions that are characteristic of the symmetry of the unit cell). We remind you again that this construction represents reciprocal space, so the length of the two vectors shown with their origins at 0,0 is equal to $1/d$ or d^*. For $c = 10$ Å and $a = 6$ Å, the length of these vectors in normal space is 2.92 Å, corresponding to $2\theta = 30.65°$, $CuK\alpha$. (Use a ruler and see if you can figure this out.)

We will calculate the intensity of diffraction from a randomly oriented powder of small crystals at one specific diffraction angle, $30.65°$ as in Fig. 10.1. A randomly oriented powder means that there are crystals suitably oriented to produce any reflection of the type $k0l$ at the proper 2θ; and because the crystals are small (the spots are large), the measured intensity will include diffraction from some spots that are just nicked by the intensity summation vector. Or, to put it another way, the summation of the intensity at $30.65°$ will catch just the edges of some of the broad Bragg reflections.

Equation (10.1) must be summed for all values of h and l that lie on the circular arc whose radius is given by the magnitude of the vector d^* or $1/d = 0.343$. Visualize the vector as rotating through the angles $0°$ to $90°$, where $0°$ is defined as coincidence with Z^*. For each angular increment, we compute h and l, substitute those into Eq. (10.1), and add the resulting intensity to the accumulating sum. Our intensity includes the integrated 103 reflection because the arc passes right through it. In addition, the arc cuts through a significant portion of the 201 maximum adding again to the sum. And finally, the integration just catches the "outside" edge of the 200 diffraction region.

We have computed the intensity at 30.65° 2θ, but we need to go on and build up an entire diffraction pattern. The summation is set to zero, the vector d^* is increased by a small amount corresponding to an increased 2θ, and the summation is repeated, giving the intensity at the next 2θ value.

The calculations are more complicated for three dimensions. The vector must be rotated about the axis Z^* for each angular change in the X^*-Z^* plane. Each rotation generates a circle of summation lying in the plane normal to Z^*. The radii, *i.e.*, d^*, of the successive circles increase as the tip of the vector is moved clockwise from Z^* to X^* (Fig. 10.2). The summation of intensity for a specific 2θ includes the sums of all portions of the spots that lie on a spherical surface whose radius is d^*. Figure 10.2 shows the construction. It illustrates the commonly used nomenclature that designates the columns of reciprocal lattice spots along Z^* as "rods." Note: a common misunderstanding is that the spherical surface shown on Fig. 10.2 is the sphere of reflection or the Ewald sphere. It is not. It is a surface such that all points on it represent hkl coordinates that have identical values of d.

The constructions of Figs. 10.1 and 10.2 are worth exploring. Suppose that a calculated 00l or basal diffraction pattern is required. Then the orientation of the vector is fixed coincident with Z^* and is increased in length

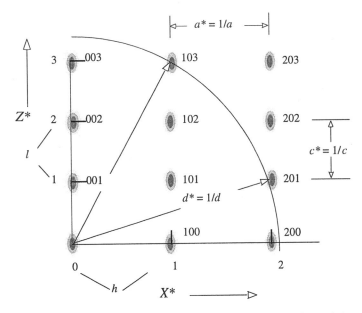

Fig. 10.1. Reciprocal lattice representation of a crystal of finite size and the summation vector for calculating the intensity of diffraction from a randomly oriented powder.

incrementally for a calculation of the intensity at each 2θ. How would a powder of very large crystals appear? The spots would be exceedingly small dots, and the only intensity recorded at 30.65° would be from the 103 reflection—and perhaps even that would be missed and show up at a different vector length corresponding to, say, 30.60°. What if the crystals were smaller in both the X and Z directions? Well, it depends on how much smaller, but perhaps the calculated intensity would then contain additional components from the 003 and 202 reflections. What if all the crystals were so

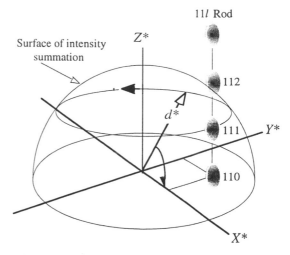

Fig. 10.2. Intensity summation in three-dimensional reciprocal space for the intensity of diffraction from a set of 11l planes in a randomly oriented powder.

small that they consisted of separate unit cells (N=1)? There would be no spots. They would merge into a uniform field of non-Bragg scattering, and the diffraction pattern would then consist of broad maxima like that produced by a melted molecular substance. The character of the scattering pattern would be controlled entirely by F, which we have not considered in this discussion. Finally, what if there were some disorder in the crystallite assemblage? Depending on the nature of the disorder and the symmetry of the structure, some spots would be extended in one or more of the three dimensions, others would be unaffected, and additional diffuse spots would appear between the spots in some of the rods. The amounts of these changes would depend on the extent of the disorder. What if the crystals were reasonably large in the X and Y dimensions, but were only one unit cell thick along Z? Then, the vertical arrays of spots shown by Figs. 10.1 and 10.2 would be drawn out into rods or pillars of continuous intensity along Z^*, modulated only by slow changes in F along Z^*. This is the turbostratic condition, which is discussed next.

TURBOSTRATIC DISORDER

Theory

This type of disorder is present in almost all smectite minerals. The 2:1 silicate layers are stacked along the normal to d(001) so that no layer is tilted with respect to that line, but the layers are displaced from each other in the X-Y plane by random amounts and are rotated about the normal by random amounts. Imagine carelessly tossing playing cards one by one onto a pile on a

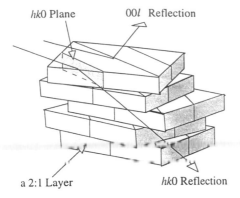

hk0 Plane   00l Reflection

a 2:1 Layer   hk0 Reflection

Fig. 10.3. Turbostratic stacking of 2:1 layers.

flat surface. If the cards represent the 2:1 layers, the pile is a turbostratic "crystal," or at least it acts like a crystal with respect to the basal X-ray reflections. Figure 10.3 shows such an arrangement. This structure is not possible with the micas or chlorites because the interlayer ions or hydroxide structure key adjacent layers into repeatable and fixed positions. But the weakly bonded exchangeable cations in the smectites evidently force no such ordering.

Let's rotate the turbostratic crystal of Fig. 10.3 about an axis perpendicular to the layers and try to get diffraction from a set of prismatic planes like the hk0 (Fig. 10.3). Diffraction will take place if the layer on the top is aligned suitably with respect to the incident beam. But the other four layers cannot diffract from this same plane because their orientations do not coincide with that of the top layer. In fact, in this example, no orientation of the crystal can produce hk0 (or hkl) reflections from more than one layer. The other layers might as well not be there, as far as the diffraction process is concerned. With respect to the hk0 plane, we say that the layers are optically incoherent with respect to each other, and a randomly oriented aggregate of such crystals will produce hkl diffraction only from single layers. To calculate diffraction from the hk0 plane, you would set $N_3 = 1$ in Eq. (10.1). Note that the planes of the 00l series are optically coherent. Diffraction from them, given the proper angle of incidence, would be described by $N_3 = 5$ for Fig. 10.3.

Return to Fig. 10.1 and single out one of the vertical rows of spots, say the 20l. We remarked earlier (p. 334) that the intensity of, for example, an h0l reflection falls to zero along Z* at the values $l_0 \pm 1/N_3$. This means that the intensity of the 202 spot falls to zero at the position of the 203, and the 203 falls to zero at the position of the 202 spot. In between their positions they are additive, producing an intensity continuum between these two and by induction, between all spots in a row. In short, the spots of Fig. 10.1 are drawn out into vertical rods along Z* whose vertical intensity changes slowly with l according to Eq. (10.2), as has been noted earlier. The diameter of the rods, i.e., their X-Y cross section, is controlled by the X-Y dimensions of the crystals [N_1 and N_2 in Eq. (10.1)].

X-ray diffraction requirements for three-dimensional optical coherence are very strict, and disorder need not be nearly as drastic as that shown by Fig. 10.3. For crystals that average a few hundred Å in the X and Y directions, random rotations of only half of a degree produce fully turbostratic diffraction patterns.

The reciprocal space construction is the easy way to visualize the origin of the two-dimensional diffraction band that identifies turbostratic disorder. Our development is taken from Brindley (1980). You might want to read his treatment if we leave you with any confusion. Figure 10.4 serves as the basis of the discussion. Shown are the familiar (by now) reciprocal space coordinates and a single rod for a turbostratic structure. The series of concentric arcs depicts a sequence of values of d^* (or 2θ) over which the intensity is summed. These are numbered to establish correspondence with the diffraction profile of Fig. 10.4B, which is

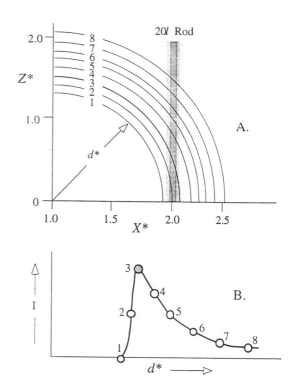

Fig. 10.4. Summation arcs in reciprocal space (A) and the development of a two-dimensional diffraction band (B).

diagrammatic only. Start with arc 1 and imagine integrating the intensity along its length. Only the left edge of the rod is intercepted, so the intensity is very weak for this value of d^*. Arc 2 cuts down through a significant portion of the rod, and its path almost intersects the center, where the intensity is greatest. The resulting intensity is sharply increased, causing the characteristic steep rise on the low-angle side of the developing two-dimensional reflection or band. Arc 3 cuts most of the way through the rod and has the longest path length through it. It produces the maximum intensity. The remaining arcs produce continuously diminished intensity because they pass through the rod at angles that are progressively nearer to 90°, which would give the minimum arc in F.

The shape of the band is modified if changes are taken into account along Z^*. In extreme cases, a rapid diminution of F along Z^* causes the asymmetric tail of the band to be reduced, giving rise to a quite symmetrical peak that appears to be a legitimate three-dimensional reflection. Such an example is the 02; 11 band for the micas at about $d = 4.5$ Å. Brindley (1980) shows numerous calculated examples of the effect of F on band shapes.

Two-dimensional bands such as the one illustrated by Fig. 10.4B occur at diffraction angles corresponding to *h*00, 0*k*0, and *hk*0 reflections. Three-dimensional peaks (*h, k,* and *l* ≠ 0) are missing, and their positions lie in the tails of the two-dimensional bands. A word of caution: The intensity maximum of the band does not necessarily coincide with *d* for any specific reflection such as the 020, and this can cause difficulties in the measurement of *d*(060), which is commonly used to estimate the *b* dimension.

Random powder XRD studies of smectite, I/S, and illite can be profitably restricted to the angular range 2θ = 16° to 44° because this range contains all

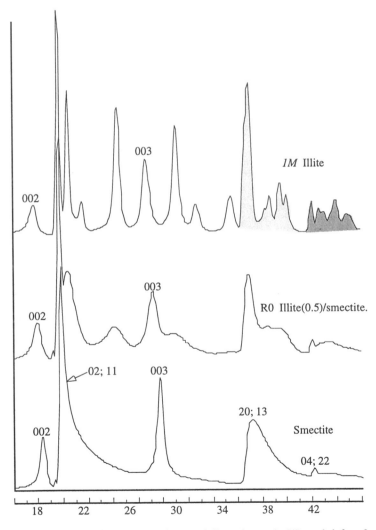

Fig. 10.5. Fully turbostratic smectite, partially turbostratic I/S, and defect-free *1M* illite. Unshaded regions contain peaks for which *k* ≠3*n*, *k* = 3*n* reflections are lightly shaded, and dark shading depicts mixed peak types.

the strong nonbasal reflections, with the exception of the 060, that collectively provide the most diagnostic information on the kinds and extents of disorder present. Remember, basal peaks will be present too; so in the case of I/S, it is desirable to simplify the basal pattern so as to minimize interferences from the 00l series. This is best done by dehydration, and that procedure was described in Chapter 6.

Smectite

A diffraction pattern for a turbostratic smectite is illustrated in Fig. 10.5. The crystallite dimensions are $N_1 = 60$, $N_2 = 30$, and $N_3 = 1$ (Eq. 10.1) for the hk pattern and a distribution of values for $N_3 = 1$ to 30 for the basal series assuming a dehydrated preparation [$d(001) = 9.7$ Å]. The two-dimensional diffraction bands are labeled with hk because for $N_3 = 1$ there can be no integral l component in the diffraction peaks. The calculated pattern includes the effects of the structure factor, and these cause the markedly different shapes of the three bands. Examination of the smectite pattern of Fig. 10.5 leads to the conclusion that the 20; 13 band is the best to use for detecting (and quantifying, as we will see later) turbostratic disorder. This band is intense compared to the 04; 22 band and much more asymmetrical than the 02; 11. Also shown in Fig. 10.5 for comparison is a *1M* structure containing a significant portion of turbostratic disorder (middle trace) and a defect-free *1M* illite (top trace).

Illite/Smectite

Mixed-layered illite/smectite contains 2:1 illite layers that are in perfect juxtaposition because the surface structure of the layers is keyed on the interlayer K. Experimental evidence published by Reynolds (1992) and other unpublished data indicate that the expandable, or smectitic interlayers, are not keyed; they are sites of turbostratic defects. Figure 10.6A is a cartoon of a few MacEwan crystallites of the many that make up a crystalline aggregate that we call a specific I/S. These are shown diagrammatically in random orientation. The question is, how do they diffract? The 00l series is the well-known irrational pattern described and discussed in Chapter 8. But the three-dimensional reflections are identical to those produced by a random arrangement of separated illitic crystallites (Fig. 10.6B). This behavior is caused by the nature of turbostratic defects that break the optical coherence of the MacEwan crystallites and cause the nonturbostratic fragments to diffract like separate crystallites. You might say that quartz and feldspar crystals in a powder aggregate are separated by turbostratic-like interfaces and that is why you get the simple sums of their diffraction patterns, and not a mixed-layered quartz/feldspar pattern. These illitic fragments are the fundamental particles of Nadeau et al. (1984) and, to use their terminology, the fundamental particles show interparticle diffraction for the 00l series, but not for three-dimensional reflections.

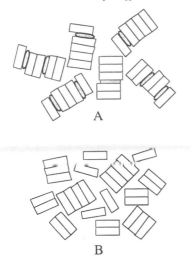

A

B

Fig. 10.6. Randomly oriented MacEwan crystallites (A) and their illite particles (B).

I/S has an incidence of turbostratic defects that is proportional to the percent expandability or smectite content of the species (Reynolds, 1992). The I/S series thus separates the end members of (1) pure illite with no turbostratic defects, and (2) pure smectite with turbostratic defects separating each 2:1 layer. The middle trace of Fig. 10.5 represents R0 illite(0.5)/smectite, and here is how it was calculated. The statistical distribution of MacEwan crystallites for this species was calculated, and the numbers of 1, 2, 3, etc. unit-cell-thick illite particles were summed for all the crystallites in the assemblage, producing a synthetic particle-size histogram of the type measured by Nadeau et al. (1984), who used transmission electron microscopy. We assumed here that the internal stacking was $1M$, and the three-dimensional diffraction pattern was computed by this distribution of illite crystallites. The excellent agreement that this procedure gives with respect to real samples, for which the illite content is known from the 00l patterns, is shown by numerous examples in Reynolds (1993) and McCarty and Reynolds (1995).

If you compare the three traces in Fig. 10.5, you can see the deterioration of the three-dimensional peaks caused by increased expandability. Figure 10.7 shows a calculated series of profiles that demonstrate the effects of increased turbostratic character on the 20l; 13l diffraction region. You could estimate percent expandability by analyzing the shapes of the 20l; 13l profiles from experimental samples, but it is easier and more accurate to base your interpretation on the 00l series.

The illitic fundamental particles in I/S could be internally stacked according to any of the mica polytype schemes, but the evidence so far suggests that they are always stacked in the $1M$ or $1Md$ arrangement. You might think, "So, I/S is a $1M$ illite structure that contains some turbostratic defects, and that is what is meant by the term $1Md$." Unfortunately, many would agree with that reasoning. Such a mineral is indeed a disordered $1M$ structure, but that is not the meaning of the term $1Md$ as used by, for example, Smith and Yoder (1956), Méring (1975), and Bailey (1988). We need help here from the International Committee on Nomenclature! In this book, we follow the authors cited who used the term "$1Md$" to denote the $1M$ structure disordered by randomly distributed $n120°$ and $n60°$ rotations. Turbostratic stacking, small crystallite size (which is really the same thing—the lower limit being crystallites one unit cell thick), mixed-layering, and poorly identified "poor crystallinity" do not qualify for our use of the "d" in $1Md$.

Rotational Disorder in Illite and Illite/Smectite

Layer rotations that are integral multiples of 60° ($n60°$) or 120° ($n120°$) affect only some of the *hkl* reflections. Distortions of the surface oxygen planes in the dioctahedral micas produce a plane with trigonal symmetry; and for the trioctahedral types, the octahedral cation sheet is also trigonal. Trigonal symmetry has a 3-fold axis of rotation, so rotations of 120° or 240° have no effect on some of the atomic patterns. Other aspects of the mica structure are not trigonal, so $n120°$ and $n60°$ rotations do cause changes in a crystal's diffraction pattern. Méring (1975) has used the term *semiordered* for mica structures with randomly distributed $n120°$ rotations. The prefix *semi* means that this type of disorder destroys only some of the *hkl* reflections; planes unaffected by $n120°$ rotations may continue to diffract; for these, there is no disorder.

The point comes up: How *d* is *1Md*? You can visualize a complete series that covers the range between a pure *1M* structure and one that is totally disordered; and indeed, members of such a series are present in nature, although they have been previously identified only as either *1M* or *1Md*. We define the location of a structure in that series by the variable P_0, which means the probability of a zero-layer rotation with respect to the layer below it along *Z*. For the case of $n120°$ rotations, and defining P_{120} and P_{240} in a fashion identical to P_0,

$$P_0 + P_{120} + P_{240} = 1 \qquad (10.3)$$

and

$$P_{120} = P_{240}$$

i.e., rotations of 120° and 240°, with respect to the underlying layers, are crystallographically equivalent operations so that there is no tendency for one to be more likely than the other. If $P_0 = 1/3$, then all rotations are equally probable, and we take this as the extreme case of disorder. $P_0 = 1$ describes the end-member *1M* structure.

The disordered case brings up an interesting point. One could look at *1Md* as disordered $2M_1$ ($2M_{1d}$) because plus and minus 120° rotations are as common as any other pair. Perhaps it is a *3Td* because the spiral axis caused by 120°-120°-120°-etc. rotations are also represented. Bailey (1988)

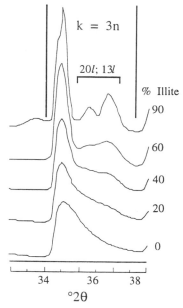

Fig. 10.7. The 20*l*, 13*l* diffraction region and the affects of turbostratic stacking that accompanies expandability in I/S.

suggested that we call the totally disordered case simply Md because the layers are certainly monoclinic and they are stacked in a disordered fashion. We like this idea but retain the older nomenclature because this change has not been formally sanctioned.

We must delve deeply into the crystallography of the micas to see why these strictly defined layer rotations produce the diffraction results that they do. The shape of the illite (or mica) unit cell is very nearly orthohexagonal. This means that the lattice has only a slightly deformed hexagonal pattern. But the symmetry of the unit cell is monoclinic. The shape and dimensions of the unit cell dictate at what diffraction angles the hkl reflections occur, and their intensities are determined by the occupants of the atomic positions that fix the symmetry of the cell and manifest themselves in the structure factor F_{hkl}. Perfect orthohexagonality is present if the following relations hold.

$$b = a\sqrt{3} \tag{10.4}$$

and

$$c\cos\beta = -a/3 \tag{10.5}$$

where a, b, and c are the unit cell dimensions and β is the monoclinic angle. The relation of Eq. (10.5) means that the top oxygen plane of the 2:1 layer is translated in the $-X$ direction with respect to the upper oxygen plane of the underlying layer by an amount that is equal to 1/3 of the a dimension. The ideal monoclinic angle is $\beta = 99.8°$ for a $d(001)$ value of 10 Å, characteristic of the micas, and $a \approx 5.2$ Å and $b \approx 9.0$ Å, which are also characteristic of micas and almost always perfectly orthohexagonal. If we compare ideal β with the muscovite or illite value of $\beta = 101.3°$, we see that there is some overshoot that has displaced the top oxygen plane a bit more than is ideal. This is mostly caused by the larger octahedral site at the vacant cation location that lies on the M1 position in the X-Z plane. Consequently, the muscovite-like illite layer is not perfectly orthohexagonal, but it is close enough to make the following discussion applicable.

Figure 10.8 shows two illite unit cells, the uppermost of which has been rotated by 120° with respect to the bottom one. The bold arrows indicate the $-X$ directions for each cell. The axis of rotation is normal to $d(001)$ and passes through the center of the K atom that lies in the interlayer plane that separates the two cells. This figure is not a cartoon—it has been accurately scaled. The lower cell has marked on it the (200) plane, and the upper cell has a trace of a member of the ($\bar{1}31$) set. Note that orientation of the (200) plane of the lower cell is coincident with the ($\bar{1}31$) plane of the upper cell. Therefore the 2 planes are in optical continuity. Furthermore, $d(200) = d(\bar{1}31)$, and the intensities $|F(200)|^2$ and $|F(\bar{1}31)|^2$ are equal except for multiplicity. In short, the intensity of the 200 reflection from a crystal is unaffected by layer rotations of 120° (or 240°).

Consider now a different reflection like the 020 (Fig. 10.9). A 120° rotation has indeed aligned the projections of the (11*l*) and (020) planes on the *X-Y* interface, but no integral or near-integral value of *l* produces coincident orientation in three-dimensional space between the (020) and any (11*l*) plane. Optical continuity between the two unit cells is lost for this arrangement, and there would be no 020 or 11*l* reflections.

The different geometries demonstrated by Figs. 10.8 and 10.9 can be extended to a general rule. All reflections for which $k = 3n$ are unaffected by 120° or 240° rotations *(k = 0* counts as a 3*n* reflection). Reflections for which $k \neq 3n$ will be missing or broadened, depending on the incidence of rotations. The situation for *n*60° rotations is more complicated.

Figure 10.10 shows a calculated powder diffraction pattern for a *1M* illite. The top trace is the entire pattern, and the bottom one shows the high-angle region in more detail. The peaks are labeled, and their indices demonstrate that for the *1M* illite space group (*C2/m*) reflections are limited to those for which $h+k$ is an even number. Ignore the 00*l* series, which gets in the way but tell us nothing about rotational disorder, and you can see that the pattern can be divided into three regions. Between 19° and 34° 2θ, all the peaks are of the type $k \neq 3n$. Specifically, they are the 11*l* and 02*l* reflections. Between 34°

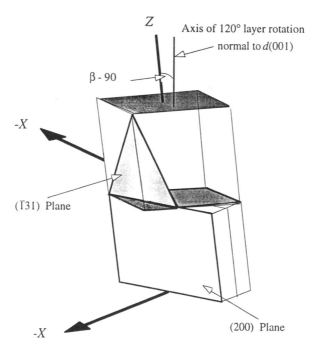

Fig. 10.8. Coincidence of the (200) and ($\bar{1}$31) planes caused by a layer rotation of 120° (-*X* axis of upper unit has rotated 120° from that of the lower -*X* axis).

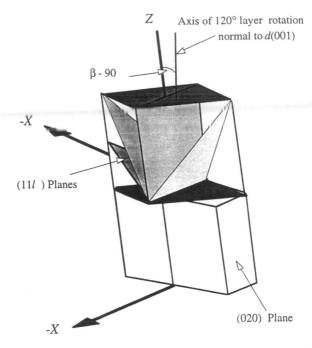

Fig. 10.9. A rotation of 120° produces no coincidence between the (020) plane with any member of the (11l) series.

and 39°, they are $k = 3n$ types (20l and 13l). From 39° 2θ to 44°, the reflectiontypes are mixed. From what you have learned, you can conclude that the low-angle region will be affected by rotational disorder, and the intermediate 2θ range will not. The higher-angle range, which fortunately contains only weak peaks anyway, will show complicated changes if rotational disorder is increased.

Turbostratic stacking affects *all* the peaks, so it is a simple matter to distinguish it from rotational disorder. Figure 10.11 demonstrates the principles described above. In both patterns, the 11l; 02l reflections are weak and broad, attesting to some kind of disorder. The I/S (upper trace) shows a poorly modulated 20l; 13l profile because interstratification with smectite has introduced many turbostratic stacking defects. The lower trace represents an illite in which rotational disorder has broadened the 11l; 02l peaks so that they resemble those in the I/S pattern; but the $k = 3n$ angular region has well-developed peaks because 120° and 240° rotations do not affect them. The similarities of the $k \neq 3n$ peaks for the two are remarkable.

We need to discuss rotational disorder and the $k \neq 3n$ peaks—a subject that penetrates to the heart of the meaning of such disorder. We know that turbostratic defects resulting from mixed-layering cause the $k \neq 3n$ spots, as well as the $k = 3n$ spots, to be extended toward the shape of pillars of nearly

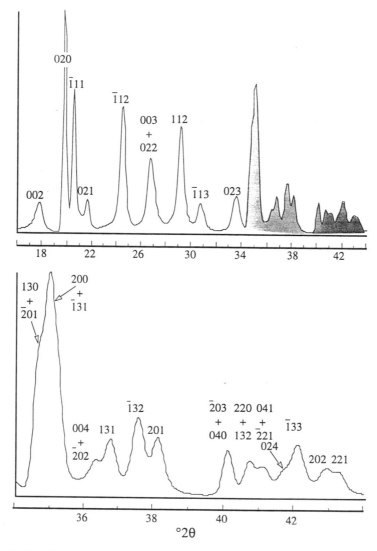

Fig. 10.10. Calculated random powder diffraction pattern of *1M* illite. Unshaded regions contain peaks for which $k \neq 3n$, $k = 3n$ reflections are lightly shaded, and dark shading depicts mixed peak types. The lower tracing is an expansion of the high-angle region of the upper tracing.

constant intensity along Z because if the crystal is only one unit cell thick, there can be no three-dimensional Bragg diffraction. Is the same principle op-erating for the rotationally disordered condition? No, the $k \neq 3n$ rods are essentially pillars of uniform intensity for a very different reason. No single crystallite can represent a rotationally disordered illite. There is no such thing as a representative crystallite. The diffraction pattern from a *1Md* specimen is the sum of the *intensities* of very many crystallites with different stacking sequences whose abundances are determined by the statistics of the composition. If $P_0 = 1/3$, then all rotations are equally probable, and the crystallite assemblage contains individuals that consist of *1M, 2M₁, 3T*, and other polytypes not found in nature such as *3M*, in addition to sequences too long to repeat within the crystallite thickness. These produce reciprocal lattice spots all along the $k \neq 3n$ rods at values of *l* different from the *1M* positions, and there are so many of them that the rods approach a condition of constant intensity along *Z**. For example, a *2M₁* stacking sequence produces spots midway between the *1M* spots, and a *3T* generates two spots between each *1M* reciprocal lattice node.

Figure 10.12 illustrates the character of the diffraction patterns for the range *1M* to *1Md*. The $k \neq 3n$ reflections behave as expected, but the $k = 3n$ region shows some changes despite what we said earlier, namely, that the $k = 3n$ reflections should not be affected by $n120°$ rotations. The changes in these

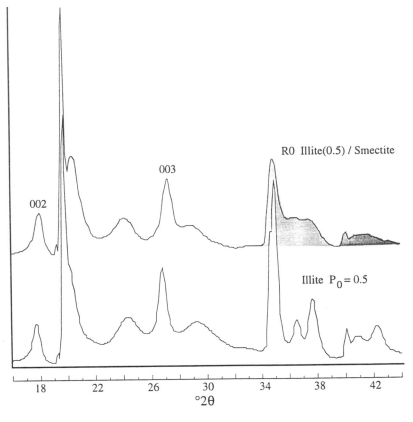

Fig. 10.11. Two kinds of disorder, rotational ($P_0 = 0.5$) and turbostratic (mixed-layering), and their manifestations in powder diffraction patterns. Unshaded regions contain peaks for which $k \neq 3n$, $k = 3n$ reflections are lightly shaded, and dark shading depicts mixed peak types.

diffraction profiles occur because the *IM* layer is not perfectly orthohexagonal; its monoclinic angle is 101.3° instead of the ideal value of 99.8°. These reflections provide a very important diagnostic parameter; if you can see any separation of the two peaks between 36° and 38° into four components, as in the top tracing, the structure has very little disorder and rates the name *IM*. An examination of the literature suggests to us that earlier studies would have categorized the top three traces as *IM* despite the disorder present. Clay mineralogists have yet to correlate decreasing rotational disorder with any geologic parameter.

Bailey (1975) has observed that *n*60° rotations are uncommon in the

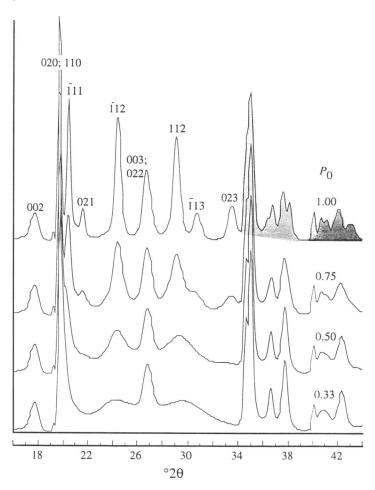

Fig. 10.12. Calculated diffraction patterns for the series *tv-1M/tv-1Md* illite. Unshaded regions contain peaks for which $k \neq 3n$, $k = 3n$ reflections are lightly shaded, and dark shading depicts mixed peak types. P_0 is the probability of a zero-layer rotation with respect to the layer below it along Z.

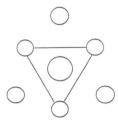

A. Upper layer rotated 60°

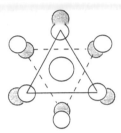

B.   Upper layer rotated 120°

Fig. 10.13   Ditrigonal (A) and octahedral (B) oxygen coordination about interlayer potassium in illite. Shaded atoms represent oxygen atoms in the lower plane.

dioctahedral micas, the best-known example of which is the rare $2 M_2$ polytype. He suggests that the reason is that the oxygen coordination about the interlayer potassium is energetically unfavorable. Figure 10.13 demonstrates the differences between 60° and 120° rotations. Diagram A shows that the upper and lower atomic planes are exactly superimposed, leading to ditrigonal coordination of oxygen about potassium. This arrangement brings the upper and lower oxygen atoms in close proximity, causing repulsive forces to destabilize the structure. Figure 10.13B depicts a 120° rotation. Shaded atoms represent oxygen atoms in the lower plane. The lower and upper oxygen triangles form octahedral coordination about potassium and maximize the separation of oxygen atoms in the upper and lower planes, leading to a more stable configuration.

McCarty and Reynolds (1995) found $n60°$ disorder in detectable amounts in Paleozoic K-bentonites (I/S). Correlations with chemical composition suggest that $n60°$ disorder increases with octahedral substitution of Mg for Al (and less Al for Si in the tetrahedral sites), such as occurs in the mica end member celadonite. Interesting partial corroboration of this conclusion is provided by the results of Sakharov et al. (1990), who reported that disorder in glauconite in dominated by $n60°$ rotations. The common wisdom is that glauconite tends to be more celadonitic than illite.

Rotations of $n60°$ and $n120°$ have different effects on diffraction patterns. For the former, reflections of the type $k = 3n$ have no coincident planes across the interfaces between such rotations, unlike the case for $n120°$ rotations. So we expect, and do see, changes in the region between 34 and 39° 2θ that accompany $n60°$ disorder, changes greater than those shown in Fig. 10.12. Figure 10.14 shows calculated patterns for partially disordered *1M* structures ($P_0 = 0.7$) in which the proportion of $n60°$ rotations varies from none ($P_{60} = 0$ and therefore the remaining 0.3 of rotations are all $n120°$ rotations) to the end member case in which the proportions of 0°, 60°, 180°, and 300° rotations are equal ($P_{60} = 1$). The tendency of this type of disorder is to coalesce the two "double" reflections into a single maximum. The most sensitive indicator for $n60°$ disorder is the lack of separation of these two peaks.

The effects of $n60°$ rotations on the $k \neq 3n$ diffraction pattern are complicated because 60°, 180°, and 300° rotations produce dissimilar results, and each influences some but not all the reflections. Because some of them are unaffected (like the case of $n120°$ rotations on the $k = 3n$ peaks), the $k \neq 3n$ diffraction region will appear to be somewhat less disordered than would be the case for rotations of $n120°$ (as in Fig. 10.12).

Cis-Vacant Illite and Interstratified Cis- and Trans-Vacant Illite/Smectite

We have described in Chapter 4, Fig. 4.3, two kinds of illite unit cells, the centrosymmetric and well-known *trans*-vacant (*tv*) variety and the newly (in illite) recognized *cis*-vacant (*c*v) noncentric structure. Our knowledge of *c*v structures is due to Russian scientists, particularly V. A. Drits and his colleagues. Tsipursky and Drits (1984) reported that many dioctahedral smectites have this structure, and Drits et al. (1984) derived the unit cell and atomic parameters for *c*v illite and published calculated diffraction patterns for it. Zvyagin et al. (1985) first reported a *c*v illite, and Reynolds and Thomson (1993) and Drits et al. (1993) described other

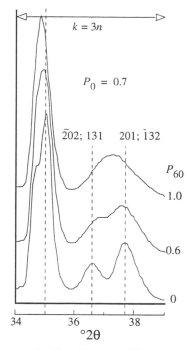

Fig. 10.14. The effects of $n60°$ rotations on the $k = 3n$ reflections 34 to 39° 2θ.

occurrences. McCarty and Reynolds (1995) showed that many Paleozoic K-bentonite I/S minerals have illitic portions that consist of randomly interstratified *c*v and *tv* layers. They are not simply I/S, but interstratified *tv*-illite/*c*v-illite/smectite. Quantification of such structures requires comparisons of experimental with calculated diffraction patterns such as those produced by WILDFIRE© or one of the Russian programs.

Figure 10.15 shows experimental diffraction patterns of *tv* and *c*v illite. The *tv* specimen is a hydrothermal sericite provided by D. D. Eberl, and the *c*v example is the illite from the Potsdam sandstone described by Reynolds and Thomson (1993). Reflections from the same *hkl* planes are at measurably different diffraction angles because the *c*v variety has β = 99.3° and for the *tv* cell, β = 101.3°. β = 99.3° also gives a "cleaner" $k = 3n$ diffraction profile because it is close to the ideal orthohexagonal value of 99.8°. You should have no trouble telling these two structures apart. The four 11*l* peaks of almost equal intensity are the diagnostic *c*v illite signature. We are confident that many occurrences of *c*v-illites will be documented if you students out there will take the trouble to achieve random orientation for your samples and get high-quality powder diffraction patterns from the many illites that you will see as clay scientists. The resulting database will (we hope, with the usual scientific optimism) finally make it possible to assign specific geologic conditions to the occurrences of this "new" structure. Analyses of more than 100 K-bentonites, including some published by McCarty and Reynolds (1995), the report of Lee et al. (1995) of large amounts of *cv IM* illite in sandstones of poor reservoir quality, and other as yet unpublished work, lead us to conclude that for these rocks, the *cv* variety is far more abundant than the *tv* species.

There is a disquieting circumstance in this otherwise optimistic picture. Recall that the *cv* variety is almost perfectly orthohexagonal. Well, the *3T* mica polytype is indeed hexagonal; consequently, both structures produce almost identical peak positions. By nasty chance, even the intensities of the reflections from the two are very similar. Reynolds and Thomson (1993) were able to resolve this difficulty by showing that, within the accuracy of measured peak positions, their data fit the monoclinic *cv* structure better than they accommodate the *3T* type. But if there is rotational disorder, or interstratification with *tv* layers, you might not be able to measure peak positions with sufficient accuracy to confidently exclude the *3T* interpretation. The differentiation would be simple by single-crystal methods, but they are not likely to be of much help with clay-sized crystallites. Discrimination can be accomplished by oblique-texture electron diffraction, but despite its long-

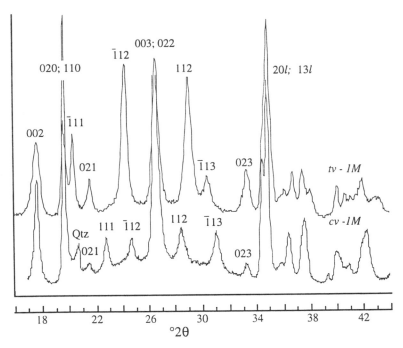

Fig. 10.15. Experimental powder diffraction patterns of *cv*- and *tv*-illite. Note the diagnostic (111) peaks.

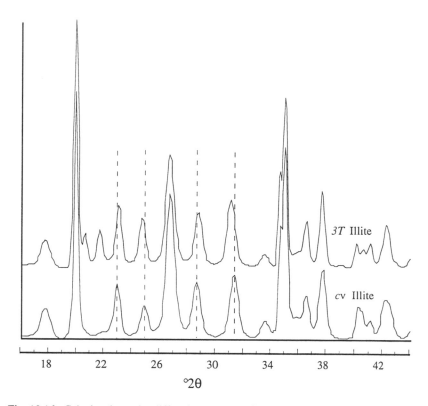

Fig. 10.16. Calculated powder diffraction patterns of *3T tv*-illite and *1M cv*-illite.

term use and development by Zvyagin and his colleagues, it is a relatively unfamiliar technique outside of Russia. The problem is illustrated by Fig. 10.16. Where does this leave us? There is a lot to do on the subject of *cv* illite. But we will assume that the "new" structure is the *cv* variety, not the *3T* polytype, as we go on to discuss its diffraction characteristics in disordered and interstratified structures.

A series of calculated patterns for interstratified *cv/tv* illite is portrayed in Fig. 10.17. The end members have been omitted to simplify the graphics, but that omission is not critical because you would be hard pressed to confidently identify 20% or less of either component in the other. We define *Pcv* as the decimal fraction of *cv* layers in the interstratification. Diagnostic features for the interpretation of *cv/tv* interstratification are (1) a shift in the positions of the 112 and $\bar{1}$13 reflections with changes in *Pcv*; (2) appearance of the 111 peak at higher values of *Pcv*; and (3) at large values of *Pcv*, four peaks of nearly equal intensity distributed in two pairs on opposite sides of the 003 illite reflection. The occurrence of the 111 is the crucial criterion. If you can see it at all, the structure contains at least 50% *cv* layers (*Pcv* ≥ 0.5). Just to allay your suspicions that we are too carried away with synthetic data, we show on Fig. 10.18 some "real" examples taken from McCarty and Reynolds

(1995). These are I/S minerals from Paleozoic K-bentonites analyzed in the dehydrated condition. Note that they compare favorably with the calculated patterns on Fig. 10.17 except that some rotational disorder is present ($P_0 < 1$) and small amounts of smectite interstratification have introduced some turbostratic disorder. The $k = 3n$ region near 35° 2θ on the top trace shows some evidence of $n60°$ rotational disorder (compare to Fig. 10.14).

A physical mixture of *cv* and *tv* types is easy to distinguish from an interstratified *cv/tv* structure for illites with little expandability and free from rotational disorder. The mixture will produce partially resolved reflections from each type, whereas the interstratification will yield peaks at intermediate

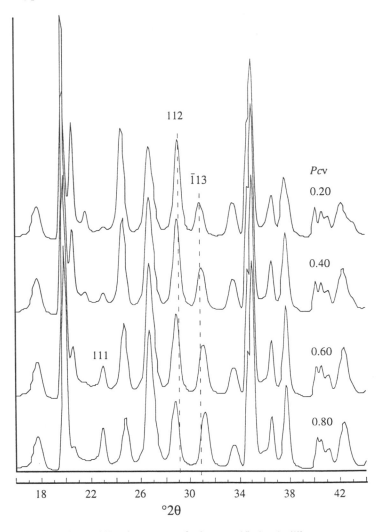

Fig. 10.17. Calculated diffraction patterns for interstratified *cv/tv* illite.

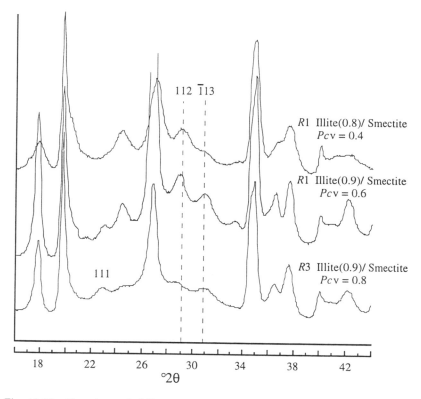

Fig. 10.18. Experimental diffraction patterns of I/S from Paleozoic K-bentonites (McCarty and Reynolds, 1995).

positions. V. A. Drits (personal communication, 1993) has pointed out that the 020; 110 doublet at about 19.9° 2θ is different for the *tv* and *cv* structures, as the calculated data in Table 10.1 demonstrate. The importance of this observation is twofold. A peak at 19.72° identifies the *tv IM* polytype, and a peak at 19.91° is diagnostic for the *cv IM* structure. Note that two peaks near 19.9° conclusively demonstrate the physical admixture of the two types of layers. Unfortunately, as rotational disorder increases, this discrimination cannot be made, and at some point it is impossible to distinguish a mixture from an interstratification. For the completely disordered case, $P_0 = 1/3$, there is little difference between diffraction patterns of the two cases. Figure 10.19 shows such a pattern for $Pcv = 0.6$, which is typical for many nonexpandable illites from shales. We have messed this pattern up a bit to better simulate experimental diffraction data. Some noise has been added, and a small degree of preferred orientation was introduced into the calculations. Does this common type of diffraction pattern mean that many *1Md* illites are interstratifications of *cv* and *tv* layers? Tune in in a few years and maybe we will know.

Figure 10.20 shows the effects of increasing *n*120° disorder on the

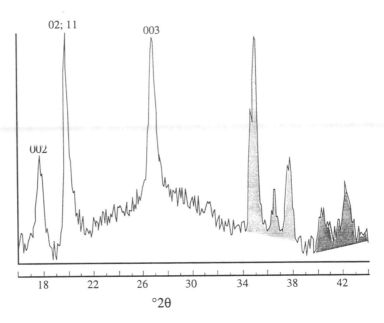

Fig. 10.19. Calculated diffraction pattern for completely disordered ($P_0 = 1/3$) interstratified *cv/tv* illite ($Pcv = 0.6$). Unshaded regions contain peaks for which $k \neq 3n$, $k = 3n$ reflections are lightly shaded, and dark shading depicts mixed peak types.

diffraction patterns of *cv 1M* illite. As you would expect, the results are similar to those shown on Fig. 10.12 for the disordered *tv 1M* series.

We close this chapter with an overview of the process of interpreting disorder by means of three-dimensional powder XRD patterns. Figure 10.21 illustrates the way to go about it. Figure 10.21A shows the diffraction angles that contain the $k \neq 3n$ reflections. The weak, broad Ī11, Ī12 and 112 reflections and the absence of the 111 peak should tell you that:

1. The layer structure type is *trans*-vacant *1M*, and

2. Disorder is present as evidenced by the breadths of the reflections.

The disorder could be turbostratic, caused by interstratification with smectite. Or it could be rotational stacking disorder due to $n120°$ or $n60°$

Table 10.1   Intensities and diffraction angles for the 020 and 110 reflections.

Peak	Intensity	d	2θ
tv 110	<1	4.44	20.01
tv 020	77	4.50	19.72
cv 110	79	4.46	19.91
cv 020	9	4.50	19.72

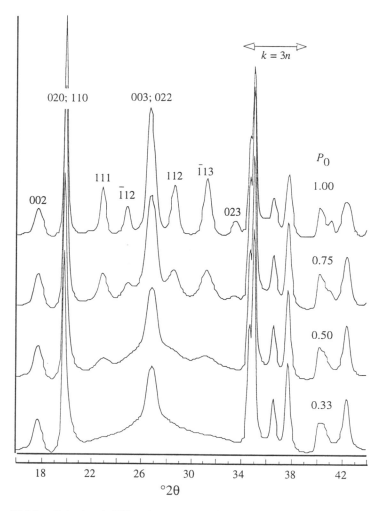

Fig. 10.20. Calculated diffraction patterns for the series *cv-1M/cv-1Md* illite illustrating increasing *n*120° disorder.

rotations. It might be a combination of all these. Consider now the diffraction profiles for the $k = 3n$ reflections 130, $\bar{2}01$, 200, $\bar{1}31$, $\bar{2}02$, 131, $\bar{1}32$, and 201 (see the high-angle region of Fig. 10.10) that lie between 34° and 39° 2θ (Fig. 10.21 B, C, and D). If the pattern for this diffraction region looks like Fig. 10.21B, then the structure is disordered by turbostratic defects. If it resembles Fig. 10.21C, randomly distributed *n*120° rotations along *Z* are responsible for the disorder evident in the pattern of Fig. 10.21A. Alternatively, the $k = 3n$ diffraction profile may be best matched by Fig. 10.21D, in which case rotational disorder of the *n*60° kind dominates. You will have to "read between the lines" if a real experimental pattern shows diffraction effects intermediate among these more-or-less end member examples. To do better

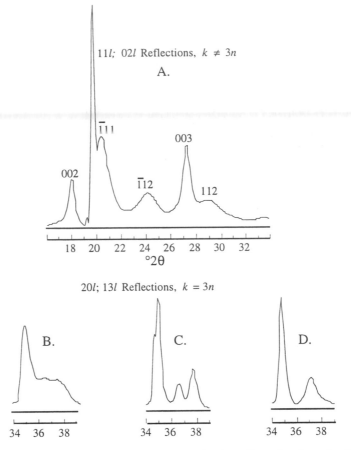

Figure 10.21. An example of the interpretation of a three-dimensional powder XRD pattern for a disordered *tv 1M* illite.

than that, the only recourse is to model the diffraction pattern by computer calculations, and adjust the disorder parameters until a good fit is obtained between the calculated and experimental results.

CONCLUSIONS

1. Turbostratic defects that occur across smectite interlayer positions in I/S diffuse all the three-dimensional reflections. In the limit where scattering is restricted to single 2:1 layers, the three-dimensional scattering pattern is marked by asymmetrical bands whose heads roughly correspond to 2θ angles of "prism" reflections like the 110, 020, 130, 200, 040, 220, etc.

2. Rotational disorder is caused by randomly distributed layer rotations of $n120°$ or, more rarely, $n60°$. The $n120°$ rotations broaden and weaken $k \neq 3n$ reflections like the $11l$ and $02l$ sequences, but have only slight effects on the $k = 3n$ peaks such as the $20l$ and $13l$ series. It is this type of disorder that produces illites for which the term *1Md* should be restricted.

3. Disorder of the $n60°$ kind (n is an odd number) can be identified by a tendency of the peaks at $\approx 36.6°$ and $\approx 37.7°$ 2θ to merge into a single peak at about $37°$ 2θ.

4. Rotational disorder produces a continuous series that separates the pure *1M* polytype from the totally disordered species that has been called *1Md*. Modern nomenclature is inadequate to deal with this series, and in most of the literature, structures are identified as *1M* even though they contain significant amounts of disorder.

5. The well-known *trans*-vacant illite or mica unit cell is common in illite, but so is the newly identified *cis*-vacant variety. These illite types can occur as physical mixtures or as components of interstratifications. They are distinguished by the positions and intensities of the $11l$ and $02l$ reflections.

6. Many or even most specimens of I/S will display all the types of disorder discussed in this chapter: mixed-layering, rotational disorder, and *cis/trans* interstratification.

7. These types of disorder can be identified and quantified if the time is taken to prepare proper samples for X-ray analysis and if attention is paid to diffractometer operating parameters—particularly, the use of long count times in the step-scan procedures. To quote the phrase often used by athletes, "we'll have to step up to the next level," to sort these structures out and assign geological meaning to them.

REFERENCES

Bailey, S. W. (1975) Cation ordering and pseudosymmetry in layer structures: *Am. Mineral.* **60**, 175-87.

Bailey, S. W. (1988) X-ray diffraction identification of the polytypes of mica, serpentine, and chlorite: *Clays and Clay Minerals* **36**, 193-213.

Brindley, G. W. (1980) Order-disorder in clay mineral structures: in Brindley, G. W., and Brown, G., editors, *Crystal Structures of Clay Minerals and Their X-Ray Identification*, Monograph **5**, Mineralogical Society, London, 125-95.

Brindley, G. W., and Méring, J. (1951) Diffraction des rayons X par les structures en couches desordonnees: *Acta Cryst.* **4**, 441-47.

Drits, V. A., Plançon, B. A., Sakharov, B. A., Besson, G., Tsipursky, S. I., and Tchoubar, C. (1984) Diffraction effects calculated for structural models of K-saturated montmorillonite containing different types of defects: *Clay Minerals* **19**, 541-61.

Drits, V. A., and Tchoubar, C. (1990) *X-Ray Diffraction by Disordered Lamellar Structures*: Springer-Verlag, New York, 371 pp.

Drits, V. A., Weber, F., Salyn, A. L., and Tsipursky, S. I. (1993) X-ray identification of one-layer illite varieties: Application to the study of illites around uranium deposits of Canada: *Clays and Clay Minerals* **41**, 389-98.

Grathoff, G.H., and Moore, D.M. (1996) Illite polytype quantification using WILDFIRE™ calculated patterns: *Clays and Clay Minerals*, **44** (in press).

Lee, M., Lewandowski, J., and Kinzel, M. (1995) Reservoir heterogeneity due to fault related diagenesis in N.W. Germany: *Abstracts with Program*, **27** no. 6, Geological Society of America, New Orleans, A-251 (abstract).

Méring, J. (1975) Smectites. in Giescking, J.E., editor, *Soil Components, Vol 2. Inorganic Components*: Springer-Verlag, New York, 97-119.

Méring, J., and Oberlin, A. (1967) Electron-optical study of smectites: *Clays and Clay Minerals*, 17th Nat. Conf., Pergamon Press, 3-25.

McCarty, D. K., and Reynolds, R. C., Jr. (1995) Rotationally disordered illite/smectite in Paleozoic K-bentonites: *Clays and Clay Minerals* **43**, 271-84.

Nadeau, P. H., Wilson, M. J., McHardy, W. J., and Tait, J. M. (1984) Interstratified clays as fundamental particles: *Science* **225**, 923-25.

Reynolds, R. C. (1992) X-ray diffraction studies of illite/smectite from rocks, and <1μm oriented powder aggregates: The absence of laboratory-induced artifacts: *Clays and Clay Minerals* **40**, 387-396.

Reynolds, R. C. (1993) Three-dimensional powder X-ray diffraction from disordered illite: Simulation and interpretation of the diffraction patterns: in Reynolds, R.C., and Walker, J.R., editors, *Computer Applications to X-Ray Diffraction Methods*: Clay Minerals Society Workshop Lectures, Vol. 5, 44-78.

Reynolds, R. C. (1994) *WILDFIRE©, A Computer Program for the Calculation of Three-Dimensional Powder X-Ray Diffraction Patterns for Mica Polytypes and their Disordered Variations*: R. C. Reynolds, 8 Brook Rd., Hanover, NH.

Reynolds, R. C., and Thomson, C. H. (1993) Illite from the Potsdam Sandstone of New York: A probable noncentrosymmetric mica structure: *Clays and Clay Minerals* **41**, 66-72

Sakharov, B. A., Besson, G., Drits, V. A., Kamenava, M. Yu, Salyn, A. L. and Smoliar, B. B. (1990) X-ray study of the nature of stacking faults in the structure of glauconites: *Clay Minerals* **25**, 419-35.

Smith, J. V., and Yoder, H. S. (1956) Experimental and theoretical studies of the mica polymorphs: *Mineral. Mag.* **31**, 209-35.

Tsipursky, S. I., and Drits, V. A. (1984) The distribution of octahedral cations in the 2:1 layers of dioctahedral smectites studied by oblique-texture electron diffraction: *Clay Minerals* **19**, 177-93.

Zvyagin, B. B, Robotnof, V. T., Sidorenko, O. V., and Kotelnikov, D. D. (1985) Unique mica with noncentrosymmetric layers: *Izvestiya Akad. Nauk SSSR, Geol.* **35**, 121-24 (in Russian).

Appendix:
Modeling One-Dimensional X-Ray Patterns

We describe a model for the calculation of one-dimensional X-ray diffraction patterns for the clay minerals. It applies to mixed-layered as well as to simple structures. There is not enough information in this section to allow you to sit down and write your own program for mixed-layered clay minerals, but if you study the key references (Reynolds, 1980; James, 1965; Bethke and Reynolds, 1986), you should be able to do just that. You should be able to construct a model like this for the simple clay minerals by using Eqs. (3.9), (3.13), (3.14), and (3.15) as a start and then adding refinements discussed in this Appendix. A program called NEWMOD© that does all the things discussed below is commercially available (Reynolds, 1985).

Calculated diffraction patterns are useful for both instruction and research. It takes a very long time for the beginning clay scientist to encounter and interpret diffraction patterns from a wide range of simple and mixed-layered clay minerals. But they can be computed and studied, and this is a very fast alternative. Experiments with compositional variations reinforce intuition or teach the effects of these variations on the patterns. For example, you can quickly answer the questions: What is the effect of increasing Fe substitution on the intensity ratio of the 002/005 reflections for illite? What happens to the basal diffraction pattern of a mica if large amounts of Li are assumed to be in octahedral coordination? For the mixed-layered clay minerals, such questions are almost endless, and answering at least a few of them by means of calculated patterns will greatly increase the effectiveness of the investigator.

At the research level, calculated patterns are invaluable for estimating the composition of simple clay minerals; for identifying, working out the composition, and identifying ordering types for mixed-layered clay minerals; and for generating standard intensities for quantitative standardization (Chapter 9). The limit is not yet in sight. As computers become faster and memory is increased, existing models will be refined further and will run fast enough to be usable in truly interactive modes. Modeling will lead the way toward the identification and quantification of more complicated structures such as mixed-layered clay minerals that contain more than two components.

THE INPUT VARIABLES

Simulating the Instrument

A realistic model of a diffraction pattern requires that you specify all the important mineral structural data, and that you know the values and settings that describe the optical configuration of the diffractometer being simulated. There are a good many instrumental variables. They are listed in Table A.1.

Table A.1. Instrumental variables

Lambda	1.5418
Divergence slit	1.0
Goniometer radius	20
Soller slit 1	6.6
Soller slit 2	2.0
Sample length	3.6
Quartz reference intensity	25,000

Lambda (λ) is the wavelength of the incident radiation in Ångstrom units. $CuK\alpha$ is most commonly used in the United States, and $CoK\alpha$ is favored in Europe, but other wavelengths can be used. The divergence slit controls the beam divergence expressed in degrees for the primary, incident X-ray beam. This quantity is needed, in conjunction with values for the goniometer radius and the sample length, in order to correct for the intensity loss at low diffraction angles caused by an illumination spot whose length exceeds that of the sample. Values for the goniometer radius and the sample length are expressed in centimeters. Angular values in degrees are shown for the two Soller slits or, if only one is used, for the effective aperture or axial divergence of the radiation tunnel that has no slit. The latter is 6.6° for the Siemens D-500, and that value is probably acceptable for other instruments with one "missing" Soller slit. Soller slit 1 is the primary or incident beam slit, and Soller slit 2 is the diffracted beam slit. Calculation of the Lorentz factor requires values for the Soller slits and the preferred orientation of crystallites in the sample.

The quartz reference intensity denotes the peak intensity in counts per second for the reflection from a randomly oriented quartz powder at $d = 3.34$ Å. This measurement must be made on the goniometer with which the results from the program are to be compared, and the goniometer setup must be the same as that described by the parameters for the calculation. The quartz reference intensity is required for the calculation of absolute intensities.

Describing the Clay Mineral

A useful inventory of clay mineral types for a diffraction model such as NEWMOD© consists of illite, biotite, kaolinite, serpentine, tri-tri-chlorite, two-water-layer di-smectite, two-water-layer tri-smectite, two-glycol-layer-di-smectite, two-glycol-layer tri-smectite, one-water-layer di-smectite, one-water-layer tri-smectite, two-water-layer Mg-vermiculite, and one-glycol-layer Na-vermiculite. You might also consider adding di-di-chlorite, di-tri-chlorite, and tri-di-chlorite. The compositions of all these, save kaolinite, are variable, and the model must have a provision for defining their compositions.

The items in Table A.2 describe the variables needed for modeling X-ray diffraction patterns of clay minerals. The default values are good averages, and the ranges are the most common. Use of these values will produce line breadths that realistically simulate those of many natural samples. We offer a few words about each value. The exchange capacity for smectites is required, and it is conveniently expressed as the number of univalent exchangeable cations per $(Si,Al)_4$. Options should be available for specifying the exchange cation in smectites, and Na, K, Mg, Ca, and Sr are useful choices. Ca is a good default ion. The Fe content must be specified for the octahedral sheets in dioctahedral and trioctahedral minerals, and for chlorites; Fe must also be specified for the hydroxide interlayer. The number of atoms of K is needed for modeling illite, and, as described later, this value can be manipulated to simulate Na for paragonite and NH_4 for ammonium illite. Modeling of the diffraction patterns for pyrophyllite and talc requires the value of zero here. Smectites and vermiculites have interlayer complexes of exchangeable cations and water or ethylene glycol. These complexes have a definite atomic structure that must be built into the calculation by extension of the unit cell formulation for G (Eq. 3.9) beyond the hexagonal oxygen surface plane of the silicate layer.

The quantity μ^* is the mean sample mass absorption coefficient, and 45 is a realistic average value for low-Fe clay minerals irradiated by $CuK\alpha$. μ^* controls only absolute intensities, and σ^* is the standard deviation of a Gaussian orientation function for the crystals in the powder aggregate. It may be measured for the most accurate work, such as quantitative analysis or one-dimensional structure determinations for the clay minerals (Reynolds, 1986). The σ^*, in conjunction with the Soller slits, controls both absolute intensity and angle-dependent intensity through the Lorentz factor. High N and low N refer, respectively, to the number of unit cells stacked in the Z direction that make up the largest and smallest crystallites. A range of N is used here to eliminate the spurious ripples between peaks of the interference function (see Fig. 3.14).

The quantity $q(N)$ describes the crystallite size distribution. In other words, if, within the range of N specified, we want crystallites 9 unit cells thick to constitute 22% of the crystallites, then $q(9) = 0.22$. For routine calculations of diffractograms or quantitative standardization, set all values of $q(N)$ to the same value [see Eq. (3.15), p. 92]. Note, however, that $q(N)$ must be normalized to unity by dividing each initial value of $q(N)$ by the sum of the initial values for $q(N)$. If peak shape is to be modeled, then you can experiment with the defect-broadening option that requires a value for δ, the mean defect-free distance.

Table A.2. Default chemical, structural, and sample variables

Exchange capacity	0.35
Exchange cation	Ca
Fe atoms per 2 or 3 octahedral sites	0-2, 0-3
K per 12-fold site	0-1
Interlayer complex	1-water, 2-glycol layers, etc.
μ^*	15
σ^*	12
Low N	3
High N	14
$q(N)$	1
δ	Variable
Proportion of component 1[a]	0-1
Reichweite[a]	0, 1, 2, 3

[a]Applies to mixed-layered clay minerals.

The distance δ is the average distance in a crystal that is not interrupted by a stacking fault that breaks the crystal into two or more X-ray coherent scattering domains. The defect-broadening option produces a special kind of particle-size distribution that, as you will see when you try it, generates very realistic looking peak shapes, particularly for pure minerals. More on this subject later.

Mixed-layered clay minerals require two parameters: a compositional parameter that fixes the relative amounts of each of the two mineral types in the interstratification, and a numerical statement that describes the ordering rules for the sacking sequence along Z (the Reichweite). Let the two types of layers be A and B. Only one of these need be specified by a decimal fraction between 0 and 1 because the sum of the decimal fractions of the two is equal to unity ($P_A + P_B = 1$). Reichweite values that seem to have natural significance at the present time are 0, 1, 2, and 3.

THEORY

Earlier published work by Reynolds (e.g., Reynolds, 1980; Reynolds and Hower, 1970) used a simplification in the mathematics that limited the algorithm to mixed-layered clay minerals that have unit cells centrosymmetric on projection to Z. To handle additional clay mineral types such as kaolinite and serpentine, we need a general formulation of the intensity equation such as that given by Eqs. (A.1) and (A.2):

$$I = Lp\sum_{S} G_j^* G_k \sigma_s(\cos \phi S + i \sin \phi S) \tag{A.1}$$

$$\phi = (4\pi \sin\theta) / \lambda \tag{A.2}$$

The G subscripts identify the layer types that are separated by the spacing S. The layer transforms G originate at the center of the lowermost oxygen plane of the tetrahedral sheet (Fig. A.1). Lp is the Lorentz-polarization factor described later. S is the value of a spacing, in Ångstroms, defined by the number of layers stacked along Z (see Fig. A.1). G_j^* is the complex conjugate of the layer transform of layer A or B, and G_k is the transform of layer type A or B.

By way of review, let some arbitrary variable

$$U = X\cos a + i\,Y\sin b; \quad \text{then } U^* = X\cos a - i\,Y\sin b$$

where $i = \sqrt{-1}$. If you do the multiplications of Eq. (A.1) correctly, all the imaginary components will cancel out, i.e., all occurrences of i will be gone. The summation of Eq. (A.1) is taken over all values of S including the limit $S = 0$ (single layers). σ_S is the frequency of occurrence of the spacing S. It is this quantity that greatly complicates the calculation of diffraction patterns for mixed-layered clay minerals. We cannot go into this sufficiently here to allow you to do the calculations, but a few comments should give you the main ideas. The interested reader is referred to Reynolds (1980) and Bethke and Reynolds (1986) for details.

The frequency term consists of two parts. The first part consists of the number of times that a given array fits into a crystallite that has N layers. The array shown by Fig. A.1 contains five layers, and it will fit only once into a five-layered crystallite. It will fit twice into a six-layered crystallite, three times in a seven-layered crystallite, etc. The rule, then, is to multiply the probability of occurrence by $N + 1$ minus the number of layers in the array. The probability is the complicated part. Figure A.1 shows only one array out of all the possible combinations and permutations. Suppose that $N = 10$. Then there are 2^{10} such arrays containing 10 layers, 2^9 containing 9, etc., and 2^5 (32) that contain 5 layers. We consider one of the 32 in Fig. A.1. The other arrays in this example are ignored that contain 2, 3, 4, 6, 7, 8, 9, and 10 layers. Let the proportion of type A in the assumed mixed-layered clay mineral be 0.7 (for example); then $P_A = 0.7$ and $P_B = 0.3$ because $P_A + P_B = 1$. If the interstratification is random (i.e., $R = 0$), then the probability of occurrence is simply P_A^3 times P_B^2, where the numbers in the exponents refer, respectively, to the numbers of A and B layers in the array. σ_S for this term of the summation is $5 \times (0.3^2 \times 0.7^3) = 0.1544$. Suppose now that $R = 1$. Then we need to compound the conditional or transitional probabilities for this array. Proceeding upward from the bottom, we have

$$\sigma_S = 5(P_B\,P_{B.A}\,P_{A.A}\,P_{A.B}\,P_{B.A}) \tag{A.3}$$

To make this calculation, however, we need values for the conditional probabilities $P_{B.A}$, $P_{A.A}$, etc. Read these as "the probability of an A following a B," and "the probability of an A following an A." We have touched on this subject in Chapter 5 (Box 5.3, p. 173), but more detail is needed here. If $R = 1$ and the composition of the mixed-layered mineral is greater than 50% A, then a B-type layer cannot follow a B-type layer. In other words, $P_{B.B} = 0$. Note that $P_{B.A} + P_{B.B} = 1$ because something must follow a B. It follows then that $P_{B.A} = 1$. So far, we have the transitional probabilities $P_{B.A}$ and $P_{B.B}$, and we need $P_{A.B}$ and $P_{A.A}$ to finish the calculation. The probability of occurrence of the *pair AB* is equal to the probability of an A times the probability of a B following an A, or $P_A P_{A.B}$. The probability of occurrence of the *pair AB* is equal to that of the *pair BA*, and the latter is equal to $P_B P_{B.A}$. So we write

$$P_A P_{A.B} = P_B P_{B.A}$$

or

$$P_{A.B} = P_B P_{B.A} / P_A = 0.3 \times 1 / 0.7 = 0.4286$$

and

$$P_{A.B} + P_{A.A} = 1$$

so

$$P_{A.A} = (1 - 0.4286) = 0.5714$$

We can now calculate σ_S from Eq. (A.3):

$$\sigma_S = 5 \times 0.3 \times 1 \times 0.5714 \times 0.4286 \times 1 = 0.3674$$

Notice that σ_S is larger for the *R1* structure than for the *R0* type. The reason is that we have described an array that has three interfaces of the types $A.B$ and $B.A$, and these types are favored by the condition of nearest-neighbor

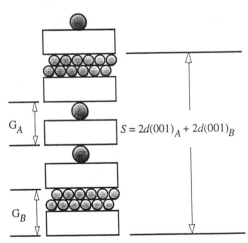

Fig. A.1. One of the possible spacings S formed by a core of two A type layers and one B type layer, with a B on one end and an A on the other.

ordering, which is the $R = 1$ structure. A different array (reading from bottom to top) A-B-B-A-A also contains three A-type layers and two B types. For $R = 0$ it has the same frequency of occurrence as the example described before, but for $R = 1$, $\sigma_S = 0$. See if you can figure out why this statement is correct.

The calculation of σ_S is more involved for cases of $R > 1$ because you must compound long-range transitional probabilities such as $P_{AB.B}$ and $P_{ABA.A}$. If you wish to follow through on these cases, we refer you to the literature cited earlier.

The Lorentz-polarization factor is given by

$$Lp = \frac{(1 + \cos^2 2\theta)\,\Psi}{\sin\theta} \tag{A.4}$$

where the cosine expression is the polarization factor, $\sin\theta$ is the single-crystal Lorentz factor, and ψ is the powder ring distribution factor. [We must avoid confusion here. In Eq. (3.12), the denominator was given as $\sin 2\theta$ for single crystals. This quantity is correct for the *integrated* intensity. $\sin\theta$ applies to the point-by-point intensity of a diffraction profile.] For a randomly oriented powder, ψ is proportional to $1/\sin\theta$; and for a single crystal, ψ is constant. Oriented clay aggregates present values of ψ that are intermediate to these special cases. The development of the theory is given by Reynolds (1986). The computation of ψ requires the root-mean-square angular divergence (in degrees) of the primary and diffracted beam Soller slits and a value for σ^*, which is the standard deviation of a Gaussian distribution that describes the frequency of occurrence of crystallites that are tilted by a given angle (ω) from the mean orientation, which is the orientation parallel to the surface of the powder aggregate. The powder ring distribution factor ψ controls angle-dependent intensity as well as absolute intensity. For this reason, well-oriented clay aggregates produce much stronger diffraction intensities than do poorly or randomly oriented aggregates. The absolute intensity at moderate and high angles is approximately proportional to $1/(\sigma^*)^2$. The derivations of the relevant equations and their applications are given by Reynolds (1986).

The calculated intensities are put on an absolute basis (I_0) by multiplying the calculated intensity values (I) by the quotient shown in Eq. (A.5) (Reynolds, 1983); $\bar{d}$ is the weighted mean value of $d(001)$ for the two components; $\bar{V}$ is the weighted mean volume of the unit cell; ρ is the mean density; μ^* is the sample mass absorption coefficient; and $\bar{N}$ is the weighted mean of the crystallite size distribution, given by the two additional equations. *The constant K is a collection of physical constants, an empirical constant, and the quartz reference intensity. This collection accounts for instrument operating conditions.* The quartz reference intensity is the peak intensity, in counts per second, for the quartz (101) reflection at $d = 3.34$ Å. The measurement is made on a pure quartz powder or on one of the polycrystalline quartz aggregates commonly supplied by diffractometer manufacturers. The measurement must be made under the same operating conditions as those used for the clay mineral diffraction patterns that are to be compared with the calculated results. Overall errors in the accuracy of I_0 amount to about 10% if μ^* and σ^* are measured (Reynolds, 1989).

$$I_0 = I \frac{K'd}{\bar{N}\bar{V}^2 \rho \omega^*}$$ (A.5)

where

$$\bar{N} = \sum_N q(N)N$$

and

$$\sum_N q(N) = 1$$

The calculated intensities are corrected for the angle-dependent effects of a short sample and/or a long beam projection on the sample. These factors, coupled with the goniometer radius, cause progressively greater intensity losses at lower diffraction angles, and there is some angle above which such effects are absent. The length of the beam "footprint" on the sample, which is controlled by the primary divergence slit, is

$$LB = R_0 \, \alpha \, (\pi/180) \, / \sin\theta$$ (A.6)

where R_0 is the goniometer radius in centimeters, α is the angular divergence of the primary divergence slit in degrees, and $\pi/180$ is the radian conversion. The calculated intensities are multiplied by the sample length in centimeters and divided by LB [Eq. (A.6)], only for angles at which LB exceeds the sample length.

Structures of the Component Layers

The layer transforms G are Fourier transforms of the atomic structure along the Z direction. Transform means change, of course, and these transforms effect a change from (1) the coordinates for the position of electrons in space to (2) the amplitude of scattering as a function of the angle of diffraction. Unit cell volumes are based on the half-unit cell, $(Si,Al)_4O_{10}$, except for the 1:1 minerals, which are based on Si_2O_5. Atomic scattering factors are calculated according to Wright (1973). Half-ionized atomic scattering factors are used for Si, Al, and silicate O, and fully ionized values are used for the other cations. Neutral atom (i.e., not ions) factors obtain for ethylene glycol. For water and OH$^-$ the scattering factor is the sum of the factors for H^+ and O^{2-}.

Debye-Waller temperature corrections for atomic vibrations are computed using $B = 1.5$ for cations and $B = 2$ for anions. $B = 11$ is used for ethylene glycol molecules and for one of the water molecule sites in the two-layered water-smectite structure. B is a measure of the mean square displacement of an atom from its ideal place in the structure. It is used to adjust the atomic scattering factor in this form (the higher the temperature, the greater the displacement, the less efficient the scattering)

$$f = f_0 \exp(-B \sin^2\theta / \lambda^2) \qquad (A.7)$$

Equation (A.7) shows how B is applied. The quantity f_0 is the scattering factor for an atom or ion at rest, and f is the temperature-corrected value.

NEWMOD© uses three 2:1 silicate layer structures. The dioctahedral layer is based on unpublished work by Bridges and Reynolds on illite and is used for all the dioctahedral components except smectite, which is taken from Reynolds (1965). All trioctahedral 2:1 layers are defined by a structure published by Mathieson and Walker (1954) for vermiculite. Kaolinite atomic coordinates are those of Zvyagin (1960), and serpentine has the generalized 1:1 trioctahedral structure described by Bailey (1969). The two-layered ethylene glycol structure is taken from Reynolds (1965) and the one-layer Na ethylene glycol structure from Bradley et al. (1963). The two-layered water structure (smectite) is based on unpublished work by Hower and Reynolds, and the one-layered water structure (smectite) assumes that exchangeable cations and water molecules are located at the center of the interlayer space. The water content of the latter is based on the data of Mooney et al. (1952). Atomic coordinates for the trioctahedral chlorite hydroxide sheet are given by Brindley (1961a), and the dioctahedral chlorite hydroxide interlayer has the Al-OH separation of a generalized kaolinite structure (Brindley, 1961b). The two-layered water vermiculite interlayer structure is taken from Mathieson and Walker (1954).

Some of the interlayer structures are only educated guesses. However, small errors in atomic positions cause only small discrepancies in intensities at intermediate and low diffraction angles where most clay mineral work is done. But if you use these data to calculate intensities at angles as high as 50° 2θ and beyond (CuKα), be advised that the intensity results are accurate only if the calculations involve the better-known components such as illite, biotite, two-layered glycol smectite, two-layered water vermiculite, tri-tri-chlorite, and kaolinite.

Table A.3 shows the atomic coordinates for the silicate 2:1 skeletons with respect to an origin at the center of the octahedral sheet (Fig. 3.20, p. 98). Atomic coordinates for interlayer complexes have as their origin ($Z = 0.00$) the center of symmetry on projection to Z for the interlayer structure. Only half of each structure is given, and for the sites not on the center of symmetry, the numbers of atoms in each are doubled. This location of the origin gives a

Table A.3.   One-dimensional atomic structures of silicate layers; coordinates are in Ångstroms measured perpendicular to $d(001)$ along Z

2:1 Di-silicate[a]		Di-smectite		2:1 Tri-silicate	
3.300	6 O	3.27	6 O	3.27	6 O
2.720	4 Si	2.70	4 Si	2.75	4 Si
1.065	4 O + 2 OH	1.06	4 O + 2 OH	1.06	4 O + 2 OH
0.000	1.7 (Fe, Al) + 0.3 Mg	0.00	0.7 (Fe, Al) + 0.3 Mg	0.00	3 (Fe,Mg)

1:1 Tri-silicate		Kaolinite	
4.32	3 OH	4.369	1 OH
3.31	3 (Fe,Mg)	4.311	2 OH
2.27	2 O + OH	3.396	1 Al
0.58	2 Si	3.382	1 Al
0.00	4 O	2.288	1 O + 1 OH
		2.274	1 O
		0.651	1 Si
		0.636	1 Si
		0.150	1 O
		0.143	1 O
		0.000	1 O

[a]Except for dioctahedral smectite.

coordinate reference that is most useful for calculating diffraction patterns using the approach demonstrated in Chapter 3 (Fig. 3.20 and associated discussion). To calculate G for any of the centrosymmetric minerals, calculate G for the silicate skeleton (G_S) and G for the interlayer material (G_I). G for the skeleton plus the interlayer is given by $G = G_S + G_I \cos(4\pi \, [d(001) / 2] \sin\theta / \lambda)$. The coordinates are given in the present form for simplicity, but you will have to transform them if you deal with mixed-layered clay minerals and use the noncentrosymmetric origin shown in Fig. A.1. The kaolinite and serpentine zero coordinates lie at the center of the basal oxygen sheet or oxygen atom of the 1:1 layer. Table A.4 lists the atomic coordinates and atomic compositions of the interlayer structures dealt with by NEWMOD©.

ADVANCED TECHNIQUES

The model represented by NEWMOD© has applications that might not be obvious from an examination of the input parameters. Some of these are described here.

Pure Minerals

Diffraction patterns of pure minerals are produced by interlayering a component with itself. The proportions and Reichweite (R) are irrelevant, but roundoff errors may be minimized if you select 50/50 proportions and $R = 0$.

Table A.4 One-dimensional atomic structure of interlayer material, mineral names, and $d(001)$; coordinates are in Ångstroms measured perpendicular to $d(001)$ along Z

Illite $d(001) = 10.0$		Biotite $d(001) = 10.0$		Tri-tri-chlorite $d(001) = 14.2$	
0.00	0-1 K	0.00	0-1 K	1.02	6 OH
				0.00	3 (Fe, Mg)

2 EG; $d(001) = 16.9^a$		1 EG; $d(001) = 12.9^a$		2-Water; $d(001) = 15.0^a$	
2.33	1.7 CH_2OH^c	0.95	0.35 Na	1.20	1.4 H_2O
1.38	1.7 CH_2OH^c	0.45	2 CH_2OH^c	1.06	0.69 H_2O
0.51	1.2 H_2O^b			0.35	0.69 H_2O
				0.00	nM^+

1-Water; $d(001) = 12.4^a$		Mg-Vermiculite; $d(001) = 14.32$	
0.00	2 $H_2O + nM^+$	1.06	4.32 H_2O
		0.00	0.32 Mg

[a]Interlayer material for smectite. [b]Substitute exchangeable cations for water here. [c]Temperature factor, $B = 11$.

Compositional Superstructures

The algorithm described here and used by NEWMOD© allows the specification of different chemical compositions for each of the layer types. Consequently, you can proceed in the fashion described here for a pure mineral and specify, for example, low Fe in one "component" and high Fe in the other. The calculated pattern will contain the weak compositional superstructure reflections between each of the normal reflections if $R = 1$. If $R = 0$, the result is identical to that of a pure mineral with the average Fe content in each component. Higher Reichweite designations may produce bizarre results whose natural significance should not be dismissed until we begin to look for structures like these.

Layer Types Not Specifically Included

Some imagination is required for the simulation of layer types that are not included in the menu. The examples given here illustrate the principles, but they do not exhaust the possibilities. Suppose you wish to calculate the pattern for interstratified talc/saponite. To simulate the talc, select biotite and enter zero for the K and Fe contents. Then change $d(001)$ for biotite from 10 to 9.33 Å. Pyrophyllite can be treated similarly, except that illite is selected and $d(001)$ is changed to 9.2 Å. You might worry that talc and pyrophyllite do not have tetrahedral Al substitution, whereas biotite and illite do. But such substitution has only a small effect on intensity because the scattering factors are very similar for $Al^{1.5+}$ and Si^{2+}.

Atom Types Not Incorporated in the Model

The atomic scattering factor at low and intermediate diffraction angles is proportional to the number of electrons in the atom or ion. To simulate, for example, ammonium illite, divide 10 by 18 to get 0.56, because NH_4^+ has 10 electrons and K^+ has 18. Select illite, enter 0.56 for K, and increase $d(001)$ to 10.32 Å. Paragonite is handled in identical fashion because, by chance, Na^+ also contains 10 electrons, except that in this case $d(001)$ is decreased to 9.66 Å. This method is useful for dealing with some exotic micas and for extending the range of exchangeable cations in smectites so that it includes Cs, Rb, Pb, Zn, etc. For exchangeable cations, the charge on the modified cation must be identical to that of the cation selected from the standard menu, or else you will need to alter the exchange capacity.

Defect Broadening

Defect broadening refers to diffraction line broadening caused by randomly distributed stacking faults along the Z direction. The theory is described by Ergun (1970), who applied it to carbons. The statistical occurrence of such defects is given by δ, which is the mean defect-free distance, given here by the number of unit cells that are contiguous without interruption. $\delta = 10$ means that, on the average, coherent scattering domains are 10 unit cells thick, smaller domains are more abundant, and larger ones less so. The probability of an optically coherent length δ is given by

$$q(N) = \exp[- (N\text{-}1) / \delta] \qquad (A.8)$$

The largest crystallites (high N) are assumed to be very large compared with δ, but the exponential term becomes so small at large N that for practical purposes the calculations need not extend beyond high $N = 7\delta$. Realistic values for clay minerals are from 6 to 10, so computations can require values of high N as large as 50. Such a calculation for mixed-layered clay minerals is feasible only because of the efficiency of the very fast statistical algorithm developed by C. Bethke (Bethke and Reynolds, 1986). Experience with this model for line broadening has shown that very realistic results are obtained for clay minerals by setting high $N = 5\delta$.

Defect broadening produces a special kind of particle-size distribution [Eq. (A.8)] that produces diffraction lines that have a Cauchy or Lorentzian shape, in contrast to the more Gaussian line shapes produced by the condition $q(N) = 1$. If you try the defect-broadening option, we believe that you will agree that the realistic line shapes produced (particularly for non-mixed-layered clay minerals) suggest that real clay minerals consist of fairly large crystallites that have a mosaic substructure. As high N is diminished from 7δ, the line shape becomes more and more Gaussian due to the presence of finite-crystallite-sized effects.

REFERENCES

Bailey, S. W. (1969) Polytypism of trioctahedral 1:1 layer silicates: *Clays and Clay Minerals* **17**, 355-71.

Bethke, C. M. and Reynolds, R. C., Jr. (1986) Recursive method for determining frequency factors in interstratified clay diffraction calculations: *Clays and Clay Minerals* **34**, 224-26.

Bradley, W. F., Weiss, E. J., and Rowland, R. A. (1963) A glycol-sodium vermiculite complex: *Clays and Clay Minerals*, Proc. 10th Nat. Conf., Austin, Texas, Pergamon Press Inc., 117-22.

Brindley, G. W. (1961a) Chlorite minerals: in Brown, G., ed., *The X-Ray Identification and Crystal Structures of Clay Minerals*: Mineralogical Society, London, 242-96.

Brindley, G. W. (1961b) Kaolin, serpentine and kindred minerals: in Brown, G., ed., *The X-Ray Identification and Crystal Structures of Clay Minerals*, Mineralogical Society, London, 51-131.

Ergun, S. (1970) X-ray scattering by very defective lattices: *Phys. Rev. B* **131**, 3371-80.

James, R. W. (1965) *The Optical Principles of the Diffraction of X-Rays*: Vol. II of the Crystalline State: Series edited by Sir Lawrence Bragg, Cornell University Press, 664 pp.

Mathieson, A. McL., and Walker, G. F. (1954) Crystal structure of Mg-vermiculite: *Amer. Minerl.* **39**, 231-55.

Mooney, R. W., Keenan, A. C., and Wood, L. A. (1952) Adsorption of water vapor by montmorillonite: *Jour. Amer. Chem. Soc.* **74**, 1367-74.

Reynolds, R. C., Jr. (1965) An X-ray study of an ethylene glycol-montmorillonite complex: *Amer. Minerl.* **50**, 990-1001.

Reynolds, R. C., Jr. (1980) Interstratified clay minerals: in Brindley, G.W., and Brown, G., editors, *Crystal Structures of Clay Minerals and Their X-Ray Identification*, Monograph No. 5, Mineralogical Society, London, 249-303.

Reynolds, R. C., Jr. (1983) Calculation of absolute diffraction intensities for mixed-layered clays: *Clays and Clay Minerals* **31**, 233-34.

Reynolds, R. C., Jr. (1985) *NEWMOD* © *a Computer Program for the Calculation of One-Dimensional Diffraction Patterns of Mixed-Layered Clays*: R. C. Reynolds, Jr., 8 Brook Rd., Hanover, NH.

Reynolds, R. C., Jr. (1986) The Lorentz factor and preferred orientation in oriented clay aggregates: *Clays and Clay Minerals* **34**, 359-67.

Reynolds, R. C., Jr. (1989) Principles and techniques of quantitative analysis of clay minerals by X-ray powder diffraction: in Pevear, D. R., and Mumpton, F. A., editors, *Quantitative Mineral Analysis of Clays*, CMS Workshop Lectures, Vol. **1**, The Clay Minerals Society, 4-36.

Reynolds, R. C., Jr., and Hower, J. (1970) The nature of interlayering in mixed-layered illite-montmorillonites: *Clays and Clay Minerals* **18**, 25-36.

Wright, A. C. (1973) A compact representation for atomic scattering factors: *Clays and Clay Minerals* **21**, 489-90.

Zvyagin, B. B. (1960) Electron diffraction determination of the structure of kaolinite: Soviet Physics, *Crystallography* **5**, 32-42.

Index

Remember that the Table of Contents also can help you find things.